Lecture Notes in Earth Sciences

Editors:
S. Bhattacharji, Brooklyn
G. M. Friedman, Brooklyn and Troy
H. J. Neugebauer, Bonn
A. Seilacher, Tuebingen and Yale

Springer
Berlin
Heidelberg
New York
Hong Kong
London
Milan
Paris
Tokyo

Horst J. Neugebauer   Clemens Simmer (Eds.)

Dynamics of Multiscale Earth Systems

With approx. 150 Figures and 50 Tables

 Springer

Editors

Professor Horst J. Neugebauer
Universität Bonn
Institut für Geodynamik
Nussallee 8
53115 Bonn

Clemens Simmer
Meteorologisches Institut
der Universität Bonn
Auf dem Hügel 20
53121 Bonn

"For all Lecture Notes in Earth Sciences published till now please see final pages of the book"

ISSN 0930-0317
ISBN 3-540-41796-6 Springer-Verlag Berlin Heidelberg New York

Cataloging-in-Publication Data applied for
Bibliographic information published by Die Deutsche Bibliothek.
Die Deutsche Bibliothek lists this publication in the Deutsche Nationalbibliographie; detailed bibliographic data is available in the Internet at <http://dnb.ddb.de>.

This work is subject to copyright. All rights are reserved, whether the whole or part of the material is concerned, specifically the rights of translation, reprinting, re-use of illustrations, recitation, broadcasting, reproduction on microfilms or in any other way, and storage in data banks. Duplication of this publication or parts thereof is permitted only under the provisions of the German Copyright Law of September 9, 1965, in its current version, and permission for use must always be obtained from Springer-Verlag. Violations are liable for prosecution under the German Copyright Law.

Springer-Verlag Berlin Heidelberg New York
a member of BertelsmannSpringer Science+Business Media GmbH

http://www.springer.de

© Springer-Verlag Berlin Heidelberg 2003
Printed in Germany

The use of general descriptive names, registered names, trademarks, etc. in this publication does not imply, even in the absence of a specific statement, that such names are exempt from the relevant protective laws and regulations and therefore free for general use

Typesetting: Camera ready by editors
Printed on acid-free paper   32/3141 - 5 4 3 2 1 0

Preface

In many aspects science becomes conducted nowadays through technology and preferential criteria of economy. Thus investigation and knowledge is evidently linked to a specific purpose. Especially Earth science is confronted with two major human perspectives concerning our natural environment: *sustainability* of resources and *assessment* of risks. Both aspects are expressing urgent needs of the living society, but in the same way those needs are addressing a long lasting fundamental challenge which has so far not been met. Following on the patterns of economy and technology, the key is presumed to be found through a development of feasible concepts for a *management* of both our natural environment and in one or the other way the realm of life. Although new techniques for observation and analysis led to an increase of rather specific knowledge about particular phenomena, yet we fail now even more frequently to avoid unforeseen implications and sudden changes of a situation. Obviously the improved technological tools and the assigned expectations on a management of nature still exceed our traditional scientific experience and accumulated competence. Earth- and Life-Sciences are nowadays exceedingly faced with the puzzling nature of an almost boundless network of relations, i. e., the *complexity* of phenomena with respect to their variability. The disciplinary notations and their particular approaches are thus no longer accounting sufficiently for the recorded context of phenomena, for their permanent variability and their unpredictable implications. The large environmental changes of glacial climatic cycles, for instance, demonstrate this complexity of such a typical phenomenology. Ice age cycles involve beside the reorganisation of ice sheets as well changes of the ocean-atmosphere system, the physics and chemistry of the oceans and their sedimentary boundaries. They are linked to the carbon cycle, and the marine and terrestrial ecosystems and last not least the crucial changes in the orbital parameters such as in eccentricity, precession frequency and tilt of the planet during its rotation and movement in space. So far changes of solar radiation through the activity of the sun itself have not yet been adequately incorporated. The entire dynamics of the climate system has therefore the potential to perform abrupt reorganisation as demonstrated by sedimentary records. It becomes quite obvious, in order to reveal the complex nature of phenomena we evidently have to reorganise our own scientific perspectives and our disciplinary bounds as well.

Reflecting assessment of environmental risks from a perspective of complexity, we may discover quickly our traditional addiction to representative averages and the idea of permanence which we equate with *normality* in contrast to exceptional *disasters*. A closer look onto those extremes, however, gives ample evidence that even our normality is based onto ongoing changes which are usually expressed through the variance around the extracted means. Within the concept of complexity the entire spectrum of change has to be considered, whatever intensity it may have. In such a broader context, variability is usually not at all representative for the particularity of phenomena. We might be even more astonished being confronted with an inability of predicting not only the extreme disasters, but equally well the small changes within the range of variance about averages, what we called previously the case of normality.

Encouraged by the potential of new technologies, we might again become tempted to "search for the key where we may have some light, but where we never have lost it". That means, the idea of management without improving our understanding of complexity of Earth-systems will just continue creating popular, weak and disappointing approaches of the known type. It is definitely nor sufficient and neither successful to search for some popular compact solutions in terms of some kind of index-values or thresholds of particular variables for the sake of our convenience. Complexity of phenomena is rather based on ongoing change of an extended network of *conditional relations*; that means, the change of a system can generally not be related to one specific cause. What we therefore need for the search of appropriate concepts of sustainability is primarily an improved *complex competence*. At that state we are facing in the first order very fundamental scientific problems ahead of the technological ones. We encounter a class of difficulties which exceed by far any limited disciplinary competence as well as any aspect of economical advantage. The difficulties we do encounter when investigating complexity are well expressed synonymously by the universal *multi-scale* character of phenomena. We might thus strive for an enhanced discussion of the expanding scale concept for the investigation of geo-complexity.

Scales are geometrical measures and quite suitable to express general properties or qualities of phenomena such as length, size or volume. Even so space-related features, as for instance intensity or momentum might be addressed by scales as well. The close relationship of scales to our perception of space justifies their universal character and utility. An entire *scale-concept* might be established and practised through characterising objects and phenomena by their *representative scales*. In this sense we might even comprehend a complex phenomenology by analysing it with respect to its inherent hierarchy of related scales. The consideration of scales might give us equally well an idea about the conditional background of phenomena and their peculiar variability. Complexity of change of entire systems such as for instance fracture pattern, slope instabilities, distribution of precipitation and others are progressively investigated by means of a statistical analysis of scales through the fit of characteristic *scaling laws* to large data sets. At present, it appears to be the most efficient way of revealing the hidden nature of complexity of phenomena which are depending on a large context. Insight comes thereafter mostly from the conditions, necessary for reproducing statistical probability distributions through modelling. However, our traditional addiction to the common background of linear deterministic relations gives us a hard time establishing quantitative models on complex system behaviour. Representing complex change quantitatively requires the interaction between different variables and parameters at least on a local basis. Consequently, there are numerous attempts to build representative complex models through gradually adopting non-linearity up to the degree of self-organisation in the critical state. Conceptual as well as methodological problems of both analysing and modelling complexity have been assigned to the scale-concept for the convenience of a universal standard for comparison. Within this context a key question becomes addressed implicitly throughout the various measurements, techniques of analysis and complex modelling which will be discussed: *do our considerations of natural phenomena yield sufficient evidence for the existence of representative averages.*

About ten traditional disciplines, among them applied mathematics, computer science, physics, photogrammetry and most of the geo-sciences formed a united power of competence in addressing the complexity of *Earth-systems* on different manifestations. Related to a limited area of mountain and basin structures along a section of the Lower Rhine, we studied two major aspects: the historical component of past processes from Earth history by means of structural and sedimentary records as well as temporary perspectives of ongoing processes of change through recent observations. Our common matter of understanding was the conceptual aim of performing complementary *quantitative modelling.* Therefore the presented volume gives in the beginning a comprehensive presentation on the different aspects of the scale-concept. According to the addressed phenomena the scale related representation of both spatial data sets and physical processes will be discussed in the chapters two and three. Finally some structural and dynamical modelling aspects of the complexity of change in earth systems will be presented with reference to the scale-concept.

After a period of ten years of interdisciplinary research, plenty of scientific experience has been developed and accumulated. Naturally, even a long record for warm acknowledgements arose from a wide spectrum of support which made this unusual scientific approach possible. Therefore the present book will give some representative sampling of our themes, methods and results at an advanced state of research work. We will take the opportunity to gratefully acknowledge the substantial funding of our scientific interests within the Collaborative Research Centre 350: *Interaction between and Modelling of Continental Geo-Systems* — by the German science foundation — Deutsche Forschungsgemeinschaft — enclosing the large group of critical reviewers. In the same sense we like to thank our head board of the Rheinische Friedrich-Wilhelms-University at Bonn. Large amounts of data have been kindly provided from various sources: a peat mining company, water supply companies and various federal institutes and surveys. We are very grateful for the numerous occasions of generous help. Finally we like to thank our colleagues from two European research programs in the Netherlands and Switzerland. The editors are indebted to Michael Pullmann and Sven Hunecke for their valuable help in preparing this book.

Last not least, we like to acknowledge the assistance of the reviewers with the contributing papers with gratitude.

Horst J. Neugebauer & Clemens Simmer

Table of Contents

I Scale Concepts in Geosciences

Scale Aspects of Geo-Data Sampling 5
 Hans-Joachim Kümpel

Notions of Scale in Geosciences 17
 Wolfgang Förstner

Complexity of Change and the Scale Concept in Earth System Modelling . 41
 Horst J. Neugebauer

II Multi-Scale Representation of Data

Wavelet Analysis of Geoscientific Data 69
 Thomas Gerstner, Hans-Peter Helfrich, and Angela Kunoth

Diffusion Methods for Form Generalisation 89
 Andre Braunmandl, Thomas Canarius, and Hans-Peter Helfrich

Multi-Scale Aspects in the Management of Geologically Defined Geometries 103
 Martin Breunig, Armin B. Cremers, Serge Shumilov, and Jörg Siebeck

III Scale Problems in Physical Process Models

Wavelet and Multigrid Methods for Convection-Diffusion Equations 123
 Thomas Gerstner, Frank Kiefer, and Angela Kunoth

Analytical Coupling of Scales — Transport of Water and Solutes in Porous Media 135
 Thomas Canarius, Hans-Peter Helfrich, and G. W. Brümmer

Upscaling of Hydrological Models by Means of Parameter Aggregation Technique 145
 Bernd Diekkrüger

Parameterisation of Turbulent Transport in the Atmosphere 167
 Matthias Raschendorfer, Clemens Simmer, and Patrick Gross

Precipitation Dynamics of Convective Clouds 186
 Günther Heinemann and Christoph Reudenbach

Sediment Transport — from Grains to Partial Differential Equations 199
 Stefan Hergarten, Markus Hinterkausen, and Michael Küpper

Water Uptake by Plant Roots — a Multi-Scale Approach 215
  Markus Mendel, Stefan Hergarten, and Horst J. Neugebauer

IV Scale-Related Approaches to Geo-Processes

Fractals – Pretty Pictures or a Key to Understanding Earth Processes? ... 237
  Stefan Hergarten

Fractal Variability of Drilling Profiles in the Upper Crystalline Crust 257
  Sabrina Leonardi

Is the Earth's Surface Critical? The Role of Fluvial Erosion and Landslides 271
  Stefan Hergarten

Scale Problems in Geometric-Kinematic Modelling of Geological Objects .. 291
  Agemar Siehl and Andreas Thomsen

Depositional Systems and Missing Depositional Records 307
  Andreas Schäfer

Multi-Scale Processes and the Reconstruction of Palaeoclimate 325
  Christoph Gebhardt, Norbert Kühl, Andreas Hense, and Thomas Litt

A Statistical-Dynamic Analysis of Precipitation Data with High Temporal
Resolution 337
  Hildegard Steinhorst, Clemens Simmer, and Heinz-Dieter Schilling

List of Authors

BRAUNMANDL, Andre, Bonner Logsweg 107, 53123 Bonn, abraunma@de.ibm.com

BREUNIG, Martin, Prof. Dr., Institut für Umweltwissenschaften, Hochschule Vechta, Oldenburger Straße 97, 49377 Vechta, mbreunig@iuw.uni-vechta.de

BRÜMMER, Gerhard W., Prof. Dr., Institut für Bodenkunde, Rheinische Friedrich-Wilhelms-Universität Bonn, Nußallee 13, 53115 Bonn, bruemmer@boden.uni-bonn.de

CANARIUS, Thomas, Dr., Hochstraße 71, 55128 Mainz, thomas.canarius@db.com

CREMERS, Armin Bernd, Prof. Dr., Institut für Informatik III, Rheinische Friedrich-Wilhelms-Universität Bonn, Römerstraße 164, 53117 Bonn, abc@cs.uni-bonn.de

DIEKKRÜGER, Bernd, Prof. Dr., Geographisches Institut, Rheinische Friedrich-Wilhelms-Universität Bonn, Meckenheimer Allee 166, 53115 Bonn, b.diekkrueger@uni-bonn.de

FÖRSTNER, Wolfgang, Prof. Dr., Institut für Photogrammetrie, Rheinische Friedrich-Wilhelms-Universität Bonn, Nußallee 15, 53115 Bonn, wf@ipb.uni-bonn.de

GEBHARDT, Christoph, Meteorologisches Institut, Rheinische Friedrich-Wilhelms-Universität Bonn, Auf dem Hügel 20, 53121 Bonn, c.gebhardt@uni-bonn.de

GERSTNER, Thomas, Institut für Angewandte Mathematik, Rheinische Friedrich-Wilhelms-Universität Bonn, Wegelerstraße 6, 53115 Bonn, gerstner@iam.uni-bonn.de

GROSS, Patrick, Steinmetzweg 34, 64625 Bensheim, kapahogro@t-online.de

HEINEMANN, Günther, Prof. Dr., Meteorologisches Institut, Rheinische Friedrich-Wilhelms-Universität Bonn, Auf dem Hügel 20, 53121 Bonn, gheinemann@uni-bonn.de

HELFRICH, Hans-Peter, Prof. Dr., Mathematisches Seminar der Landwirtschaftlichen Fakultät, Rheinische Friedrich-Wilhelms-Universität Bonn, Nußallee 15, 53115 Bonn, helfrich@uni-bonn.de

HENSE, Andreas, Meteorologisches Institut, Rheinische Friedrich-Wilhelms-Universität Bonn, Auf dem Hügel 20, 53121 Bonn, ahense@uni-bonn.de

HERGARTEN, Stefan, Dr., Lehrstuhl für Geodynamik, Geologisches Institut, Rheinische Friedrich-Wilhelms-Universität Bonn, Nußallee 8, 53115 Bonn, hergarten@geo.uni-bonn.de

HINTERKAUSEN, Markus, Dr., Birkenweg 16, 71679 Asperg, markus@geo.uni-bonn.de

KIEFER, Frank, Dr., Institut für Angewandte Mathematik, Rheinische Friedrich-Wilhelms-Universität Bonn, Wegelerstraße 6, 53115 Bonn, kiefer@iam.uni-bonn.de

KUNOTH, Angela, Prof. Dr., Institut für Angewandte Mathematik, Rheinische Friedrich-Wilhelms-Universität Bonn, Wegelerstraße 6, 53115 Bonn, kunoth@iam.uni-bonn.de

KÜHL, Norbert, Dr., Institut für Paläontologie, Rheinische Friedrich-Wilhelms-Universität Bonn, Nußallee 8, 53115 Bonn, kuehl@uni-bonn.de

KÜMPEL, Hans-Joachim, Prof. Dr., GGA-Institut, Stilleweg 2, 30655 Hannover, kuempel@gga-hannover.de

KÜPPER, Michael, Lehrstuhl für Geodynamik, Geologisches Institut, Rheinische Friedrich-Wilhelms-Universität Bonn, Nußallee 8, 53115 Bonn, kuepper@geo.uni-bonn.de

LEONARDI, Sabrina, Dr., Geologisches Institut, Rheinische Friedrich-Wilhelms-Universität Bonn, Nußallee 8, 53115 Bonn, sabrina@geo.uni-bonn.de

LITT, Thomas, Prof. Dr., Institut für Paläontologie, Rheinische Friedrich-Wilhelms-Universität Bonn, Nußallee 8, 53115 Bonn, t.litt@uni-bonn.de

MENDEL, Markus, Dr., Lehrstuhl für Geodynamik, Geologisches Institut, Rheinische Friedrich-Wilhelms-Universität Bonn, Nußallee 8, 53115 Bonn, mendel@geo.uni-bonn.de

NEUGEBAUER, Horst J., Prof. Dr., Lehrstuhl für Geodynamik, Geologisches Institut, Rheinische Friedrich-Wilhelms-Universität Bonn, Nußallee 8, 53115 Bonn, neugb@geo.uni-bonn.de

RASCHENDORFER, Matthias, Am Lindenbaum 38, 60433 Frankfurt, matthias.raschendorfer@dwd.de

REUDENBACH, Christoph, Dr., Fachbereich Geographie, Universität Marburg, Deutschhausstraße 10, 35037 Marburg, c.reudenbach@mailer.uni-marburg.de

SCHÄFER, Andreas, Prof. Dr., Geologisches Institut, Rheinische Friedrich-Wilhelms-Universität Bonn, Nußallee 8, 53115 Bonn, schaefer@uni-bonn.de

SCHILLING, Heinz-Dieter, Prof. Dr., Meteorologisches Institut, Rheinische Friedrich-Wilhelms-Universität Bonn. Heinz-Dieter Schilling was one of the driving forces of the research program. He died on November 22th, 1997, on a research expedition in the table mountains of Venezuela.

SHUMILOV, Serge, Institut für Informatik III, Universität Bonn, Römerstraße 164, 53117 Bonn, shumilov@iai.uni-bonn.de

SIEBECK, Jörg, Institut für Informatik III, Universität Bonn, Römerstraße 164, 53117 Bonn, siebeck@iai.uni-bonn.de

SIEHL, Agemar, Prof. Dr., Geologisches Institut, Rheinische Friedrich-Wilhelms-Universität Bonn, Nußallee 8, 53115 Bonn, siehl@uni-bonn.de

SIMMER, Clemens, Prof. Dr., Meteorologisches Institut, Rheinische Friedrich-Wilhelms-Universität Bonn, Auf dem Hügel 20, 53121 Bonn, csimmer@uni-bonn.de

STEINHORST, Hildegard, Dr., Abteilung ICG-1 (Stratosphäre), Institut für Chemie und Dynamik der Geosphäre, Forschungszentrum Jülich, 52435 Jülich, h.steinhorst@fz-juelich.de

THOMSEN, Andreas, Institut für Informatik III, Rheinische Friedrich-Wilhelms-Universität Bonn, Römerstraße 164, 53117 Bonn, thomsen@uni-bonn.de

Part I

Scale Concepts in Geosciences

Scale Concepts in Geosciences

This introductory chapter is meant to guide the reader into the broad wealth of scales in geoscience from three different aspects. We start with the all-important observations, the marks put on them by the measurement process itself and its sampling characteristics. It follows the necessity to stay back and critically ask ourselves what the observables really are and what kind of information they might provide in relation to the processes under scrutiny. Then we detach ourselves from the observation standpoint and approach the concept and problems of scales in geoscience from the data modelling viewpoint. Finally, we combine the aspect of scales in observations, data models, and physical model development into a self-consistent geo-system model environment.

When observing spatial or temporal scales with measuring devices, we first have to acknowledge, that measurements itself comprise processes with their own dynamics and thus own statistics and errors, and also scales. Kümpel shows in his contribution the many implications which follow from this aspect. Besides the necessarily incomplete measuring process itself, which can never provide a complete image of the processes under consideration, data selection by sampling in space or time, or both, can never be avoided, and thus puts its own trace marks on the data. So, apart from the scales and scale interactions of the observed system networks, we also have to deal with artefacts, which might mix spatial and temporal scales, mimic real processes and thus put us easily on wrong tracks. The problem will be largely aggravated, when we try to detect temporal scales from geological records, which already are the result of recording processes not controlled by ourselves, processes itself not yet fully understood. Since the geological records may and will contain missing time slices due to erosion processes, the situation can easily become quite intricate.

Putting observations into hard data in the form of data models intended for process analysis is yet another process dealt with in the contribution by Förstner. Creating data models might seem unproblematic at first glance, but this process also leaves its own marks and has its own pitfalls. At this stage at latest we have to ask ourselves: What are scales? How do we characterise scales? Do we mean a map scale, or sizes or lifetimes of objects? The question of scales comes in here in a very general way, especially when we cross the borders between disciplines. We observe that scales are always attributed to the objects of disciplinary interest, thus already driven by disciplinary views of processes. Objects often even define scales, like e. g. midlatitude cyclones in the atmosphere define the so-called synoptic scale in meteorology. We must be aware of these internal ways of thinking and consequently very critically reflect disciplinary scale concepts when heading for the broader the whole geo-system encompassing pathway of knowledge.

The degree of our knowledge of complex geo-systems can best, probably only, be quantified by our ability to predict geo-system behaviour with process models. Neugebauer stresses in his contribution that prediction ability is an indispensable component of our knowledge building process. First principles (e.g. the conservation laws of mass, energy and momentum) can only build the skeleton of such models or its components because of the multi-scale problem inherent in any geoscience endeavour. The individual disciplines have dealt with their own scale problems in a wide variety of ways, which will be discussed in different chapters of the book. In order to extend both knowledge and predictability beyond disciplinary achievements, we have to connect disciplinary process models to form more complete geo-system models. Crossing the disciplines we have to acknowledge the disciplinary achievements in intra-medium multi-scale interaction modelling. But at the same instant we are forced to strive for more general and also more complete methods in order to deal with so far underrepresented and even more extreme multi-scale cross-media interactions.

Scale Aspects of Geo-Data Sampling

Hans-Joachim Kümpel[*]

Section Applied Geophysics, Geological Institute, University of Bonn, Germany

Abstract. Sampling data values in earth sciences is all but trivial. The reason is the multi-scale variability of observables in nature, both in space and in time. Sampling geo-data means selecting a finite number of data points from a continuum of high complexity, or from a system of virtually infinite degree of freedom. When taking a reading a filtering process is automatically involved, whatever tool or instrument is used. The design of adequate sampling strategy as well as proper interpretation of geo-data thus requires a-priori insight into the variability of the target structure and/or into the dynamics of a time varying observable. Ignoring the inherent complexity of earth systems can ultimately lead to grossly false conclusions. Retrieving data from nature and deducing meaningful results obviously is an expert's task.

Earth systems exhibit an inherent, multi-scale nature, hence, its observables too. When we want to sample geo-data, e. g. by taking soil samples or readings from a sensor in the field, we have to keep in mind that even the smallest soil sample is an aggregate of a myriad of minerals and fluid molecules; and that the sensor of any instrument, due to limited resolution, renders values that are integrated over some part of the scanned medium. Any set of measurements extracts only a tiny piece of information from an Earth system.

In this book, an Earth system classifies a complex, natural regime on Earth. It could be a group of clouds in the atmosphere, a network of fissures in a rock, a hill slope, a sedimentary basin, the Earth's lithosphere; or it could be a specific geographical region as, for instance, the Rhenish Massif, or a geological province like the Westerwald volcanic province, a part of that Massif. Earth systems are coupled systems, in which various components interact; and they are open systems, as there is always some interaction with external constituents. What Earth systems have in common, besides of their multi-scale structures in space and their complex dynamics in the time domain, is a large, virtually infinite number of particles or elements, and consequently of degrees of freedom.

The problem of adequate data sampling in an Earth system obviously exists on both the spatial and the time scale: Taking a reading may need seconds, hours, months. Can we be sure that the dynamics of the target parameter can be neglected during such time spans? We wouldn't have to bother if structural variabilities were of comparatively large wavelengths, or if temporal fluctuations exhibit rather low frequencies. This will, however, generally not be the case.

---

[*] kuempel@gga-hannover.de [now at: Leibniz Institute for Applied Geosciences (GGA), Hannover, Germany]

When observables present a multi-scale variability, the resolution of readings will actually often be less than the nominal resolution of the sensor or of the measurement technique we apply.

How do we classify sampled data in terms of accuracy, precision, resolution? The values we call 'readings', are they average values? What kind of averaging is involved? If — on the scale of sampling — a multi-scale variability is significant, should we name the readings 'effective' values, since we can but obtain a rather limited idea about their full variability? What would be the exact understanding of an effective value?

It may be useful to recall some basic definitions of limitations that always exist in sampling. By 'resolution' we mean the smallest difference between two readings that differ by a significant amount. 'Precision' is a measure of how well a result was determined. It is associated with the scatter in a set of repeated readings. 'Bias' is the systematic error in a measurement. It can be caused by a badly calibrated or faulty instrument and corrupts the value of a reading that may have been taken with high precision. 'Accuracy' is a measure of correctness, or of how close a result comes to the true value (Fig. 1). However, the true value is usually unknown. Scientists need to estimate in some systematic way how much confidence they can have in their measurements. All they have is an imagination, or model, of what they believe is the true value. An Earth system itself is often imagined as a grossly simplified model. 'Error' is simply defined as the difference between an observed (or calculated) value and the true value. And 'dynamic range' of an instrument or technique denotes the ratio of the largest difference that can be sensed (i. e. its measuring range) and its resolution.

Readings are typically corrupted by statistical (or random) errors that result from observational fluctuations differing from experiment to experiment. The statistical error can be quantified from the observed spread of the results of a set of representative measurements, i. e. by testing how well a reading is reproducible (if repetition is possible). A bias is more difficult to identify and in many cases is not even considered. It is repeated in every single measurement and leads to systematic differences between the true and the measured value. A bias can be assessed by application of standards, or by comparison with redundant observations obtained by using a different technique. An accurate reading has both a low statistical and a low systematic error. Bias and precision are completely independent of each other. A result can be unbiased and precise, or biased and imprecise, but also unbiased and imprecise, or biased and precise. In the two latter cases the results may not be satisfactory. By the way, fuzzy logic is a useful technique to handle imprecise or even indistinct problems — yet at the expense of resolution.

To be more specific on problems that are particular to geo-data sampling, let us first consider cases for limitations on the spatial scale. In all geoscience innumerable kinds of sampling methods do exist. A few examples are given below, mainly from geophysics. They can give only a brief glance of the broadness of conflicts existing with geo-data sampling. Apart from that, in view of the multi-scale nature of observables, we may differ between two aspects: Limited representative character of a single observation, e. g. due to the extended geometry of an instrument or a sampling technique; and limited representative character of several readings, taken of a quantity that presents a certain variability in space, for instance, along a profile.

A frequent problem in geoscience that is common to both aspects is 'undersampling'. In an Earth system, we will never have sufficient readings to unambiguously describe a structure or a process in all its details. Imaging of details on all scales is impossible due to the limited dynamic range of any scanning technique. There is always a 'bit' or 'byte' (1-D), a 'pixel' (2-D), or a 'voxel' (3-D), i.e. a smallest element bearing information, and a maximum total number of elements that can be retrieved and stored. To code the structure of a surface of one square metre at micrometre resolution in black and white (or that of a square kilometre at resolution of a millimetre) would require a dynamic range of 4×10^{12} or four million Megapixels. Common data loggers have a dynamic range of 2048 bits, broadband loggers offer 8 million. Aliasing occurs if short wavelength fluctuations are undersampled leading to appearance of these fluctuations in the long wavelength part of the archived data set (e. g. Kanasewich,

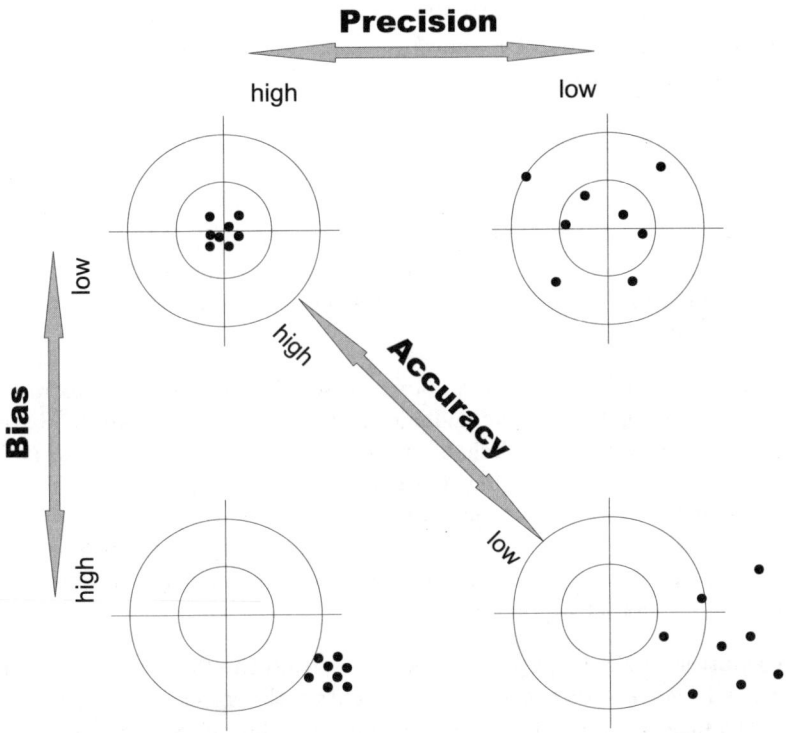

Fig. 1. Precision, bias and accuracy (inspired by Wagner, 1998).

1981). Unless short wavelength variability is suppressed through filtering prior to sampling, a certain degree of aliasing is unavoidable in geoscience. However, by the process of filtering, original data is lost. How can we still get along in geoscience?

As for the aspect of limited representative character of single readings, we may e. g. think of distance measurements, for instance through an extensometer. This is a geometrically extended instrument, two points of which are coupled to the ground. Instrument readings are changes of the distance between the two points divided by the instrument's overall spread, called the baselength. When compared to previously taken readings, data values indicate the bulk shortening or lengthening of the rock formation between the instrument's grounded points, yielding one component of the surface deformation tensor (Fig. 2a). In practice, the baselength may range from several centimetres to hundreds of metres. Sophisticated extensometers have a resolution of one part in ten billion (10^{-10}), yet each single reading represents an integral value over all shortenings or lengthenings that may occur anywhere between the grounded points (Figs. 2b, c). What can we learn from a measurement which is both so accurate and so crude?

On a larger scale, a similar problem exists with measurements called Very Long Baseline Interferometry (VLBI). These yield distance data over baselines of hundreds or thousands of kilometres, namely between two parabola antennae that receive radio signals from a distant quasar (e. g. Campbell, 2000). A single reading when processed is the change in the distance between two VLBI stations. It can be resolved with millimetre precision. However, from the multitude of point displacements occurring at the Earth's surface, like local shifts of ground positions, regional size displacements along faults, or plate tectonic movements on the global scale, only changes in the bulk distance between the VLBI stations are extracted. Any thinkable configuration of VLBI systems will largely undersample the full spectrum of ground movements. Do the measurements make sense?

A different situation is met with observation of potential fields, as in gravity, magnetics, or geoelectrics. Here, a single reading reflects a certain field strength at the location of the sensor element. When interpreting the reading one has to be aware that the observed value represents a superposition of contributions of masses, magnetic dipoles, electric charges from the closest vicinity up to — in principle — infinite distance. Any reading is actually the integral over a very large number of contributors in the sensor's near and far field. Is there any hope to reduce such inherent ambiguity?

In the latter case it helps when we know — from the theory of potential fields — that the larger the distance to a contributor, the lower will be its influence; or, a contributor at larger distance needs either to be more extended or to be of higher source strength to provoke the same effect as a close by contributor. For instance, spheres of identical excess density $\Delta\rho$ provoke the same change Δg in gravity when their distances r_i to the gravity sensor scale to their radii R_i with

Scale Aspects of Geo-Data Sampling

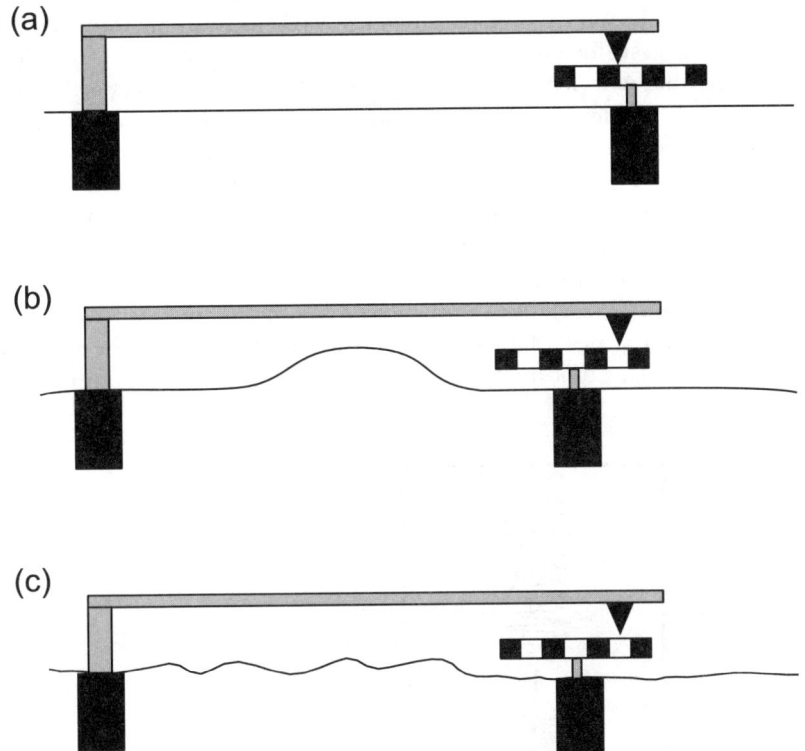

Fig. 2. Principle of extensometer (a) and impossibility to distinguish between length fluctuations of long (b) and short wavelength (c).

$R_i^3/r_i^2 = $ const. (because from Newton's law,

$$\Delta g = \frac{\Delta m_i}{r_i^2} = \frac{\Delta \rho V_i}{r_i^2} = \frac{4}{3}\pi \Delta \rho \frac{R_i^3}{r_i^2}$$

with Δm_i, V_i denoting the excess mass and the volume of a sphere; Fig. 3). Knowing basic physical laws thus allows us to differ between more and less dominating constituents, and those that are negligible — without being able to specify the shape of an anomalous body in the subsurface.

Knowledge of basic geoscientific principles, similarly, helps to understand extensometer readings or VLBI data. The concept of plate tectonics has given the clue to understand why extended areas of the Earth's surface are comparatively rigid (and therefore called plates) and large movements occur mostly at plate boundaries (e. g. Kearey and Vine, 1996). In fact, significant variations in the length of VLBI baselines can largely be attributed to plate boundary regions rather than to processes of deformation within the plates; displacements generally do not occur at arbitrary locations in between VLBI stations. This also holds for smaller scales: Extensometer readings are well understood from insight into the distribution of stiffer and weaker portions of a probed rock formation, and into the nature and wavelength of stresses that lead to shortening or lengthening of the baselength (Agnew, 1986). How representative and informative individual readings are can often be estimated from a-priori knowledge or an expert's assumption of the most influential processes being involved. This applies to many sampling procedures: A-priori knowledge leads to the imagination of models; good models are fundamental for every sampling strategy.

How well is the signature of a geo-signal represented in our data if we have sampled a whole series of values — along a profile, in an area, or in 3-D? As variability exists on all scales, the smaller the sampling intervals, the more details are resolved (Fig. 4). Some methods allow sampling of a large number of data

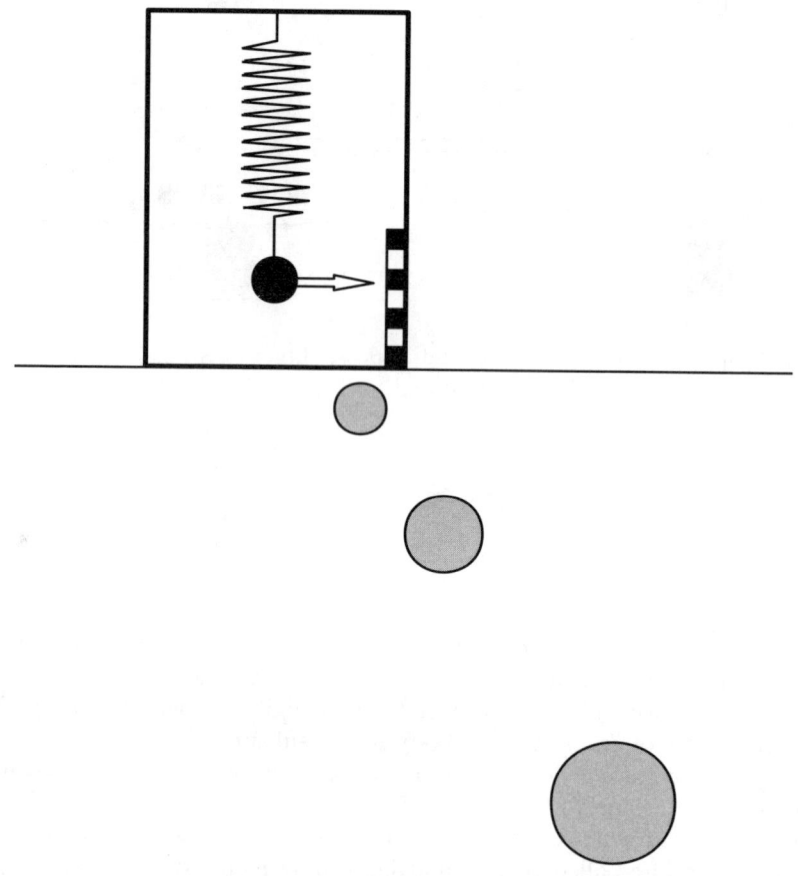

Fig. 3. Principle of gravimeter and identical attraction effect of anomalous masses at closer and larger distances.

values, others do not; or we may be able to obtain densely sampled data along one axis in space, but not along another. Rock cores recovered from boreholes or data from geophysical borehole-logging, for example, provide a wealth of information along the borehole axis (see Leonardi; this volume), with centimetre, for core material even sub-millimetre resolution. The limited number of boreholes drilled in an area, however, may restrict lateral resolution to wavelengths of tens or hundreds of kilometres, which does often not allow to trace alterations and boundaries of geological formations over respective distances. Or, whereas monitoring of in-situ deformation using extensometers — due to logistic constraints — can be operated at a few locations only, areal coverage by e.g. a geomagnetic field survey may be rather dense. Still, spatial resolution of the latter technique is far of being sufficient if we want to resolve details of e.g. the structure of pores to understand colloid transport on the micro-scale.

When surveying potential field quantities, useful information can be retrieved if we have recorded a series of values over some area and apply the principles of potential theory. For example, iterated smoothing of signal amplitudes through application of a technique called 'upward continuation' allows to define the maximum burial depth of a contributor, provided the embedding material has uniform source strength (e.g. Telford et al., 1976). Analysis of signal gradients and solution of boundary value problems likewise reduces the ambiguity of a data set. Yet, as long as we have no direct access to the hypothetic contributor, we are unable to verify its exact shape, position and strength. Again, only when we have some a-priori knowledge of the most likely petrological regime and of a geologically plausible distribution of anomalous sources in the subsurface, we can make adequate use of such observations and assessments.

In seismics and ground penetrating radar, wave or pulse reflection techniques are successfully applied to scan the morphology of subsurface structures. Although signals reflected from subsurface targets can be tracked with dense areal coverage, resolution of these techniques is proportional to the shortest wavelength that is preserved in the recorded signal. As only signals of rather long wavelength 'survive' when travelling over large distances, gross alterations in the morphology of structures or the bulk appearance of sedimentary layers is all what can be resolved from greater depth. However, even the shortest wavelength is larger than the finest structure in the target medium. How signal averaging takes place for bundles of thin layers, i.e. how seismic or radar signals are related to the real structure in the ground is still a matter of research (Ziolkowski, 1999). Experience from inspection of outcrops at the Earth's surface or an expert's imagination about structures and layering may still allow meaningful interpretation beyond the resolution of the measurement technique (Schäfer, this volume; Siehl and Thomsen, this volume).

Focusing on temporal effects, time series in geosciences often include a broad spectrum of variability. Signals may be very regular or highly irregular. Diurnal and semidiurnal constituents of tidal forces acting on the Earth, for instance, are well known from the ephemeridae (coordinates) of the influential celestial bodies, i.e. the moon and the sun. They can be computed with high accuracy,

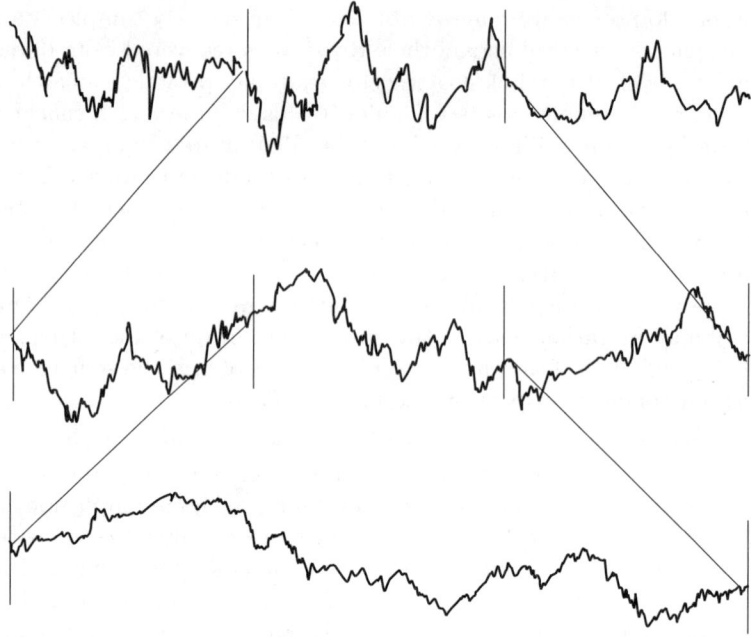

Fig. 4. Multi-scale variability of geo-data signals.

also for the future (Wenzel, 1997). The occurrence of earthquakes, on the other hand, is very uncertain. Mean recurrence periods of large events can statistically be assessed from the seismicity history. Such periods can last tens, hundreds, or thousands of years, and probably longer. Reliable earthquake catalogues report historical events for the last several hundred years, at best. Accordingly, stochastic assessment of an earthquake cycle, if existing, is rather uncertain. In no case, the available data allows to predict the date, magnitude, and location of the next earthquake (Geller, 1997).

The time scales that matter in geoscience are enormous. For VLBI measurements, correlation of fluctuations in the radio signals of quasars are carried out in the 10^{-11} s range. A pumping test to assess the transmissivity and hydraulic performance of a well-aquifer-system may last several hours, days, or weeks. Sedimentation and rock alteration processes take place over millions of years. And some of the radioactive isotopes that contribute to the Earth's heat balance have decay rates of billions of years (i.e. in the 10^{17} s), thus compete with the age of the Earth.

In geology and paleontology, sampling often means identification of markers and fossils, structural elements or petrological parameters over very different ranges of time. Depending on the presence or absence of markers, continuous sedimentary sequences may allow stratigraphic interpretation with time resolution as short as years, yet sometimes only to as much as millions of years.

Growth patterns of certain intertidal bivalves or tidal bundles in laminated sediments reveal features that are linked to the succession of high and low tides, so allow resolution of semidiurnal cycles. Such a pattern may have formed several hundred million years ago; the absolute age may still be uncertain by several million years (Scrutton, 1978).

When sampling data values of time varying signals, equivalent problems exist as in the space domain. A single value may be some type of average of a quickly changing observable. Meteorologists, for instance, take air temperature readings from devices attached to ascending balloons. During the recording process the outside temperature is permanently changing. As thermal adjustment needs time — due to the heat capacity of the sensor — only smoothed temperature values and persistent signal parts are obtained. Similarly, super-Hertz fluctuations of the Earth's magnetic field, induced by solar-ionospheric effects, are not sensed by magnetometers that are in common use for geomagnetic surveys. To what extent such fluctuations can be tolerated? Geoscientists have experienced that many time varying signals present some type of power law characteristic, meaning that high frequency variability occurs with much lower amplitudes than low frequency variability. Knowing the power law characteristic of a sequence one may neglect the existence of high frequency fluctuations for many purposes. Power law characteristics are also common in the space domain, e.g. in the morphology of the Earth's surface, of rock specimen, of fractures, in borehole data, etc. (Turcotte, 1992; Hergarten, this volume; Leonardi, this volume).

Sampling of a series of values over time, e.g. by repeatedly taking measurements at the same site, may be feasible at high sampling rate for one technique, but not for another. For instance, ground shaking can be recorded up to kilohertz resolution using a suitable seismograph; a single pumping test to check hydraulic transmissivity of an aquifer could last days or weeks. Due to advanced technology, instrumental performance allows recording of observables mostly at — what is believed — sufficiently high sampling rates. For geological records, however, undersampling in the time domain is more the rule than the exception because sedimentary sequences and rock alteration processes are often disrupted, distorted, eroded or overprinted.

All these examples show that averaging, integration, interpolation, and extrapolation are extensively used in geoscience. Researchers are obliged to train themselves to regularly ask: In how far are we allowed to interpret a single reading which has been taken from data exhibiting a certain degree of variability? To what extent does a set of readings represent the full range and variability of the observables? Depending on the type of measurement, the technique applied, and the inherent nature of accessible parameters, the consequences are different. In most cases, more readings will allow to assess parameters more reliably; but even a billion data values may not be enough. Knowing the intrinsic variability of Earth systems, overall caution is a must. Sampling of geo-data necessarily means filtering; truly original parameter values are an illusion. That is why sampling crucially requires an expert's insight into the structural and dynamic variety of the target object, which may be called a conceptual model. Likewise, correct interpretation of imperfect data requires a sound understanding of the relevant mechanisms in the Earth system under study. Geoscientists generally deal with systems that are highly underdetermined by the available data sets. Mathematically spoken, they are ill-posed problems.

Besides of undersampling, a dilemma in geoscience is that most statistical tools have been designed for either periodic signals, or for data showing some type of white or Brownian noise, or exhibiting normal (Gaussian) scattering. That is they are actually not applicable to power law data. Parameters like the Nyquist frequency or spectral resolution describe in how far periodic signals can be resolved from discrete series. But for data sets presenting power law characteristics, such assessments become vague. Still, as suitable algorithms seem to be not available, and other parameters — like fractal dimension, Hurst coefficient, power law exponent — allow only rather general conclusions, the traditional statistical tools are continously applied.

In fact, the true variability of an Earth system's observable and the involved averaging process are often unknown. It may therefore be pragmatic and more correct to call observations or instrument readings effective values. The term effective medium is used by modellers to describe a milieu that can not be represented in all its details. In poroelasticity, for instance, — a theory to describe the rheology of porous, fluid filled formations — the shape of pores and matrix bounds is ignored. By taking a macroscopic view, the behaviour of a sedimentary rock under stress and the redistribution of internal pore pressure can still be simulated in a realistic manner and is fairly well understood. Darcy's law, describing how fluid flow depends on the gradient in hydraulic head and on the conductivity of a permeable medium, is a similar example for the widely accepted use of effective parameters. It has excessively been documented that in geoscience the use of imperfect and undersampled data, together with an expert's insight into both the structural variability of the medium and the knowledge of the fundamental governing processes, leads to plausible prognoses and important findings; hence, that effective observables are valuable — and indispensable.

References

Agnew, D. C., 1986: Strainmeters and tiltmeters. Rev. Geophys., 24, 579–624.
Campbell, J., 2000: From quasars to benchmarks: VLBI links between heaven and Earth. In: Vandenberg, N. R., and Beaver, K. D. (eds.), Proc. 2000 General Meeting of the IVS, Kötzting, Germany, 20–34, NASA-Report CP-2000-209893.
Geller, R. J., 1997: Earthquakes: thinking about the unpredictable. EOS, Trans., AGU, 78, 63–67.
Hergarten, St., 2002: Fractals — just pretty pictures or a key to understanding geoprocesses?, this volume.
Kanasewich, E. R., 1981: Time sequence analysis in geophysics. The University of Alberta Press, 3rd ed., Edmonton, 480 p.
Kearey, Ph. and Vine, F. J., 1996: Global tectonics. Blackwell Science, 2nd. ed., Oxford, 333 p.

Leonardi, S., 2002: Fractal variability in the drilling profiles of the German continental deep drilling program — Implications for the signature of crustal heterogeneities, this volume.

Schäfer, A., 2002: Depositional systems and missing depositional records, this volume.

Scrutton, C. T., 1978: Periodic growth features in fossil organisms and the length of the day and month. In: P. Brosche and J. Sündermann (eds.), Tidal friction and the Earth's rotation, Springer, Berlin, 154–196.

Siehl, A. and Thomsen, A., 2002: Scale problems in geometric-kinematic modelling of geological objects, this volume.

Telford, W. M., Geldart, L. P., Sheriff, R. E. and Keys, D. A., 1976: Applied Geophysics. Cambridge Univ. Press, Cambridge, 860 p.

Turcotte, D. L., 1992: Fractals and chaos in geology and geophysics. Cambridge Univ. Press, Cambridge, 221 p.

Wagner, G. A., 1998: Age determination of young rocks and artifacts — physical and chemical clocks in Quarternary geology and archeology. Springer, Berlin, 466 p.

Wenzel, H.-G., 1997: Tide-generating potential of the Earth. In: Wilhelm, H., Zürn, W., and Wenzel, H.-G. (Eds.), Tidal Phenomena, Lecture Notes in Earth Sciences, 66, 9–26, Springer, Berlin.

Ziolkowski, A., 1999: Multiwell imaging of reservoir fluids. The Leading Edge, 18/12, 1371–1376.

Notions of Scale in Geosciences

Wolfgang Förstner[*]

Institute for Photogrammetry, University of Bonn, Germany

Abstract. The paper discusses the notion *scale* within geosciences. The high complexity of the developed models and the wide range of participating disciplines goes along with different notions of scale used during data acquisition and model building. The paper collects the different notions of scale shows the close relations between the different notions: map scale, resolution, window size, average wavelength, level of aggregation, level of abstraction. Finally the problem of identifying scale in models is discussed. A synopsis of the continuous measures for scale links the different notions.

1 Preface

The notion scale regularly is used in geosciences referring to the large variety of size in space and time of events, phenomena, patterns or even notions.

This paper is motivated by the diversity of notions of scale and the diversity of problems related to scale which hinder a seamless communication among geoscientists. The problem is of scientific value as any formalisation establishing progress needs to clarify the relation between old and new terms used.

Each notion of scale is model related as it is just another parameter and as nature is silent and does not tell us its internal model. The notion scale therefore only seemingly refers to objects, phenomena or patterns in reality. Actually it refers to models of these objects, phenomena or patterns and to data acquired to establish or prove models about objects, phenomena or patterns.

Each formalised model needs to *adequately* reflect the phenomena in concern, thus a mathematical model used in geoscience has no value in its own. The diversity of notions of scale therefore is no indication of conflicting models but of different views of reality. Integrating or even fusing models of the same phenomenon but referring to different notions of scale requires a very careful analysis of the used notion *scale*.

As there is no a priory notion of scale, integrating models with different notions of scale may result in new models with a new notion of scale, only being equivalent to the old notions in special cases. This effect is nothing new and well known in physics, where e.g. the transition from the deterministic Newtonian model of gravity to the probabilistic model of quantum physics lead to new notions of space and time.

---

[*] wf@ipb.uni-bonn.de

We distinguish two notions of the term *model*:

1. *model of the first type*: The notion model may mean a *general model*, e. g. a mathematical model, a law of nature, thus refer to a *class of phenomena*, such as the models in physics. This type of model is denoted as *model of the first type* or mathematical or generic model, depending on the context.
2. *model of the second type*: the notion model may mean a *description* of a phenomenon being in some way similar to reality, thus refer to a *specific phenomenon*, such as a digital elevation model. This type of model is denoted as *model of the second type* or specific model.

Modeling means building a model. It therefore also has two meanings: building a model of the first type, thus finding or combining rules, relations or law of natures, e. g. when specifying a process model; and building a model of the second type, thus developing a description of a situation, e. g. the layered structure of sediments in a certain area.

Scale may directly appear in general models models, e. g. a wavelength. It seems *not* to appear in descriptions, thus models of the second type, e. g. in a digital elevation model consisting of a set of points $z(x_i, y_i)$. However, it appears in the structures used for describing a phenomenon or in parameters derived from the descriptions, e. g. the point density of the points of a digital elevation model.

As the notion scale is model related we describe the different notions of scale as they are used in the different steps of modeling in geosciences and try to show their mutual relations. The paper is meant to collect well known tools and notions which may help mutual understanding within neighboured disciplines. The paper tries to formalise the used notions as far as possible using a minimum of mathematical tools.

We first want to discuss crisp and well defined notions of scale all referring to observations. Later we will discuss notions which are semiformal or even without an explicit numerical relation to the size of the object in concern, which however, are important in modeling structures. Finally we will discuss the problems occurring with the notion scale in process models.

2 Continuous Measures of Scale

There exist a number of well definable continuous numerical measures of scale, which we want to discuss first. They all describe data, thus models of the second type. They refer to different notions of scale. But there are close functional relationships, which we collect in a synopsis.

2.1 Scale in Maps

Scale in cartography certainly is the oldest notion. Other notions of scale can be seen as generalisations of this basic notion. The reason is the strong tendency of humans to visualise concepts in space, usually 2D space, e. g. using maps, sketches or in other graphical forms for easing understanding and communication.

Maps are specific models of reality and comparable to observations or data.

The *map scale* refers to the ratio of lengths l' in the map to the corresponding lengths l in reality:

$$s = \frac{l'}{l} \; [1] \qquad (1)$$

It is a dimensionless quantity. A large scale map, e. g. 1 : 1,000 shows much detail, a certain area in the map corresponds to a small detail in reality. Small scale maps, e. g. 1 : 1,000,000 show few details, a certain area in the map corresponds to a large area in reality.

The *scale number*

$$S = \frac{1}{s} \; [1] \qquad (2)$$

behaves the opposite way.

In maps of the earth, e. g. topographic or geological maps, the scale is approximately constant only in small areas. This is due to the impossibility to map the earths surface onto a plane with constant scale.

One therefore has to expect scale to vary from position to position. Therefore scale in general is *inhomogeneous* over space. The spatial variation of scale may be described by a scale function:

$$s = s(x', y') = \frac{dl'(x', y')}{dl(x, y)}$$

This implicitly assumes the ratio of lengths to be invariant to direction. This invariance w. r. t. direction is called *isotropy*. Maps with isotropic scale behaviour are called *conformal*. Conformal maps of a sphere can be realized and are called. The Gauss-Krüger-mapping and the stereographic mapping of the earth are conformal, in contrast to the gnomonic mapping.

In general scale is not only inhomogeneous thus spatially varying but also varies with the direction. Then scale is called *anisotropic*. The ratio of a small length in the map to its corresponding length in reality then depends on the direction α

$$s = s(x', y', \alpha') = \frac{dl'(x', y', \alpha')}{dl(x, y, \alpha)}$$

e. g. the map of the geographical coordinates (λ, ϕ) directly into (x, y)-coordinates by

$$x = \lambda \qquad y = \phi$$

shows strong differences in scale in areas close to the poles as there the y-coordinate overemphasises the distance of points with the same latitude λ (cf. fig. 2 left).

The four possible cases are shown in fig. 1.

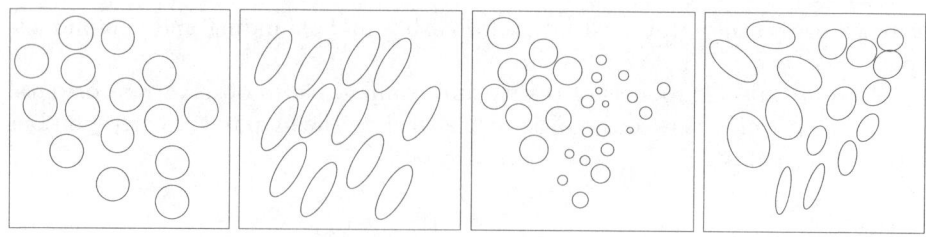

Fig. 1. Homogeneous isotropic scale, homogeneous anisotropic scale, inhomogeneous isotropic scale and inhomogeneous anisotropic scale

We will show how to formalise the concept of anisotropy and inhomogeneity when discussing scale as window size during analysis.

Here we only want to mention the so-called *indicatrix of Tissot* (cf. Tissot 1881, pp. 20 ff and Snyder 1987, pp. 20): It is the image of a differentially small circle. In the case of a mapping from one plane to another it can be determined from the mapping equations

$$\begin{pmatrix} x' \\ y' \end{pmatrix} = \begin{pmatrix} x'(x,y) \\ y'(x,y) \end{pmatrix}$$

with the Jacobian

$$\boldsymbol{J}(x,y) = \begin{pmatrix} \frac{\partial x'}{\partial x} & \frac{\partial x'}{\partial y} \\ \frac{\partial y'}{\partial x} & \frac{\partial y'}{\partial y} \end{pmatrix} \bigg|_{(x,y)}$$

thus the differentially local mapping $(dx', dy')^\mathsf{T} = \boldsymbol{J}(dx, dy)^\mathsf{T}$ Then the differentially small circle in the local (dx, dy)-coordinate system $dx^2 + dy^2 = 1$ is mapped to the differentially small ellipse, namely Tissot's indicatrix, in the local (dx', dy')-coordinate system

$$(dx'\ dy')[\boldsymbol{J}(x',y')^\mathsf{T} \boldsymbol{J}(x',y')]^{-1} \begin{pmatrix} dx' \\ dy' \end{pmatrix} = 1$$

In case of map projections from a curved surface to plane or a curved surface the equations are slightly different. Examples for this detailed scale analysis for two classical map projections are shown in fig. 2.

We will observe inhomogeneity and anisotropy also when analysing other notions of scale and formalise the concept of anisotropy and inhomogeneity when discussing scale as window size during analysis.

2.2 Resolution

Resolution often is used to refer to the scale of gridded data, representing models[2], either in time or space. It refers to the smallest unit of discourse or the least distinguishable unit.

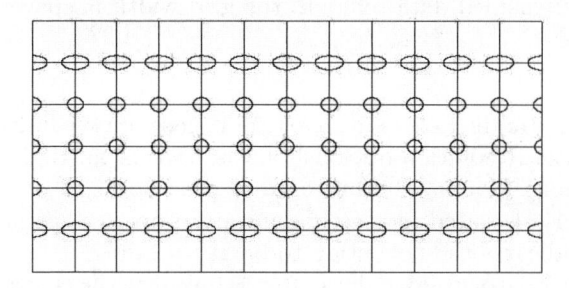

Notions of Scale in Geosciences

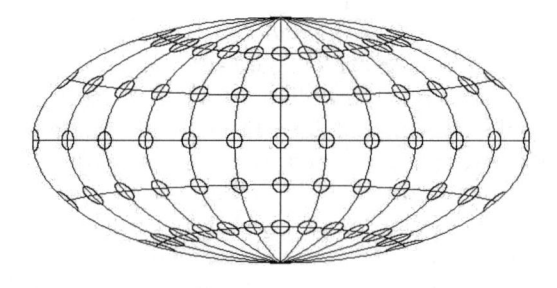

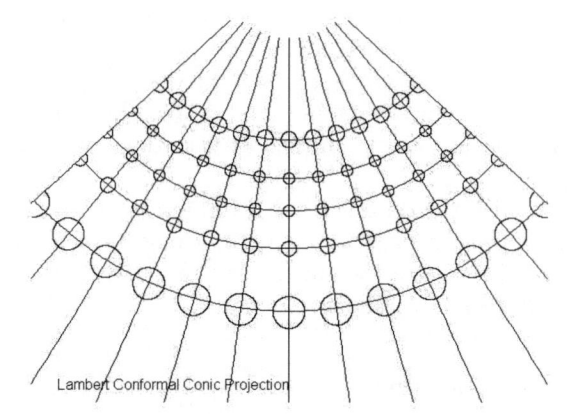

Fig. 2. Map scale for three map projections with Tissot's indicatrices indicating the local scale distortions. Top: equirectangular projection (anisotropic and inhomogeneous). Middle: Hammers projection (anisotropic and inhomogeneous). Bottom: Lambert conic conformal projection (isotropic and inhomogeneous).

It therefore also can be generalised to non spatial concepts, as e. g. when resolving differences of species of plants during their classification.

Resolution of gridded data refers to the grid width in space and time:

$$\Delta x \, [\text{m}] \quad \text{or} \quad \Delta t \, [\text{s}] \tag{3}$$

The grid size Δx standing also for Δy or Δz in case of two-dimensional or three-idimensional data. Resolution obviously has a dimension. High resolution refers to small grid widths, low resolution to large grid widths.

This notion is also used in case of data given up to a certain digit, e. g. the height 12.3 m indicates the resolution to be $\Delta z = 1\,\text{dm}$.

Resolution in case of gridded data often is linked to the envisaged application. It often reflects the accuracy of the model behind the data or for which the data are used.

This implicit assumption leads to confusion or even to severe errors in case the meta information of the data, i. e. the purpose and the instrumentation of the data acquisition is not known or given. The resolution, in the sense of the last digit, therefore should reflect the underlying accuracy, 12.4 km obviously not saying the same as 12.40 km or even 12.400 km.

The sampling theorem inherently limits the reconstruction of the underlying high resolution signal from sampled data. If the underlying signal is band-limited, i. e. is smooth to some degree measured by the wavelength of the smallest detail, the grid width should be adapted to the smallest detail, namely being not larger than half the minimum wave length $\lambda_{\min}/2$. Practically one should take the so-called Kell-factor $K \approx 0.7 - 0.8$ into account, which states that the grid size should be $K\lambda_{\min}/2$ in order to capture the smallest details, This especially is important for signals with isolated undulations, which in case of homogeneous sampling dictate the grid width.

Again resolution may be inhomogeneous and anisotropic: Geological models Siehl 1993 usually have a much higher resolution Δz in z-direction than in (x, y)-direction $(\Delta x, \Delta y)$ reflecting the different variability of the phenomena in planimetry and height. They are highly anisotropic. This can also be seen from the usually small slope, i. e. the small height variations of the terrain w. r. t. to the planar grid size.

On the other hand if the data are not given in a regular grid they may show large differences in density e. g. where undulations in the surface are locally higher. Binary, Quad- or octrees Samet 1989 or triangular irregular networks usually show a high degree of inhomogeneity.

2.3 Window Size for Smoothing

When characterising phenomena by local features or properties being below the minimum size they either are

1. of no interest or
2. aggregated from a higher resolution.

Therefore the *window size* used for determining local properties of a signal often can be a measure for scale.

We want to take the example of smoothing to show one way of formalising inhomogeneity and anisotropy already discussed before and depicted in fig. 1.

Averaging all data within a neighbourhood with a representative radius r leads to a smoothed signal with an inherent scale r. Several weighting schemes are in use: unit weight for all observations leads to a rectangular type window for one-dimensional signals, or top box or cylindrical type window for two-dimensional signals. In case points further away from the centre of the window one might use triangular ore bell-shaped weighting functions in 1D or cone-shaped or bell-shaped weighting functions in 2D. We will derive a measure, denoted as

$$\sigma_x \text{ [m]} \quad \text{or} \quad \sigma_t \text{ [s]} \tag{4}$$

in space or time signals, which is more suitable for describing the scale of a window, as it refers to the *average radius* taking the weighting scheme into account.

Fig. 3. Windows for one and two dimensional signals having the same width, thus leading to the same scale during sampling. As smoothing with box-type and bell-shaped windows with the same width leads to the same degree of smoothing, we obtain the relation between resolution, representing a box-type window, and window-size. Observe, the radius of the window, where the weight is non-zero is not a suitable measure, as it theoretical would be infinite for Gaussian windows.

In case the window is given by the weighting function $w(x)$ or the weights $w'(x)$ with the sum of all weights being 1, the scale can be computed from

$$\sigma_x = \sqrt{\int_x x^2 \, w(x) \, dx} \quad \text{or} \quad \sigma_x = \sqrt{\sum_x x^2 \, w'(x)} \tag{5}$$

depending on whether the window is continuous or discrete. Obviously the scale corresponds to the standard deviation of the function when interpreted as a probability density function.

When using a box filter, thus a filter with constant weight $1/r$ within a window $(-r/2, r/2)$ we obtain

$$\sigma_{\text{Box}} = \frac{1}{\sqrt{12}} r \tag{6}$$

When using a Gaussian bell shaped weighting function $G_\sigma(x)$ as filter the scale is identical to the parameter σ of the Gaussian function (cf. fig. 3, left).

In two dimensions the circular symmetric Gaussian is given by $G_\sigma(x,y) = G_\sigma(x)G_\sigma(y)$ It can be represented by a circle with radius s. Usually a homogeneous weighting is used to derive smoothed versions of a signal, e.g. when smoothing height or temperature data.

Inhomogeneous windowing can be achieved by spatially varying $\sigma(x,y)$. This allows to preserve finer structures by locally applying weights with smaller windows.

As soon as the window needs to be anisotropic, i.e. elliptic there is not one scalar describing the local scale. In case of different scales in x- and y-direction one may use the Gaussian: $G_{\sigma_x \sigma_y}(x,y) = G_{\sigma_x}(x)\, G_{\sigma_y}(y)$ which shows just to use different window sizes in x- and y-direction. This type of anisotropic smoothing only is adequate if the spatial structure is parallel to one of the two coordinate systems. It can be represented by an ellipse Because the anisotropy usually is not parallel to the reference coordinate system one in general needs a rotational term.

This can be achieved by generalizing the scalar value of the scale to a 2×2 matrix $\boldsymbol{\Sigma}$ representing an ellipse: $(x\ y)\boldsymbol{\Sigma}^{-1}(x\ y)^\mathsf{T} = 1$ which in case the scale matrix $\boldsymbol{\Sigma} = \mathrm{Diag}(\sigma_x^2, \sigma_y^2)$ contains the squares of the scales in x-and y-direction reduces to the special case above. Because of symmetry the scale matrix is defined by three values:

1. the larger scale σ_1
2. the smaller scale σ_2
3. the direction α in which the larger scale is valid

We then have

$$\boldsymbol{\Sigma} = \begin{pmatrix} \cos\alpha & -\sin\alpha \\ \sin\alpha & \cos\alpha \end{pmatrix} \begin{pmatrix} \sigma_1^2 & 0 \\ 0 & \sigma_2^2 \end{pmatrix} \begin{pmatrix} \cos\alpha & -\sin\alpha \\ \sin\alpha & \cos\alpha \end{pmatrix}^\mathsf{T} \qquad (7)$$

with the rotation matrix indicating the azimuth α of the most dominant scale.

We now can explicitly represent anisotropic inhomogeneous scale by a spatially varying *scale matrix*: $\boldsymbol{\Sigma} = \boldsymbol{\Sigma}(x,y)$. This makes the visualisation of the different situations of inhomogeneity and anisotropy in fig. 1 objective.

An example for an anisotropic window is given in fig. 4. Observe the form to be similar to a small hill having different width in the principle directions and not being oriented parallel to the coordinate system. This enables to link the window size and form with the intended scale of the analysis, e.g. when analysing landforms.

2.4 Average Wavelength

The notion scale often is used to indicate the average size of objects or phenomena in concern, both in space and time. Though the average size usually is not a

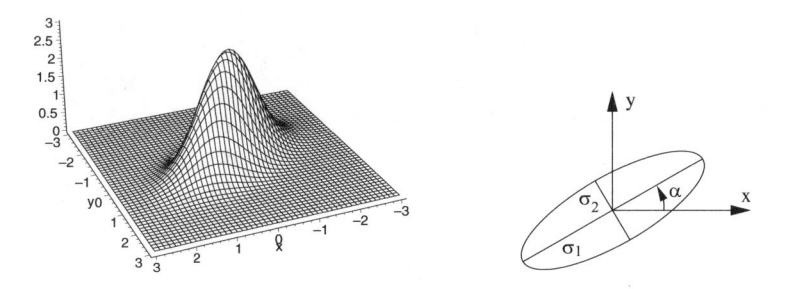

Fig. 4. Left: An anisotropic Gaussian window. It may directly related to the local form of a surface indicating its orientation and its two scales in the two principle directions. Right: Parameters for describing the anisotropic scale.

very precise value and depends on the crispness of the context the value as such can be defined well and thus related to other concepts of scale.

In case we have one-dimensional signals with a dominant wavelength we may identify the scale of that signal with the average wavelength

$$\overline{\lambda}\,[m] \quad \text{or} \quad \overline{\lambda}\,[s] \tag{8}$$

Fig. 5 and 6 show two simulated signals with dominant wavelength, one with average wavelength $\overline{\lambda} = 40$ the other with $\overline{\lambda} = 160$.

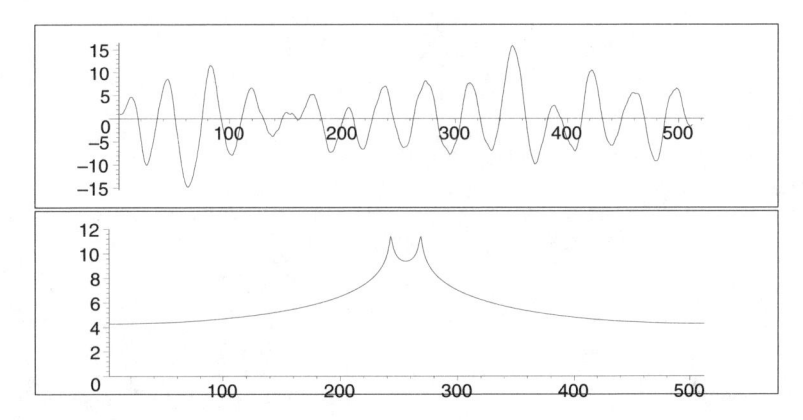

Fig. 5. A signal with short wavelength, $\overline{\lambda} = 40$ (above) together with its amplitude spectrum (below). It is a profile generated using an autoregressive process (cf. below sect. 4.1). The amplitude spectrum shows two peaks, symmetric with respect to the origin, which is in the middle of the graph. The distance between the peaks is large (compared to the following figure) indicating a large frequency, thus a short wavelength.

As the wavelength $\lambda\,[m]$ or $\lambda\,[s]$ of a periodic signal $z(x)$ or $z(t)$ is $\lambda = 1/f$, where $f\,[1/m]$ or $f\,[Hz]$ is the frequency we could also average the frequency. This should take the amplitudes of the periodic signals into account.

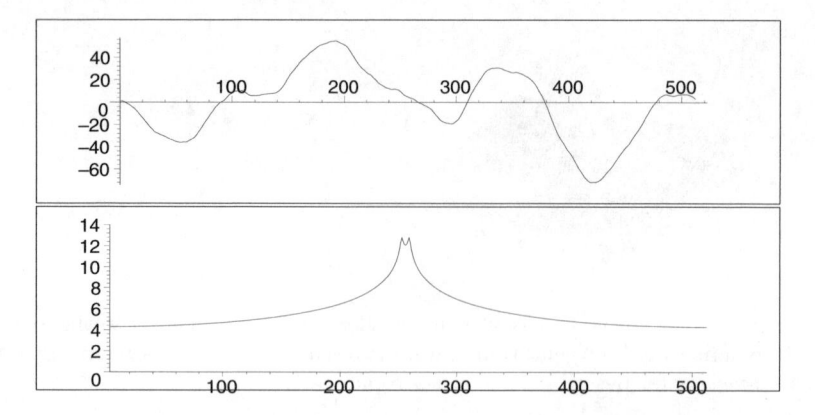

Fig. 6. A signal with long wavelength, $\overline{\lambda} = 160$ (above) together with its amplitude spectrum (below). It is a profile generated using an autoregressive process (cf. below). The amplitude spectrum shows two peaks, symmetric with respect to the origin, which is in the middle of the graph. The distance between the peaks is small (compared to the previous figure) indicating a small frequency, thus a long wavelength.

In case we take the quadratic mean, weight it with the squared magnitude $Z(f)$ of the amplitude spectrum[1], thus with the power spectrum $P_z(f) = |Z(f)|^2$ Castleman 1996 and normalise, we obtain the *effective bandwidth* Ryan et al. 1980 assuming independent and homogeneous noise

$$b = \sqrt{\frac{\int_{-\infty}^{+\infty} f^2 P_z(f)\, df}{\int_{-\infty}^{+\infty} P_z(f)\, df}}$$

It can easily be derived from the original signal *without performing a spectral analysis*:

$$b = \frac{1}{2\pi} \frac{\sigma_z}{\sigma_{z'}}$$

where

$$\sigma_z = \sqrt{\frac{1}{(N-1)} \sum_{i=1}^{N} (z(x_i) - \mu_z)^2} \quad \text{and} \quad \sigma_{z'} = \sqrt{\frac{1}{N} \sum_{i=1}^{N} (z'(x_i))^2}$$

are the standard deviations of the signal and its derivative, assuming the mean of the slope to be zero.

Thus we may derive an estimate for the *signal internal scale*, namely the average wavelength from

$$\overline{\lambda} = 2\pi \frac{\sigma_z}{\sigma_{z'}} \tag{9}$$

---

[1] We use the Fourier transform $F(u) = \int \exp(-j2\pi u x) f(x) dx$ according to Castleman 1996.

Notions of Scale in Geosciences

This is intuitive, as a signal with given amplitude, thus given σ_z, has longer average wavelength if it is less sloped. Of course this is only meaningful in case there is a dominant wavelength.

For regularly spaced data we may determine the standard deviation of the signal and the slope from:

$$\sigma_z = \sqrt{\frac{1}{N-1}\sum_{i=1}^{N}(z_i - \mu_z)^2} \qquad \sigma_{z'} = \sqrt{\frac{1}{N-2d}\sum_{i=d+1}^{N-d}\left(\frac{z_{i+d} - z_{i-d}}{2d}\right)^2} \qquad (10)$$

showing no systematic error if N is large enough, say $N > 40$. One may need to smooth the data before calculating the derivative. The distance d for calculating the slope can be used to apply some (crude) smoothing.

For two-dimensional data $z = z(x, y)$ we can also derive the data internal scale, which now may be anisotropic, as we have discussed above. We obtain a 2×2-matrix for the average squared wavelength for two-dimensional data

$$\overline{\boldsymbol{\Lambda}} = 4\pi^2 \sigma_z^2 \, \boldsymbol{\Sigma}_{z'z'}^{-1} \qquad (11)$$

with the matrix

$$\boldsymbol{\Sigma}_{z'z'} = \begin{pmatrix} \sigma_{z_x}^2 & \sigma_{z_x z_y} \\ \sigma_{z_x z_y} & \sigma_{z_y}^2 \end{pmatrix}$$

containing the variances $\sigma_{z_x}^2$ and $\sigma_{z_y}^2$ of the slopes $z_x = \partial z/\partial x$ and $z_y = \partial z/\partial x$ and their covariance $\sigma_{z_x z_y}$. The wavelength in general will be anisotropic, thus depend on the direction. The maximum and the minimum wavelength show in two perpendicular directions. Their value results from the square-roots of the eigenvalues of $\overline{\boldsymbol{\Lambda}}$. The direction of the largest wavelength can be determined in analogy to (9) from

$$\alpha = \frac{1}{2}\mathrm{atan2}(2\overline{\boldsymbol{\Lambda}}_{12}, \overline{\boldsymbol{\Lambda}}_{11} - \overline{\boldsymbol{\Lambda}}_{22})$$

being an inversion of (7).

As an example, the data inherent scale matrix has been calculated from the ripple image fig. 7. The ellipse visualises the scale in the different directions. The diameters indicate the the estimated wavelength in that orientation. The wavelength of the ripples is approximated by the *shorter* diameter, which appears to be a fairly good approximation. The wavelength in the direction along the ripples would only be infinity, in case the ripples would be perfectly straight, thus the directional variation of the ripples leads to a finite, but long estimated wavelength.

2.5 Smoothness

In contrast to the previous notions of scale which are related to the extension of the object in space or time we also use the notion scale related to the smoothness of the object.

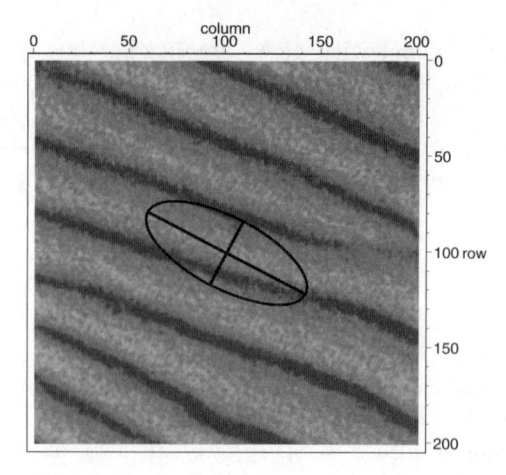

Fig. 7. Wavelengths and orientation estimated from an image of a ripple structure. The derivatives are calculated similar to (10) with $d = 5$. The wavelength of the ripples corresponds to the *shorter diameter* of the scale-ellipse. It locally is a bit larger, as it has been calculated from the complete image shown.

Objects loose crispness in case they are viewed from further away in space or are further in the past. We use the notion of a coarser scale if we talk about objects, in case we are not interested in the details, even if we do not change map scale or the grid size when presenting the object. Objects then appear generalised.

In case smoothness is achieved by an averaging process we may formalise this by linking it to the used window size Δx or Δt.

Assume a surface is rough or noisy with a certain standard deviation σ_n. Smoothing the data with a window size of width σ_x or σ_t reduces the roughness of the noise by a *factor*

$$k = \sqrt{12}\frac{\sigma_x}{\Delta x_0} \qquad \text{or} \qquad k = \sqrt{12}\frac{\sigma_t}{\Delta t_0} \qquad (12)$$

Thus the smoothing factor k is the larger the larger the smoothing is. The degree of smoothing is directly related to the scale parameter induced by the size of the smoothing window. Obviously this is a relative measure and requires to define a basic unit in which σ_x or σ_t is measured. In case of gridded data this is the basic resolution Δx_0 or Δt_0 of the data. Smoothing then would lead to data which have less resolution, thus could be represented with a coarser grid with grid width Δx or Δt.

This relation easily can be derived from analysing the effect of smoothing gridded data $z(x, y)$ with a box filter. Assuming the data to have random errors and have the same standard deviation σ_z, then smoothing with a $r \times r$-window leads to a standard deviation of the smoothed data, assuming $\Delta x_0 = 1$ for this

Notions of Scale in Geosciences

derivation

$$\bar{z}(x,y) = \frac{1}{r^2} \sum_{u=x-r/2}^{x+r/2} \sum_{v=y-r/2}^{y+r/2} z(x-u, y-v)$$

which is the average of r^2 values which therefore has standard deviation

$$\sigma_{\bar{z}} = \frac{1}{r}\sigma_z = \frac{1}{\sqrt{12}\,\sigma_x}\sigma_z$$

where the window-size $\sqrt{12}\,\sigma_x$ of the box filter from (6) is used. Thus the noise is reduced by a factor $k \approx 3.5\sigma_x$.

The relation only holds for noise in the data. In case the data are already smooth and are further smoothed the relations are more complicated.

2.6 Synopsis of Continuous Scale Measures

We now want to find the relations between these continuous measures.
We want to distinguish:

1. Map scale s as dimension-less quantity indicating the zooming factor usually being much smaller than 1, thus indicating a reduction.
2. Map scale number S as dimensionless quantity indicating the reduction factor usually being much larger than 1.
3. Resolution Δx or Δt as spacing of gridded data also the smallest distinguishable unit in the data.
4. Window-size σ_x or σ_t as average radius of function for averaging or smoothing data. The value is identical to the standard deviation of a Gaussian window.
5. Average wavelength $\bar{\lambda}_x$ or $\bar{\lambda}_t$ for signals with a repetitive structure.
6. Relative smoothness k of the data referring to smoothness of original data.

Having six measures we need at least five relations in order to be able to derive one from another. We have the following relations, which only hold in certain contexts, which are indicated in the equations:

1. Map scale and map scale number (table 1, row 3, col. 3)

$$S = \frac{1}{s}$$

This relation is generally valid.

2. Resolution and window size (table 1, row 4, col. 4)

$$\sigma_x \stackrel{s}{=} \sqrt{\frac{1}{12}} \Delta x \approx 0.28 \Delta x$$

A rectangular window with width Δx has the same effect in smoothing as a Gaussian window with parameter σ_x. The notion resolution Δ_x refers to the width of a rectangular window whereas the notion window size refers to the parameter of a Gaussian window of equivalent smoothing effect. Obviously, the *relation* between resolution and window size is made *via the effect of smoothing*, indicated by the s above the equal sign.

3. Map scale and resolution can be linked in case we define the basic resolution Δx_{map} of a map. This may e. g. relate to the printing resolution, e. g. 600 dpi, leading to a resolution of $\Delta x_{\text{map}} = 25.6\,[\text{mm}]/600 \approx 40\,\mu\text{m}$. As this resolution is on the printed map, thus at map scale we have the relation (table 1, row 5, col. 5)

$$s = \frac{\Delta x_{\text{map}}}{\Delta x}$$

where Δx now relates to the resolution in object space, e. g. for a map of scale 1:10,000 printed with 600 dpi we have $\Delta x \approx 40\,\text{cm}$.
This relation is generally valid.

4. Average wavelength $\overline{\lambda}_x$ and window-size σ_x can be linked the following way: In case we smooth *a pure noise signal* the resulting signal will be smooth. Its average wavelength is (table 1, row 6, col. 6)

$$\overline{\lambda}_x \stackrel{sw}{=} 2\sqrt{2}\,\pi\,\sigma_x \approx 9\,\sigma_x$$

The reason is: The Gaussian window function $G_\sigma(x)$ has effective bandwidth $b = 1/2\sqrt{2}\,\pi\,\sigma_x$. The graph in fig. 8 indicates this relation between window size and average wavelength to be plausible.

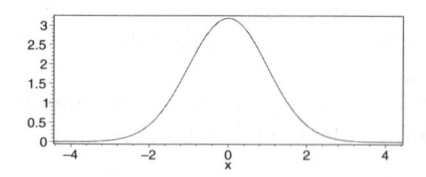

Fig. 8. Gauss function $8\cdot G_1(x)$ with $\sigma = 1$ in the range $[-\sqrt{2}\pi, \sqrt{2}\pi] \approx [-4.5, 4.5]$. It appears as one wave with length $\lambda = 2\sqrt{2\pi} \approx 9$.

Obviously, the *relation* between wavelength and window size is made *via the effect of smoothing white noise*, indicated by the *sw* above the equal sign.

5. Relative smoothness k and window size σ_x can be linked if we refer to some *basic resolution* Δx_0. It is defined as that resolution, where the data show *most detail*, thus show *highest roughness* or have original uncorrelated measurement errors. Then we have an increase k of smoothness or a reduction $1/k$ of noise. Thus we have (table 1, row 7, col. 6)

$$k \stackrel{sw}{=} \sqrt{12}\,\frac{\sigma_x}{\Delta x_0}$$

The definition of this notion of scale refers to gridded data, but can also be used for non gridded data, e. g. when reducing the density of data, then indicating the ratio of the average point distance after and before the data reduction.

Obviously, the *relation* between relative smoothness and window size is made *via the effect of smoothing white noise*.

Notions of Scale in Geosciences   31

Obviously we have three scale measures being proportional, namely resolution, window size and average wavelength, related by

$$\overline{\lambda}_x \stackrel{sw}{=} 2\sqrt{2}\,\pi\,\sigma_x \stackrel{s}{=} \sqrt{\frac{2}{3}}\,\pi\,\Delta x \quad \text{or} \quad \overline{\lambda}_x \stackrel{sw}{\approx} 9\,\sigma_x \stackrel{s}{\approx} 2.56\,\Delta x \approx 1.3\,(2\,\Delta x)$$

The factor 1.3 indicates, that waves of length λ do not need 2 samples but by a factor 1.3 more (table 1, row 6, col. 5). This corresponds to experience that the actual resolution Δx_a of gridded data is by a factor K less, thus $\Delta x_a = \Delta x/K$ than the Nyquist theorem Castleman 1996 in case of band limited signals would say, namely Δx, the Kell factor K which is between 0.7 and 0.8.

Using these five relations we can derive all possible relations between the numerical measures for scale. They are collected in the following table. The guiding relations are shown in bold face letters.

Table 1. Relations between numerical notions of scale. Each row contains the name, the symbol and the dimension of each measure of scale. In the 6 last columns each measure is expressed as a function of the others, the relations either hold generally, refer to smoothing $^{(s)}$, or refer to smoothing white noise or detail of highest roughness $^{(sw)}$. Constants: $c_1 = 1/\sqrt{12} \approx 0.288$, $c_2 = 2\sqrt{2}\pi \approx 8.89$, thus $c_1^{-1} \approx 3.45$, $c_2^{-1} \approx 0.112$, $c_1 c_2 \approx 2.56$, $c_1^{-1} c_2^{-1} \approx 0.390$.

	1	2	3	4	5	6	7	8
1	notion	dim.	s	S	Δx	σ_x	$\overline{\lambda}_x$	k
2	map scale s	1	—	$\frac{1}{S}$	$\frac{\Delta x_{\text{map}}}{\Delta x}$	(sw) $\frac{c_1 \Delta x_{\text{map}}}{\sigma_x}$	(sw) $\frac{c_1 c_2 \Delta x_{\text{map}}}{\overline{\lambda}_x}$	(sw) $\frac{\Delta x_{\text{map}}}{\Delta x_0}\frac{1}{k}$
3	map scale number S	1	$\frac{1}{s}$	—	$\frac{\Delta x}{\Delta x_{\text{map}}}$	(sw) $\frac{\sigma_x}{c_1 \Delta x_{\text{map}}}$	(sw) $\frac{\overline{\lambda}_x}{c_1 c_2 \Delta x_{\text{map}}}$	(sw) $\frac{\Delta x_0}{\Delta x_{\text{map}}} k$
4	resolution Δx	$[m], [s]$	$\frac{\Delta x_{\text{map}}}{s}$	$\mathbf{\Delta x_{\text{map}}\,S}$	—	(s) $c_1^{-1}\sigma_x$	(sw) $c_1^{-1} c_2^{-1} \overline{\lambda}_x$	(sw) $\Delta x_0\,k$
5	window size σ_x	$[m], [s]$	(sw) $\frac{c_1 \Delta x_{\text{map}}}{s}$	(sw) $c_1 \Delta x_{\text{map}}\,S$	(s) $\mathbf{c_1 \Delta x}$	—	(sw) $c_2^{-1} \overline{\lambda}_x$	(sw) $c_1 \Delta x_0\,k$
6	average wavelength $\overline{\lambda}_x$	$[m], [s]$	(sw) $\frac{c_1 c_2 \Delta x_{\text{map}}}{s}$	(sw) $c_1 c_2 \Delta x_{\text{map}}\,S$	(sw) $c_1 c_2 \Delta x$	(sw) $\mathbf{c_2 \sigma_x}$	—	(sw) $c_1 c_2 \Delta x_0\,k$
7	relative smoothness k	1	(sw) $\frac{\Delta x_{\text{map}}}{\Delta x_0}\frac{1}{s}$	(sw) $\frac{\Delta x_{\text{map}}}{\Delta x_0}\,S$	(sw) $\frac{1}{\Delta x_0}\Delta x$	(sw) $\frac{\sigma_x}{c_1 \Delta x_0}$	(sw) $\frac{\overline{\lambda}_x}{c_1 c_2 \Delta x_0}$	—

Comments We want to discuss the relevance of the other relations between the different measures of scale, this will make their dependency on the contest explicit:

- map scale/window size $s \stackrel{sw}{=} 0.28 \Delta x_{\text{map}}/\sigma_x$ (table 1, row 2, col. 6):
  In order to visualise the details in the map at the required resolution Δx_{map} and in case one has taken averages of the objects details with a window of size σ_x, then one should take a map scale of approx. s. The map resolution in this case needs not be the printing resolution, but the minimum size of details in the map to be shown separately.
  E. g. if the map should show a resolution of $\Delta x_{\text{map}} = 1$ mm and the smoothing window size at object scale is $\sigma_x = 10$ m one should choose a map scale $s \approx 0.281/10,000 \approx 1 : 35,000$).
  Vise versa, a map showing details at a separation distance of Δx_{map} can be assume to have been generated by averaging object details with a window of size σ_x

- map scale/average wavelength $s \stackrel{sw}{=} 2.56 \Delta x_{\text{map}}/\overline{\lambda}_x$ (table 1, row 2, col. 7):
  This relation is similar to the previous one: If details have wavelength $\overline{\lambda}_x$ and one wants to visualise them in the map at the required resolution Δx_{map}, then one should take a map scale of approx. s.

- map scale/smoothing factor $s \stackrel{sw}{=} \Delta x_0/\Delta x_{\text{map}} k$ (table 1, row 2, col. 8):
  In case one filters a signal with basic resolution Δx_0 with a smoothing factor k and wants to show it in a map with a resolution Δx then one should choose a scale s.
  Vices versa (table 1, row 6, col. 3), if one wants to show a signal with basic resolution of Δx_0 in a map at scale s with a resolution in that map of Δx one should apply a smoothing with factor k.

- resolution/smoothing factor $\Delta x \stackrel{sw}{=} \Delta x_0 k$ (table 1, row 4, col. 8):
  This important relations states, that a signal with resolution Δx_0 that is smoothed with a smoothing factor k, can be resampled with Δx. Vice versa (table 1, row 7, col. 5) in case one wants to decrease resolution from Δx_0 to Δx one needs to smooth with a smoothing factor k, being realizable by a smoothing filter with window size $\sigma_x = 0.28 \Delta x_0 k$ (table 1, row 5, col. 8).

Obviously, the relations hold under specific conditions, thus one needs to take the individual context of the derivation of the relations into account.

3 Discrete Measures of Scale

Taking the average size of an object as measure for scale already contains uncertainty in its definition. Though the notion of a window size is well defined the average size of objects usually is not. Moreover there are applications where even the average size of a set of phenomena is not adequate to represent the *scale* on which the phenomena are handled. Generalisation of river networks or the interpretation of sets of bore holes in no way can be *adequately* characterised by the above mentioned measures. Therefore there exist other notions of scale used in geosciences. Two are collected in the following.

Notions of Scale in Geosciences

3.1 Level of Aggregation

One of the main properties of spatial objects is the ability to partition them into parts or aggregate them into large units. In object oriented modeling these relations are represented in aggregation or is-part-of hierarchies.

Examples are manifold: The aggregation level of spatial political units as states, counties, towns or of time domains in geology.

In hydrology e. g. the scale of the modeling might be reflected in the size of catchments. Taking an average size of catchments would correspond to the window size mentioned above. However, there is no unique way to average. Taking the minimum size of the catchments as characteristics reflects the resolution. Again, there is a conceptual difference as the compartment sizes are not multiples of the minimum size, unless the basis for the definition are raster data. Also maximum sizes or the size range may be used to characterise scale of modeling.

3.2 Level of Abstraction

An even more general way to perceive scale is the property of an object to be a special case of another one or to generalise a set of objects. In object oriented modeling these relations are represented in specialisation or is-a hierarchies.

Examples again are manifold: species on biology are classified in a is-a hierarchy, phenomena in meteorology or geophysics are also classified in specialisation hierarchies. Obviously referring to a more general class of objects is using a coarser scale in reasoning, while referring to very specific objects is using a fine scale in reasoning. This type of scale selection often is used to guide modeling and may lead to specific types of mathematical relations, e. g. different types of partial differential equations.

As the different scale types — aggregation and abstraction — interfere and directly relate to the used ontology, one could use the term *ontological scale* to indicate the scale depending on the different types of hierarchies. Though it is a notion of scale with quite some degree of uncertainty in its definition, it covers all legends of maps including the abstraction hierarchies of different object types and at the same time the spatial extent of the shown objects reflected in the different signatures, enabling a differentiated description: e. g. large cities are shown with an approximation of their boundary, small are shown with a circle as symbol, principal cities are given an underlined name, other cities not underlined.

The relations between the very different notions of scale are by far not understood enough to guarantee clear specifications for data acquisition, to allow formalisation of simulation models or to enable generalisation for generating maps on demand.

4 Scale Defining Parameters in Models

The notions discussed so far all refer to description of reality either in observations or data or in models of the second type. Scale also appears in models of the first type aiming at explaining observations. Fourier techniques are the classical tool to represent and analyse scale.

The goal of this subsection is to show the difficulty of predicting scale in models.

Mathematical models, thus models of the first type, contain variables, such as pressure or velocity, which change over space and time, and parameters which are constant. Moreover, boundary conditions heavily influence the behaviour over space and time. The notion of scale of a phenomenon is depending on the parameters and the boundary conditions. Thus without a specification of the parameters or the boundary conditions a scale of the phenomenon cannot be told. In nonlinear models even small changes of the parameters may change the type of the generated process. Unfortunately there is no canonical way how a scale of the phenomenon is related to the parameters or the boundary conditions. Moreover, the notion of scale in models shows the same variety as in data.

Two examples want to demonstrate the difficulty of identifying some scale in models: both are recursive processes, where the current state x_i depends only on one or a few, say K, previous values

$$x_i = f(\{x_{i-k}\}, p_j), \qquad i = 1, \ldots, I; j = 1, \ldots, J; k = 1, \ldots, K.$$

The J parameters p_j specify the underlying the process.

4.1 Parameters in Autoregressive Models

In the first example we only deal with a linear relationship, which allows to use the powerful tools of linear system theory. Autoregressive (AR) models are one of the best understood models for one dimensional signals, such as time processes or terrain profiles. The underlying theory is found in Box and Jenkins 1976, Lücker 1980.

We want to demonstrate the ability to generate models for time series with a given spectral content and the difficulty to interpret the parameters. The models shown here explicitly contain a stochastic component to generalise specific signals.

The most simple AR-model is given by the recursive generation:

$$x_i = ax_{i-1} + \epsilon_i$$

of the sequence $\{x_i\}, i = 0, 1, \ldots$ The parameter a influences the general appearance of the process. The sequence $\{\epsilon_i\}$ of random values defines the actual form of the sequence. The distribution of ϵ may be characterised by its standard deviation σ_ϵ. Fig. 9 shows two samples of such AR-processes, one with $a = 0.95$ one with $a = 0.99$. Obviously the parameter a controls the appearance, larger a lead to longer dependencies, thus subjectively to a larger scale, thus to higher smoothness. However, the roughness of the signals does not allow to talk about a typical wavelength, i.e. representative value for the scale.

Notions of Scale in Geosciences

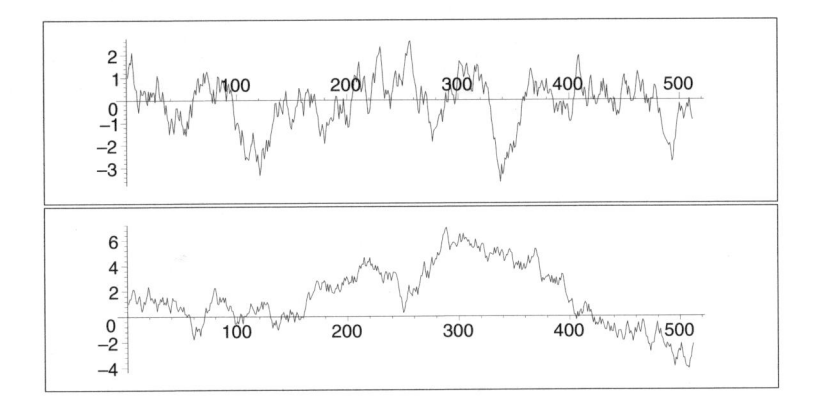

Fig. 9. Top: AR-process with $a = 0.95$. Bottom: AR-process with $a = 0.99$. In both cases $\sigma_\epsilon = 0.25$

Signals with a predominant wavelength can be achieved with higher order AR-process:

$$x_i = \sum_{k=1}^{K} a_k x_{i-k} + \epsilon_i$$

where K is the order of the AR-process. Again the parameters a_k control the behaviour.

For second order processes we are able to choose parameters a_1 and a_2 which lead to signals with one predominant scale, i.e. wavelength. In the top of the three figures 5, 6, shown above, and 10 we observe three signals with typical wavelengths 40, 160 and 40 resp.

Obviously the first two profiles differ in their dominance of a specific wavelength, the last profile is more noisy than the first. The bottom of the three profiles shows the theoretical Fourier spectrum[2]. One can easily identify two peaks. Their distance is inversely proportional to the wavelength.

This shows, that it has been quite reasonable to identify the average wavelength $\bar{\lambda} = \frac{1}{\bar{u}}$ derived from the average frequency as natural scale of a signal. In the examples, shown so far, however, we are able to predict the scale from the parameters of the underlying process.

If we now have a look at the parameters a_1 and a_2 in the following table they do not immediately show us the structure of the signals. In this special case of a linear system, one however is able to tell the properties of the signal, namely its roughness and average periodicity.[3]

---

[2] The theoretical Fourier or amplitude spectrum is only depending on the process parameters a and σ_ϵ and given by $\sigma_\epsilon/|1 - a \, \exp(j2\pi f)|$ Lücker 1980. The empirical Fourier spectrum is given by the Fourier transform of the signal.

[3] By looking at the roots of the polynomial $P(z) = 1 - a_1 z - a_2 z^2$.

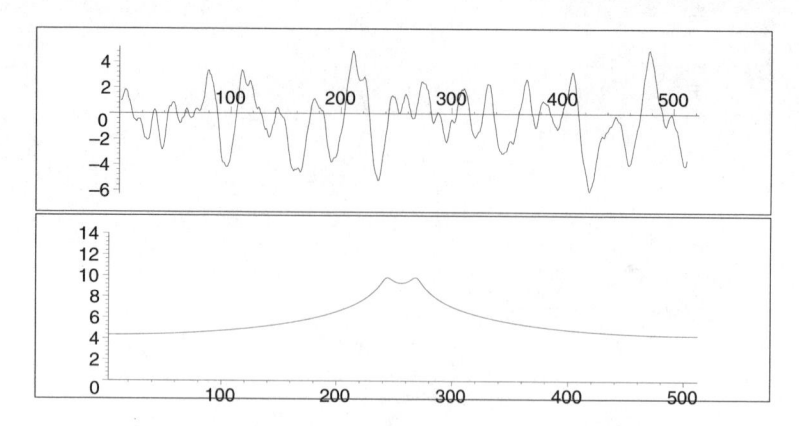

Fig. 10. Top: A rough AR-process with short wavelength. Bottom: Amplitude spectrum.

Profile	fig.	r	λ	a_1	a_2
1	5	0.99	40	1.9556229	-.98010000
2	6	0.99	160	1.9784734	-.98010000
3	10	0.95	40	1.8766078	-.90250000

We are also able to generate signals with *two* specific scales using higher order AR-processes. Figure 11 shows two signals each containing two different wavelengths, namely 10 and 40 (first two rows) and 6 and 80 (last two rows), together with their Fourier spectra. The two dominant wavelengths are clearly visible in the signals, and the Fourier spectra tell which wavelengths are present.

Again the parameters a_1, a_2, a_3 and a_4 collected in the following table do not tell anything immediately about the structure. Linear systems theory allows to predict the behaviour of the signal also in this case.[4]

Profile	fig.	r_1 λ_1 r_2 λ_2	a_1	a_2	a_3	a_4
1	11 top	0.99 40 0.99 10	3.5574765	-5.0928217	3.4866827	-.96059601
2	11 bot.	0.99 80 0.99 6	2.9684734	-3.9188887	2.9094008	-.96059601

The example intensionally is chosen such that there is *no unique natural* scale or average wavelength.

In this case theory is available to analyse the signal in detail and understand its behaviour. This theory is restricted to one-dimensional signals generated by a linear system. The theory for two-dimensional signals already is quite difficult and cumbersome.

4.2 Scale in non-linear models

The second example wants to demonstrate the inability to predict the behaviour of a nonlinear system, containing only one parameter.

---

[4] Again by analysing the roots of a polynomial $P(z) = 1 - \sum_k a_k z^k$.

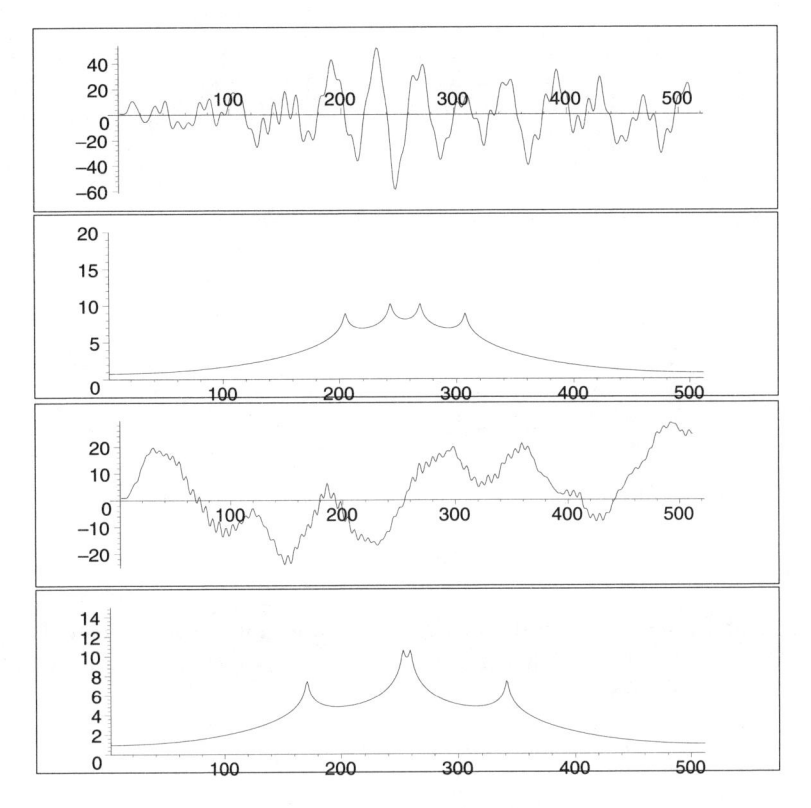

Fig. 11. Two profiles, each with two wavelengths and their theoretical Fourier spectrum

The well known logistic map Kneubühl 1995

$$x_k = a\, x_{k-1}(1 - x_{k-1})$$

is fully deterministic. It shows all types of behaviour, fixed points, periodic sequences as well as chaotic i.e. non predictable sequences for $a > 3.5699\ldots$ (cf. figure 12). Though this sequence is well studied it has the flavour of non linear systems which need to be studied individually and only in special cases allow prediction of their behaviour without simulation studies. This is in contrast to one-dimensional linear systems, which are theoretically fully understood.

The examples of this section give a clear message: In case the system can be adequately modelled as a linear system, one should exploit the power of linear systems theory. This especially holds, if reality is modelled phenomenologically or only simple linear physical laws are involved. Even in case the approximation is crude, it helps making predictions. However, all real systems are likely to be nonlinear, which makes the theory based definition of a scale very difficult.

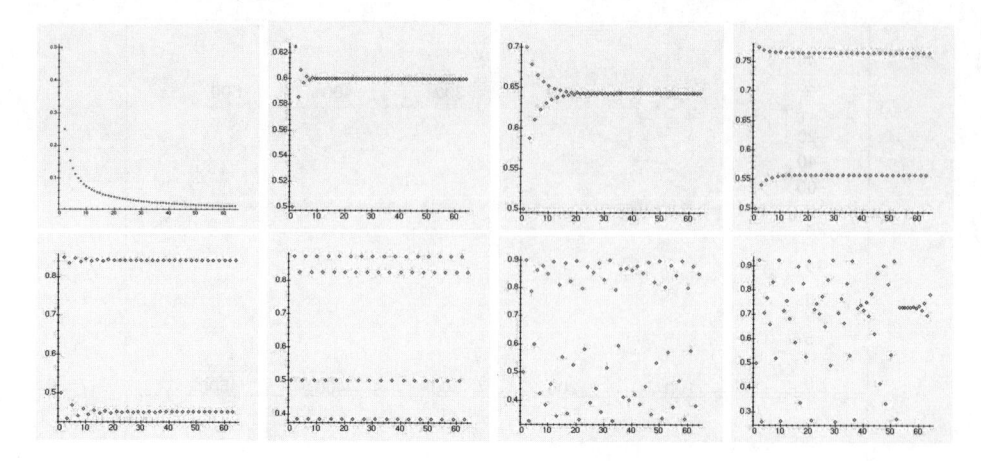

Fig. 12. Sequences of the logistic map with $a = 1$, 2.5, 2.8, 3.1, 3.4, 3.5, 3.6 and 3.7

4.3 Omni-scale models

Real data often show no dominant scale, i.e. one dominant wavelength. They often show a continuum of wavelengths, i.e. the height $z(x)$ is composed of sine waves of all frequencies:

$$z(x) = \sum_k h_k \sin(kx + p_k)$$

where the phases p_k and especially the amplitudes h_k depend on the frequency.

Terrain data have been analysed by Mandelbrot 1980. In case of terrain profiles Frederiksen et al. 1985 often found a linear dependency

$$\log h_k = c - \alpha \log k$$

between the logarithmic amplitude h_k and the logarithmic frequency k. Thus often all frequencies are present in terrain profiles, there is no specific wavelength or scale in the data.

The slope α is large for smooth terrain, e.g. in the range of $\alpha \approx 2.5$, the slope is small for rough terrain, e.g. in the range $\alpha \approx 2$ or less.

5 Conclusions and Outlook

The paper discussed the various notions of scale resulting from the diversity of the disciplines within the geosciences. They refer to data and to models and appear in several degrees of crispness.

There seem to be basically two different crisp types of notions of scale in use:

1. scale as the ratio between model and reality. The only application of this notion is the scale in *map scale*.

2. scale as a measure of the extent of an object on space or time. This holds for nearly all notions, quantitative and qualitative.

The other notions qualitatively refer to the size of the data set in concern. There are quite some close relations between the crisp notions of scale, all referring in some way to linear systems theory. Their use definitely is promising and helps getting a first insight into geoprocesses.

In contrast to this coherence, however, there are only few crisp relations between the more qualitative notions, which hinders a smooth communication. Taking the complexity of geoprocesses into account one needs to discuss how far a crisp notion of scale actually carries research and at which steps within modeling more detailed concepts need to support the description of the inherent structure of natural processes. These concepts definitely refer to the used qualitative notions of scale in some way, but they certainly will be more adequate.

References

Box, G. E. P.; Jenkins, G. M. (1976): *Time Series Analysis*. Holden-Day, 1976.

Castleman, K. R. (1996): *Digital Image Processing*. Prentice Hall Inc., 1996.

Frederiksen, P.; Jacobi, O.; Kubik, K. (1985): *A Review of Current Trends in Terrain Modelling. ITC Journal*, 1985-2, pp. 101–106.

Kneubühl, F. (1995): *Lineare und nichtlineare Schwingungen und Wellen*. Teubner Studienbücher Physik. Teubner, 1995.

Lücker, R. (1980): *Grundlagen Digitaler Filter*. Springer, 1980.

Mandelbrot, B. (1982): *The Fractal Geometry of Nature*. Freeman and C., San Francisco, 1982.

Ryan, T. W.; Gray, R. T.; Hunt, B. R. (1980): Prediction of Correlation Errors in Stereopair Images. *Optical Engineering*, 1980.

Samet, H. (1989): *The design and Analysis of Spatial Data Structures*. Addison-Wesley, Reading, Mass., 1989.

Siehl, A. (1993): *Interaktive geometrische Modellierung geologischer Flächen und Körper. Die Geowissenschaften*, 11(10-11):342–346, 1993.

Snyder, J. P. (1987): *Map Projections – A Working Manual*. US Government Printing Office, Wash., 1987.

Tissot, A. (1881): *Mémoire sur la Représentation des Surfaces et les Projections des Cartes Géographiques*. Gauthier-Villars, Paris, 1881.

Complexity of Change and the Scale Concept in Earth System Modelling

Horst J. Neugebauer[*]

Section Geodynamics, Geological Institute, University of Bonn, Germany

Abstract. The concept of scales is based on the spatial perception of phenomena. It is rather appropriate to address and to analyse the composite nature of phenomena. Consequently scales will be utilised as well for comparing various strategies of conceptual modelling, such as empirical analogues, effectual objects and relational states. This discussion becomes illustrated by means of an adaptable example and the corresponding scaling behaviour. Finally the statistical analysis of scales reveals both the limitations of phenomena and their fundamental potential for change. The universal nature of change becomes evident only through an extended perspective which is comprehending small and large-scale properties assigned to the common representative scales.

1 Nature Survey

Modelling is a principal process to access phenomena in nature. This becomes evident as soon as we reflect the various attempts of gaining knowledge about natural phenomena and from the consideration of their equivalence as well as their limited continuance. Those attempts reflect our concern with nature like an early orientation through semantic classification, the recognition of specific principles and their application for practical use, up to the desire for a prediction of risks and future developments of natural systems. It is a common and well known experience that in any case of specification or observation we rely on some kind of concept, an idea, view or reminiscence on some practical experience to begin with. That means however, whatever we sense, address or practise with respect to nature will implicitly reflect an adjacent and complementary *context* as well; it is like a specific image we take from reality; this represents a merely subtle distinction or a limited experience. Modelling has thus to do with defining, describing and comparing objects with respect to some part of the context. Therefore we are relying on an universal but concealed principle of measuring, our perception of space. The situation becomes quite simply illustrated by the phenomenon of a fault. It is an obvious example, which can be represented most strictly by the relative spatial displacement of sites. It is a fundamental convenience and of great practical use, dealing with distinguished objects instead of a non-distinguished net of potential relations. Therefore, in practice we usually discern the separated natural context of the addressed phenomena while we

---
[*] neugb@geo.uni-bonn.de

follow ideas, express ourselves in semantic terms, using graphs or performing some direct action. Although we denote objects by fixed conventional terms, our personal associations are not unique at all among us. In science we thus follow conventions for reproducible measurements and the use of comparable standards of analysis. Further on we developed alternative theories for the representation of phenomena, like for instance, the equivalent concepts of waves and particles in physics or waves and rays in seismology, which both lead to sound results, yet representing a variety of experience and perspectives with matter and energy. Last but not least, we may meet the detached context of objects again through many unexpected *re-actions* to our mental or implemented concepts. Nature is responding seldom straight and linear, but mostly rather unexpected and oblique. In this sense our technologies touch more and more the unknown context and become thus less a consequence of science but increasingly a reinforcing promoter for targets of modern sciences.

Our concern will therefore be at first to trace back the boundaries of conventional objects to their phenomenology to some extend. Spatial scales serve as a profound and powerful tool for that purpose. They should help to identify the "condensed experience" against the complementary context of natural phenomena. Scales are simply geometrical sizes or measures, such as length, area, volume or other properties attributable to them. They will be used successfully to analyse the interdependency among phenomena. For this purpose it should be sufficient to follow our well known attempts in Earth science. Clearly, the consideration of objects through the perspective of the discrimination from their dedicated context causes important consequences: For any phenomenon we find both, an associated *inner context* and the indicated *outer context*. While the inner context for instance is leading to the characteristic properties, the quality of the object; the outer one will be identified by its functioning and consequently by its possible physical origin. Under this perspective we become interested further into the mental and sensual motivation which might lead to the statement of a phenomenon. Usually a systematic comparison drawn from a large pool of observations provide the characteristic criteria of similarity in order to specify a phenomenon. This resemblance within some basic particulars serves as a general principle of spatial analogies under the signature of popular terms. The motivation for defining objects appears thus being already the most fundamental concept of modelling. Whenever phenomena are observed over sequential states, we meet the *change* of objects either in terms of a declining appearance or even as a sequence of fluctuations on a lasting sequence of states. We might again easily recognise, that considering the change of phenomena as denoted parts of a phenomenology appears to be another perspective of the principle of modelling and a model by itself. Models, respectively objects might thus come into existence either in form of semantic terms, cartoons, a sequence of numbers or even a set of mathematical equations as well.

Dealing with natural phenomena through characterising their behaviour and specific features, we face immediately the established views of scientific disciplines. What seems therefore to be distinguished from each other appears rather as mere aspect or part of our perceived phenomenology. Empirical observations and analogue experiments are as selective as for instance the so called exact representation through a differential equation, or even the performance of a statistical probability analysis. Every modelling approach has the target to attain an outermost knowledge about phenomena, to this extend we have some kind of equivalence of methods. However, when we need to know information about the possible predictability of processes in the future, for example the occurrence of an earthquake or the potential for a landslide, neither the empirical knowledge nor the deterministic quantification can lead to a proper answer independently. Modelling the statistical probability for instance, requires the conceptual and dynamical experiences as a precondition. For a better understanding of complexity of earth systems, we should therefore examine the different steps where we loose or even overlook information in our scientific approach of natural phenomena. So, discussing the purpose, the way and the implications of modelling, it will be of primary importance to reflect at first about an universal concept or measure to distinguish the complexity of pretended objects and their variability. Such a fundamental tool are the spatial scales, the basic constituents of our perception of space. On the basis of the scale concept we might be able to compare model concepts among themselves and conclude on their final benefit.

2 Abstraction of Experience

The definition of phenomena appears to be the most fundamental procedure of modelling, the formation of semantic terms. Whenever we denote particular phenomena, the introduced names represent symbols of structures, features, sets and states or even changes which have been sensed or alternatively measured. Since science is deliberate to tighten the correspondence between symbols and their meaning, we finally reach a growing diversity of the disciplinary terminology with even more specific symbols, codes and notions. Our first modelling aspect thus concerns the principle of accumulating "evidence" for a certain phenomenon around a typical dimension or scale. This procedure might concern a repeated comparison of observations, based on the correlation of distributed samples from corresponding stages of recurrent processes. This deduced basic quality of notions can easily be recognised from such simple examples as clouds, creeks, faults and others; each case resembles in some particulars between the corresponding observations but remain otherwise unlike: The procedure of naming objects is thus based on obvious similarities and a terminology has therefore the quality of analogues. The entire definition is sized implicitly through our relative perception of space. We might share this experience now by means of an example, for instance, the phenomenon and its terminus, *fault*.

The common sense of the concept is expressed by an original citation of a dictionary of geological terms: *A fault is a fracture or fracture zone along which there has been displacement of sides relative to one another parallel to the fracture.* A closer look on the statement reveals instantly the two inherent and complementary characteristics representing a fault: an outer and an inner

layer model of large-scale
master fault

100 m

fault on the reference
scale of observance

10 m

small-scale structure of
clay smear and fault
block fragmentation

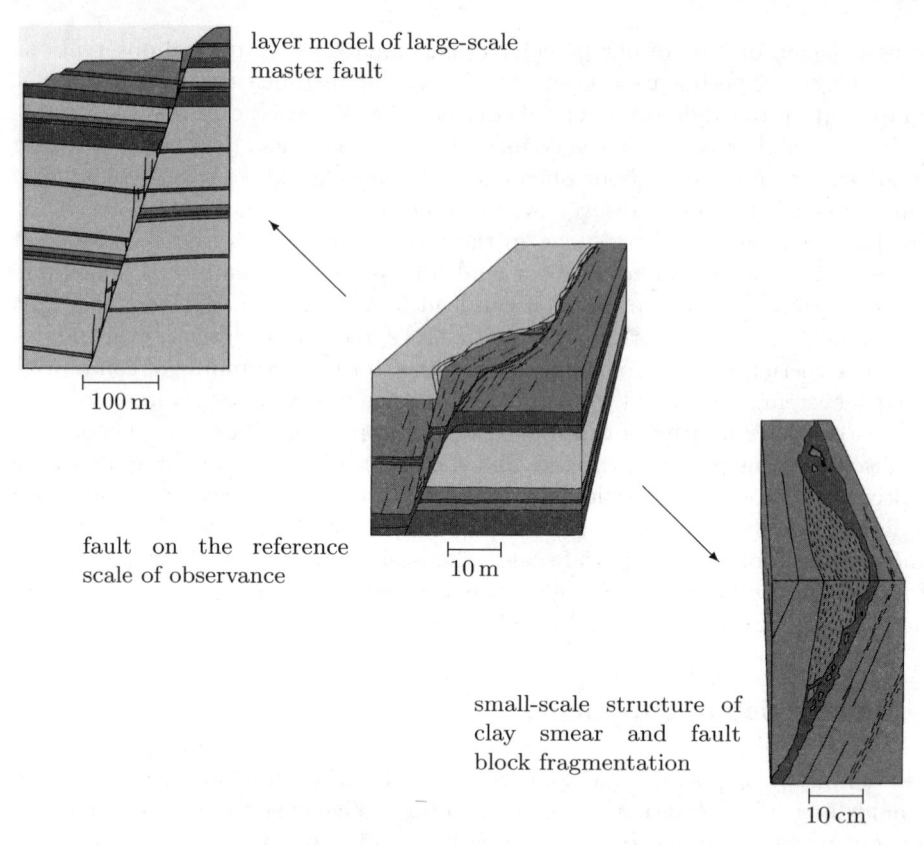

10 cm

Fig. 1. Fault structure on different levels of spatial resolution. Scale equivalent properties cover an approximate range of five orders of magnitude.

context, which together build a reciprocal spatial relation, like that illustrated in Figure 1. The inner aspect of the notation *fault* is related to the fracture itself, especially the geometrical manifestation of small-scale fragmentation and roughness as shown on the lower right side. Following the dictionary again, we distinguish a horizontally striking fault line of a certain length and shape as well as a corresponding vertically dipping fault plane as illustrated through the reference scale sketch in the central part of the cartoon. We recognise a seamless conjunction to the outer context, which appears already in the scarp and the aperture of the fracture. This outer context comprehends both material and genetic aspects of a fault. Thereafter, displacements in the past accumulated a memory indicating a historical record, the large-scale mean offset combined with the relative sense of movement as indicated on the upper right. These structural features reveal thus in terms of the relative positions of the sides the possible displacement modes like: vertically *normal* or *thrust* and horizontally *strike-slip*. Usually a fault is only an element of an even wider fault complex or a system as for instance a fault basin. Here, associated sedimentary deposits might expose another sequence of marker significant for the tectonic history of the entire fault population. To this extend the terminus fault is a comprehension of addressed structural features and physical qualities; in terms of the spatial reference of the terminology to associated scales, it is the merge of a selected hierarchical inner and outer context. With this respect, the term fault denotes already a complex semantic model or analogue of a considerably variable extent. For our consideration however, it seams still more expedient to intensify the aspect of representative spatial scales of objects into this discussion. We assumed for instance, that the mentioned features correspond approximately to about four orders of magnitude in their geometrical dimension of space. However, our experience shows an additional multitude of phenomena on respectively even smaller and larger scales associated with the definition of a fault and thus in extension to the introduced inner and outer context. So the fault is an entire concept, which is part of an external mechanical environment and the equivalent tectonic frame up to the dimension of the earth crust on a global scale. Yet, on the other end of smaller scales we have an extension of the fracture system through particular fault segments over the roughness of the fracture surface down to a significant concentration of micro-cracks and crystalline dislocations beneath the surface of the respective fault blocks. With respect to the associated scales of the phenomenology, we deal with an entire hierarchy related with the concept of a fault. Further on, we find the originally addressed phenomenon *fault* being an inherent part of an entire phenomenology, in the sense of a linked network around a reference scale of observance without obvious limitations within adjacent scales. Against this background of relations, i.e. the frame of complexity, any limitation of the phenomenology is selective, has restricted competence and is thus non-unique. However, it will illustrate the basic operation of modelling as a conceptual processing of experience within the spatial frame of perception. Modelling means therefore the condensation of our experience either in the limited analogue of semantic terms, as mentioned before. Or alternatively establishing a limited relationship between particular phenomena in order to develop a conceptual model based on their reciprocal dependence, like for instance a mathematical balance. Such conceptual models typically result from empirical correlation of repeated sequences of observations, from theoretical considerations or, especially in Earth science, from analogue experiments in the laboratory.

A conceptual model, such as our example of a fault in Figure 1, is build around a predominant reference scale of the manifestation, which is usually representing the characteristics of observance. While the corresponding phenomenology, the inner and outer context of the primary fault approach remains more or less unattended, i.e., it depends mainly on the chance for an acquisition of additional information or related ideas. So, how are we arranging ourselves with this peculiar situation, the need of deciding on appropriate bounds for the definition of objects? Obviously, this question became rarely addressed explicitly, usually we rather ask: what are the implications and where do they come from? In a common sense we escape instead towards the limitations of the causality principle. The implications are thus linked with parts of the implied inner context, while causation partly replaces elements of the outer context. Therefore small-scale phenomena are shrunk down to mere attributes expressed either in qualitative notions or as quantitative averages. The large-scale context however, is usually addressed being equivalent to the ultimate causation of a phenomenon, either through reasoning or in terms of some specific determinant. All together, this is the procedure of "sense-making", the making of an object, e. g. named fault. Although there is a mutual consent about the essentials of an object, with this procedure the scale problematic becomes concealed within the object properties and the presumed causation; it is thus changing its significance. On this fundamental level of approach, no demand is raised for an explicit integration of scales, respectively up- or down-scaling of influence. Such a condensation of the phenomenology into a causal relationship of a conceptual model will on the other side allow only a weak appraisal of quality. Modelling on a conceptual semantic level preponderates the purpose of representation and reconstruction rather than prediction. The latter requires even more specific confinements, as we will discuss later on.

Another obstacle, that impedes progress of modelling experience in this category comes from the handling of time. Our every day experience of time is adjusted to an *universal time* in the sense of Newton. Time is associated with standard cyclic changes as a scale unit as is generally known. Thereafter time is the basic variable to trace ongoing changes of phenomena typically by corresponding time series. With reference to our complex example of a fault, absolute time is missing, we only find structural evidence for the past changes in the entire context. However, those signatures are only characteristics in the sense of a relative chronological order, and there a multitude of signs appears, indicating changes all over the hierarchy of related phenomena and their assigned scales. This leaves us with the general question on the representative relevance of historical time marker in order to decipher detailed changes during the geological history. Although we might be able to distinguish between displacements of fault-sections from the integrated record of change by the sediment deposits, any proper displacement time record however remains unidentifiable and any conceptual approach appears finally to be non unique. Even for a most obvious case of a fault, we are therefore left with a complex accumulation of interfering changes on an entire zoo of spatial scales of the fault phenomenology. The predominance of spatial information and the variance of scales leads usually to a rather flexible practice of conceptual scientific cartoons

3 Consistency in Space

Let us now consequently look for some general aspects of conceptual modelling, in order to describe objects like our example of a fault quantitatively from various perspectives. In the first case, we only make use of a perceptible characteristics, i.e., its obvious *quality*, which appears to be associated with its mere *manifestation* in space. Intuitively we related the term fault to a selected scale of the perceptible phenomenology. This seems to be equivalent with the qualitative choice of an obvious scale of the system like that in Figure 1. With this specification we consequently express the phenomenon quantitatively by means of a spatial reference system. Computer science provides suitable strategies for those quantitative representations of generalised geo-objects. At that level of modelling our objects are completely represented by merely spatial elements, such as the *topological*, *metrical* and *geometrical* characters of the phenomena. All are basic attributes and sufficient for a quantitative modelling of objects. The topological relations such as the existence of a relative offset of the adjacent fault-blocks along the fracture correspond to the category of a characteristic quality. While the manifestation of the fault is expressed through a proper spatial measure, indicating the amount of relative displacement of fault blocks. In addition, the position and the orientation of the particular fault in space, according to a geometrical reference system needs to be specified. Figure 2 gives a simple example representing a fault as a spatial object: the *geometrical parameter* representation. The geometry of objects is composed by means of primitive solids and their corresponding operations. The technical design of such representations follows naturally operational aspects related to their use in optimised information systems. Generally, the required operational aspects and the systematic storage of these object representations is supported by the narrow range of the geometrical information necessary and of course, the inherent principle of similarity between the considered objects.

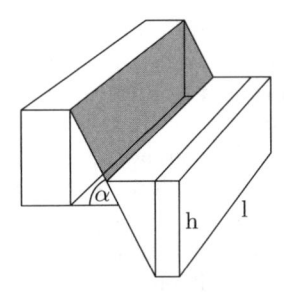

Fig. 2. Geometrical representation of a fault by the offset of fault-blocks in space.

The spatial abstraction of geo-objects serves frequently as a basic frame for numerous considerations, interpretations and even speculations. Any of these reasonings is equivalent with the immediate ascent of the context, respectively the multitude of scales involved. The idea to deduce a kind of a historical record of displacements from a spatial fault pattern seems quite attractive. However, even so a limiting state for the beginning of fault displacements might be proposed, any reconstruction of some "time-record" in the sense of the historical displacement process requires the knowledge of the entire record of the displacement rate. That means, we need to know the information on the change of the passed displacements or that of the corresponding context throughout the history. As this detailed record will usually not be available, any approach to an "inversion" of a represented spatial situation remains extremely unconstrained and appears thus rather arbitrary. The only control we might be able to perform is obviously a mere geometrical check up on the consistency of stages in space. Such a geometrically *balanced reconstruction* of an assemblage of displaced blocks needs however additional information on physical and chemical changes and the redistribution of matter. We neither have a fair chance to invert the causes of the displacements associated with faulting quantitatively from the observed spatial representation. Although we are usually dealing with a scale-limited pattern of observations, we are alternatively missing the entire scaling information of the former dynamic regime of sequential faulting. This is not a general critique of the fundamental concept, but it rather reveals the limitations of the spatial approach from a solid, quantitative viewpoint. We recognise that modelling change from the past on the spatial basis has only rather weak constraints on the complex relationship of faulting and its scaling behaviour. Thus an entire spectrum of experiences of statistical, experimental and observational nature are usually hypothetically attributed and associated to the geometrical approach. Causal relationships are therefore not yet comprehended within the spatial representation including topology and geometry.

4 Principles of Conservation

The fundamental, but rather restrictive spatial model concept became considerably improved through the incorporation of at least major aspects of causality. Only through the control of specific causes and their implications, which are both particular components of the previously addressed context, the spatial manifestation of a quality advances definitely to what we call an *object*. Whenever we like to accomplish how, for example, the displacements along a fault evolve, we have to exceed the spatial concept in many respects. Geometrically we will introduce an outer bound of space in order to make a proper statement of the question. Physically fault blocks need to be characterised by their physical state and mechanical properties, the shear zone itself by frictional parameters and others. However, it is most important to refer the model approach to a solid principle of constraints providing sufficient control on the calculated response of the object to a specific causation at any state. That means, such unique conditions are for instance met through an accurate conservation of matter, forces and energy during the entire process of displacement of the fault model throughout the whole system considered. This effort is still not generally acknowledged in Earth sciences. However, it marks clearly the transition between spatial description and interpretation on one side and the opportunity for a quantitative approximation of the caused function of phenomena by means of a reproducible analogue on the other side. Especially our abilities of approaching the responding function of an object quantitatively reveal the state of knowledge, whenever prediction is required. Returning to our fault example, quantification of rupture formation, propagation and fault-slip are common. Seismology is specially interested in the radiation of elasto-dynamic deformation according to displacement or traction discontinuities across an internal fault surface. Modelling requires thus a representative equivalent of a finite fault plane within an external space; the fault is representing some predominant spatial scale of the system, like that indicated in Figure 3. Deformed and crushed rocks of the fault gauge, shown in Figure 1,

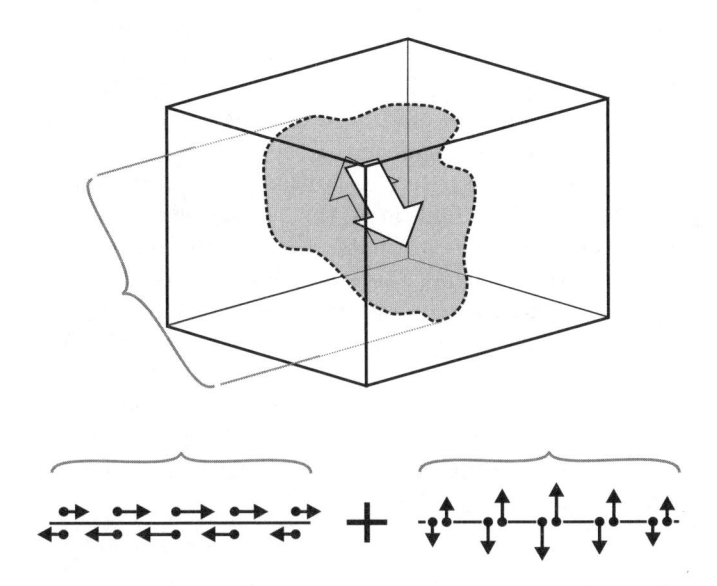

Fig. 3. Body-force-equivalent of a representative source volume for slip-motion.

exhibiting the sub-scales of the fault zone and the possibility of heterogeneous displacements, are comprehended by an effective slip within a plane of zero thickness. Because the thickness of the apparent displacement discontinuity initiating waves is almost always far less than the wavelength of the recorded seismic radiation. In the same sense the rock adjoining the fault zone, corresponding to the superior scales of the system, is described by isotropic effective elastic constants and an average density. The theoretical concept allows thus to state a constraint on the accelerations, body forces and traction or alternatively the displacements, acting throughout a representative volume, representing faulting. This approximation determines the displacement at an arbitrary point in space and time in terms of the quantities that originated the motion, the equivalent forces. This links the abstract model approach with observable features imposed through the dynamics of faulting. According to the radiated seismic wave pattern, the source of fault-slip can even be replaced by body-force equivalents for discontinuities in traction and displacement within a reference volume. The requirements necessary for the equivalence of forces are quite remarkable and contradicting to common practice in tectonics. Representative forces are usually presumed as a single couple of opposing components in the direction of fault-surface displacements, like indicated in the lower left of Figure 3. However, additional forces, single couples or linear combinations of these alternative extremes, directed normally to the fault plane are necessary in order to meet a relevant equivalence to fault-slip. It seems quite obvious now, modelling the response-function to faulting becomes a composition between the apparent spatial characteristics of a fault and the specific causation of slip by a dynamic source function. The design of the latter requires an incorporation of additional scales through the restriction to a definite finite fault plane, respectively an equivalent volume and the considered environment of the system. Herewith the object fault attains a more specific, improved quality from the related physical properties and their symmetry. Finally, the modelled function of faulting becomes more or less determined by the intensity and the time dependence of the specific source function. This is underlined by the expressed measure of fault-slip, which is called the seismic moment. It is the product of the fault area, the average slip, and the shear modulus. Even so this model concept is capable to quantify in particular dynamic faulting and seismic wave radiation, there remains still the need to access size and frequency of faulting episodes through the course of time. This peculiarity of the phenomenology is however missing in a steady-state model of fault function. This comes from the particular bounds we introduced by defining an object. As we have seen earlier, it will only be possible on the cost of the context. Looking for a representation of faulting as a particular event in space and time, we need a revision of our modelling perspective towards a transient situation of an even more complex dynamic system.

5 Return of Context?

Can we find a suitable modification of our modelling concept, which allows to follow faulting in a self-organising manner? As a sporadic appearance of limited fault-slip in an even more complex system of continuous change? Seismology and fracture mechanics indicate fluctuating areas of temporary fault-slip across larger fault zones separated by quiet intervals. Those events of faulting might thus be understood as specific states of ongoing changes in a complex system. The continuity of change has to be presumed to happen all over the hierarchy of scales, particularly within the fault zone and its neighbourhood. The temporal variation of the respective dimension of the active slip area can easiest be controlled by means of a model with a local scale approximation. This modelling concept is appropriately visualised as an aggregate of slider blocks combined through various spring elements, as shown in Figure 4. Herewith each block within the model frame represents a local scale fraction of potential state of faulting. Every local slip in the system is relaxed, that means distributed with respect to its local neighbourhood. The global driving force established on the upper boundary is opposing the frictional forces at the lower boundary, which is supporting the individual blocks. Fault failure allowing local slip motion of blocks is bound to a static or dynamical threshold, where shear stress exceeds a critical value. Principally one is repeating the balance of forces and their relaxation according to the system properties for each interval of change. This can be achieved either by means of a set of discrete differential equations or alternatively using explicit rules for the local change, which are typical for cellular

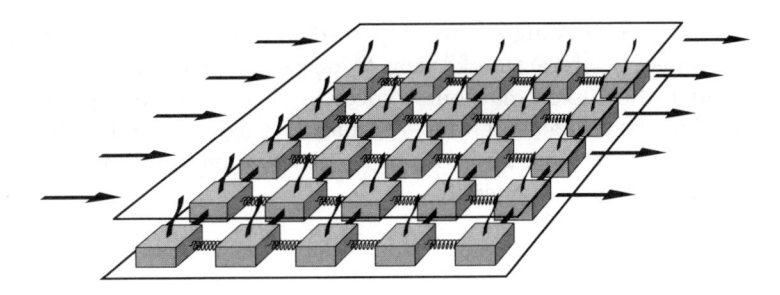

Fig. 4. Spring-block model for fluctuating fault-slip areas.

automata. The most essential and remarkable change in this modelling strategy is to assume, that the friction of an individual block recovers as soon as it comes to rest. That means, the opposing forces of external driving and internal frictional resistance become regenerated after each interval or iteration. The forces become thus interactive. Opposite to the previous concept, the momentary local state of force balance is no longer predetermined entirely from outside; it is rather evolving internally step by step for each block. The idea meets very well the observation of random obstacles to faulting like spatial roughness, dynamical friction or even long term chemical crystallisation and healing. The mutual influence between driving and frictional forces at the blocks in addition to the local relaxation rules as represented by springs and leaf springs, become thus critical for the apparent scales of the temporal slip events. Hence, every locality is now not only conditional upon the forces and their neighbours, but even so a condition for them. These features together are essential for the appearance of irregular fluctuations of slip motions over the entire model plane. That means, one of the forces is transcending the other temporarily on an respective area of fluctuating spatial scales. Turning back to our previous general discussion of the context and its implication on the change of objects, we easily recognise now: quality and manifestation of faulting undergoes progressive change, while the induced function becomes now an integral counterpart of the causation. To this extent we have overcome our assumption about the permanence of objects; we find the view of ongoing change of phenomena within their context more appropriate for a better understanding of our complex natural environment. The capability of this modelling concept appears primarily not of interest for a particular case of faulting. Although we improved the relational context of modelling, practically we will not at all be able to access the required concurrent changes during a specific tectonic situation sufficiently well. Nevertheless, the more precise rendering of the coupled phenomenology through the local scale approach enables investigations corresponding to the long term organisation of change, for instance in a complex fault system. We are placed into a position now, where we know about the fundamental background to reproduce the scaling behaviour of faulting. In particular we become able to ascertain the character of complexity from the statistics of modelled slip events; further on we might analyse the recurrence time of slip motion or study conditions for a spatial-temporal clustering. This kind of modelling approach provides a rather new insight into the role of self-organising interaction. A key to access complexity is therefore the scaling behaviour of complex relational processes. The theory of a multiple-scale approach provides basic ideas of what complexity might be and how it should be identified. An adequate analysis of proper observations could possibly reveal a necessary appraisal of the concept.

6 Encounter Complexity

In order to distinguish between different concepts of modelling and quantification, we might finally introduce an even more systematic frame of reference for the representation of phenomena compared to their context, i.e., their omnipresent scale dependence. The attendance of the scientific community to adopt the multiplicity of scale dependent constituents in the consideration of our mental and material phenomena seems not yet well established. Hence, the performance of modelling complexity is not necessarily compatible among disciplines and mostly inadequate because of the different degrees of approximation practised, as we have seen from the previous discussion. From our fault example in Figure 1, we kept in mind, that the entire hierarchy of associated scales is far from being merely an addition to an object. Scales are rather a measure equivalent to the complexity of phenomena and thus an integral part of our capacity comprehending objects. In this sense they indicate substantial modifications in the definition or modelling of objects. Therefore a comparison of different modelling concepts, their predictive abilities and implications becomes necessary. This can be achieved by the use of only few universal categories for the characterisation of phenomena, with a flexible incorporation of the corresponding context. While the context can still be taken equivalent to the addressed hierarchy of scales as before. Following our experience of similarity on different scale levels, we adopt a minimum of four conceptual modes or categories for the entire characterisation. They serve as defining devices for the degree of complexity of objects attached by accretion such as our previously discussed example of a complex fault scenario. The four particular categories as well as their changing correspondences are introduced and compiled in Figure 5. Hence, we separate describing objects in three major classes with respect to the expressed degree of complexity of a phenomenon.

In order o distinguish a phenomenon from its natural environment in a most fundamental manner, requires at least two, at the maximum four basic categories. Of prominent importance is the salient *quality* of a phenomenon, which means its addressed characteristics. We might think here either of a single quality or even a comprehensive set of qualities which remain the same in all its diverse manifestations considered. According to our fault example, the most prominent quality of a fault is the *fracture*, respectively a characteristic discontinuity of the material matrix. Associated with this characteristic is a potential performance of a specific task or the possible achievement of a concrete goal, that means its potential function. With respect to spatial and temporal aspects function might

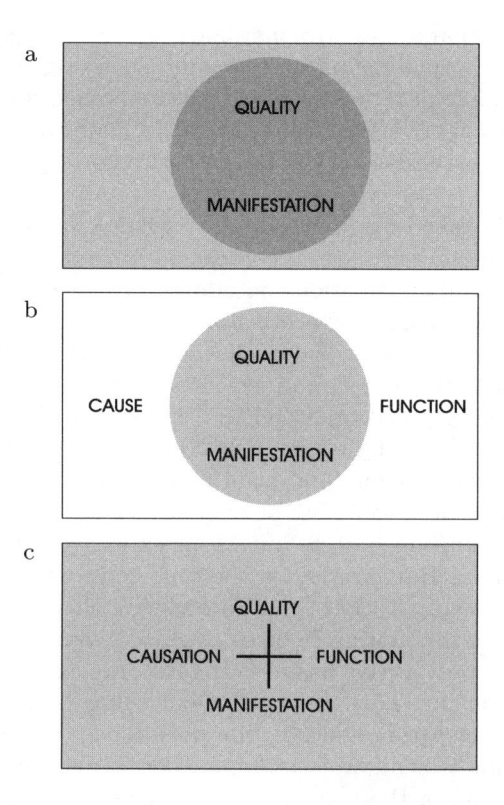

Fig. 5. Categories characterising modelling concepts: empirical analogues above, effectual objects in the middle and relational states below.

be a complete and spontaneous process, but equally well somehow distributed in space and successive in time. The function of a fault corresponds clearly to its potential performance of *displacements* along the fracture plane, which is in accord with our previous definition. Thereafter we find both, the quality and the potential function being fundamental for the third category, the *manifestation*, which is the way a phenomenon becomes evident in our spatial perception. The manifestation reflects thus the present state of a characterised object, whatever development it might have passed before. Alternatively we might understand it as the comprehension of an entire record of past stages, including the present moment of observation. From the exposition of the fault in Figure 1 we remember an entire set of spatial features of the fault structure. Together they comprise the achieved *dislocation* of blocks, representing the indicated manifestation. Finally, the function and the characteristics in association with the manifestation of a phenomenon provides the basis for any reasoning onto its *probable causation*. For our example, we might be able to derive some principal conditions upon the probable causes of the dislocation from the relative mode of displacements. For instance, it means the relative *spatial position* of exemplary fault blocks compared to the direction of gravitational body-forces. Our approach thus does not only comprise the formal definition of a fault, such as given above. It rather extends the conceptual fault representation by a comprehension of the context on the potential integration of various scales. Thus, the four categories demonstrate their capacity for an extensive review of the entire correspondence, as we will discuss later on.

We have now established a quite elementary schema for our representation of phenomena compared to their natural context on different levels of complexity. With a quality and its manifestation, we address a phenomenon spatially. We fix some reference scale in space which is linked to a certain property. Through the association of a function and its specific causation, the hitherto spatial phenomenon becomes determined as an object, having an origin and implications. The objects becomes thus accomplished by certain bounds from the context, respectively their corresponding scales. So far any modelling procedure, either semantic, conceptual, experimental or quantitative, reproduces obviously a common process of a conceptual defining of manifestations. This process not only meets the task of delimiting objects out of an extensive spectrum of internal and external conditions. But even so, we become confronted with the momentary character of the manifestations, that means with their potential for sudden change. To account for fluctuating changes of objects, we have to overcome linear determinism. Our perspective has to become reorganised with respect to the degree of relative correspondence among the four categories. We have to accept relationship between causation and function both ways, forward and backward. This of course, will not be achievable without affecting quality and manifestation simultaneously. Because of the transparency of correlation, the four categories are organising themselves. Objects become relational events of change. The introduction of the four corresponding categories is thus additionally exposing the need to compensate the obvious loss of context by denoting objects somehow. Although the problem of complexity emerged from the discussion of the multiplicity of scales assigned to a certain phenomenology, the schema of Figure 5 comprehends even so the more general aspect of inner and outer contexts with respect to the manifestation. From this point of equivalence, we might finally compare easily a variety of three fundamental families of modelling concepts qualitatively.

7 Competence of modelling

We will consider now the different traditional degrees of abstraction of phenomena: as spatial analogues, determined objects and temporary events. For this purpose we will make use of the four previous categories according to our mental and practical experience. The first modelling family, as indicated in Figure 5a, is rather dominated by the manifestation of a characteristic phenomenon, what we called a quality. While its particular function as well as the possible causation remain inherent or cryptic. This valuation of properties is emphasising the empirical nature of the approach, which is based on the fundamental similarities of numerous examples and cases, we discussed before. The result is a deficient low level of differentiation of objects, because it is lacking definitive constraints from the context. Therefore the entire process leaves plenty of space for opinions and views, it is supporting a remaining indifference of the qualified phenomena. Consequently, we meet another obstacle with the historical development, accumulated in the manifestation. As discussed previously, time appears only as an inherent historical aspect. It can hardly be reconstructed from structural features precisely; only fragmentary relation comes from assigned traces of available past physical and chemical changes within the phenomenology. It is not surprising , that under the historical aspect of time, the origination of manifestations and thus the appropriate course of the entire function remains rather hypothetical. So any unwrapping of the historical record will be strongly dependent on the examination of additional models, the consistency of the available information and the compatibility of the various sources. Basically this kind of modelling is thus solely adopting *spatial information*, selecting the major scales, as we learned from our example in the beginning. On the other hand, this entire complex spatial correspondence is a prevailing procedure for any further degree of differentiation.

The forthcoming model family becomes determined through the expressed character of the context, respectively the scales or properties addressed in addition to the spatial aspect. A phenomenon becomes thus constituted as a scale limited object through a specific correspondence between the typical function in response to a specific causation, as symbolised in Figure 5b. The explicit role of the causal relationship leads consequently towards a much more distinguished but passive quality of the manifestation. i. e. the defined object. The self-contained character of objects becomes effectuated through various sources like experiments, measurements and calculations. They are highly differentiated from their environment and therefore obey only very specific and parameterised information about their context. This procedure to configurate objects quantitatively leads to a completely *determined relationship*. Hence the time aspect is no longer a historical one. Time becomes reversed into a forward directed sequential time in response to the causation. Herewith we exceed the transition from the former flexible concepts of spatial analogues towards the precepts for the figuration of objects through a specific dynamical behaviour. We reach thus the potential for a reproducible quantification of the causal relationship. The idea of objectivity claims an unique characteristics as far as possible, i. e. an object is supposed to reproduce the same response to the same prescribed source exactly and independent of the moment of time. Obviously we have to pay for this desirable ideal of an universal characterisation. The consequences are unexpected changes of our objects due to the pretended exception from of the interacting confining context. However, this step allows a linear quantitative representation for well constraint conditions according to the limited characteristic spectrum of scales. Established laws for the conservation of mass, energy and momentum express this self-contained constraints for the consistency of the models. However, with this modelling concept in terms of definitive objects the above obstacle becomes imposed. Although, we might be able to characterise objects by calculating their response for a representative set of scales, we wont meet the experienced sudden alterations of objects during time. Hence, on this level of modelling we clearly miss, how objects come into existence and why they disappear again. If we allow "unexpected changes" of our phenomena, i. e. changes independent from the assumed causation, they will no longer be reproduced through the presented family of modelling. Too many scales with their links to the system have been eliminated by the derivation of effective parameters and constant averages.

Whenever we experience randomly fluctuating phenomena, we might easily recognise the limitations of the previous conceptual approach of objects. Although this approximation finds reasonable short term benefit in technology, it principally fails in the analysis of change. For instance, we do not find any a priori scale for rupture length on a fault or for the size of clouds during their life cycle, etc. Changing phenomena are presumably supported through an ongoing interaction of related phenomena across an entire hierarchy of scales, what we called previously their conditional context. Consequently, the relative nature of those relational events require fundamentally different modelling techniques. They should comprehend the interdependency between major system variables on the basis of the *constitutive relations* among these components. In particular the full interaction among each of the constituents, as indicated in Figure 5c, becomes necessary to succeed in reproducing changing properties, respectively scales. As indicated earlier, the correspondence of the four categories allows change in the manner of self-organisation. Clearly, the diversity of variables has to be accommodated quantitatively by suitable methods, which allow appropriate resolution of their spatial dimensions and their variability in time. While this is necessary for every variable itself, the main focus of modelling becomes now the linkage between the variables and parameters of the system. Usually a specific reproduced state of such a system, is indicated through a *critical threshold* of one variable only. It is worthwhile to notice, that there are various successful techniques approximating such coupled complex systems. Usually the rules of interaction with the respective neighbourhood are based on supporting grids, which are small compared to the scales of the commonly considered events. The entire schema is enclosed within the global frame of the model boundary conditions, affecting the various states of the system. The most straight way is the choice of a small resolution for the entire numerical schema. Whenever we have information on the scaling rules of certain phenomena, we might solve the problem separated, using variable splitting or so called nesting-techniques. Not surprisingly, the understanding of time will change again with the degree of complex coupling. While we leave behind the view of a particular dominated response by a causation, the change of the complex system becomes considerably limited to the very moment of stepwise interaction. As change occurs locally at every supporting point, the coupled system has the ability of spontaneous variation on an arbitrary size. In the limit, "time" shrinks just to the length of an interval of change in the deterministic sense. The basic approach to complexity

Complexity of Change and the Scale Concept in Earth System Modelling   57

leads thus towards a pattern of *conditions*, which originates phenomena as transient objects on a local and instantaneous supporting basis. This concept of a *relative reality* is clearly opposing the previous view of the function of stationary objects. Now the texture of phenomena appears through and through *relational*. Consequently, the presented correspondence of the categories holds generally for every local sampling point and its vicinity, thus the schema, presented in Figure 5c, resembles symbolically the entire global concept equally well. The salient features we accomplished with this concept are obviously the internal degrees of freedom, organising changes of the system through the interaction among the respective categories. Especially the scales of the modelled states of the system are no longer predetermined and the prediction of change attains a fundamentally new quality now.

8 Statistics of Scales

Finally we found that scales are generally relative features which have to do with our relative perception of space. So far we have identified scales with a representative linear measure of the spatial manifestations of a certain quality or even with a temporary state of an entire phenomenology. On the other side, modelling revealed that change of phenomena requires the peculiarity of a complex *conditional relationship* to become generated. From this viewpoint we may conclude the statistical signature of change being equally well indicative for the mere quality addressed by scales and the assignable complexity of the phenomenology. Therefore a statistical analysis of scales might give more evidence about the *nature of change*. However, as there is not a thing in nature supposed to be without change, our knowledge about change becomes critically depending on the model concepts we adopted earlier for its exploitation.

Geological structures are, for instance, fully packed with hierarchies of scales. These structures are indicating a historical record of properties from changes in the past. Yet, we are not able to decipher those scales reliably without a consistent model concept, respectively some significant statistical scaling laws. In the absence of physically sound concepts a compilation of data is the only way forward to evaluate their statistical significance. A first step in the direction of a kind of trustworthy link between observed properties and their alleged origin is the interpretation through regular change. It corresponds to the representation of phenomena according to the causal model concept, shown in Figure 5b. Somehow we are used to the concept of steady objects, which are specified spatially and through their conceivable function in response to a specific causation as described earlier. This concept leads to characteristic scaling laws. In terms of statistics, a rather constraint distribution of scales appear to be representative for those objects. As the generated phenomena are presumed to be dominated directly by specific causes, they will respond within a limited range of properties only. We found earlier the reaction being predetermined through selective limitations of the object design. This is most frequently the case for experimental testing devices in the laboratory or with our practical applications in technology and their theoretical representations. Quantitatively the spectrum of scales is completely determined by the presumed constant parameters of the corresponding differential equations. The assigned phenomena are thus indicated statistically as *scale-dependent*. Under the addressed relationship between specific causes and consequences the function of objects appears predictable within the limits of the restricted concept, e. g. as long as the variability of the natural system parameters remains negligible compared to the determined one. We call that the presumable existence of characteristic or *representative scales*, such as length, size or volume. Following this perspective, we will expect that any typical statistical scaling law of analysed properties is thus reflecting a mean value and possibly an assigned limiting variance. The normal or *lognormal* probability distribution of frequencies is a very prominent scale-limited law beside others, for instance, the *exponential* distribution. The range of sampled properties should at least cover more than one order of magnitude, as evident from Figure 6. Otherwise one would not be able to distinguish noticeably between different distribution functions.

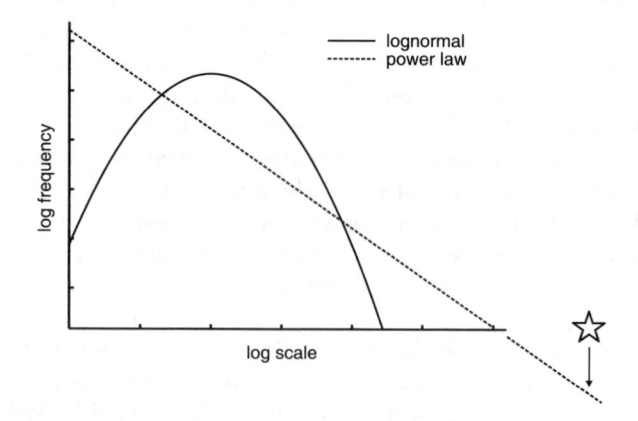

Fig. 6. Statistical distribution functions in logarithmic representation. Indication of an extraordinary sample property or scale size by star symbol.

Especially in Earth science our attention becomes progressively extended towards the perspective of irregular change as already demonstrated. When change occurs more or less unexpected, we search for both the significance of irregularities in the analysis of apparent scales as well as for the concomitant constituents of a system, allowing instantaneous mutability. Thus we are running straight into an omnipresent paradigm, called *complexity* or even more precise for our discussion: *geo-complexity*. Our previous sections on the third model category indicated already, that sudden change is most likely embedded within a complex process of *permanent change*. Therefore we have concluded that continuously changing systems need to be supported by at least more than a single causal condition. This led finally to the expectation, that irregular fluctuating phenomena are backed up by a communicating system of interfering conditions, which clearly in turn set up the reasons for an enhanced spectrum of scales. Like the spatial nature of scales itself, the analysed phenomena are frequently indicated by only a single variable or property associated with a pretended critical threshold. Thus ignoring the latent complexity of conditions may consequently lead to the false impression of *sudden changes*. In this sense, scales are mostly quantities, which comprehend varying properties of particular complex systems on a distinct level. For example, scaling fault slip, the statistical property of the slip-area is defined by a fixed threshold criterion compared to many small-scale details of distributed local slip across this area. As we are not able to reveal the multiple-scale nature from a single event of change, we make use of an equivalence we addressed already in the beginning. We sample the sequence of fluctuating changes of a specific system instead of the non-accessible complexity of single object. Therefore we postulate implicitly the equivalence among the *conditional relations*, i. e., the network of processes required to generate a certain quality temporarily. Along this line of equivalence the statistical analysis of representative scale samples appears to be synonymous for the complexity of the generating system.

Numerous statistical studies at various scales and properties and in different earth systems have shown that their probability distribution clearly follows the *power law*. This forms a straight line for a logarithmic scaling of both axis, as shown in Figure 6. The power law distribution of scales is equally well reproduced from the statistics of slip-areas through the simulations of the slider-block model, given in Figure 4. Opposite to the scale-related distribution functions, the power law has the important implication, that there is no indication of any characteristic or representative scale, state or property, like length, size or area of displacement. Therefore it is called a *scale-invariant* or *fractal distribution*. The main distinctive characterisation comes from its exponent, which is a measure for the slope and corresponds to a *fractal dimension*. Yet, with respect to our limiting perception of space the power laws have to be limited as well by physical length scales that form the upper and lower bound to the scale range of observation. Therefore, it has been appreciated that resolution effects imposed on the power law properties can be rather misleading and result in a lognormal distribution. Alternatively, fractures with scales smaller than the distribution mode are probably insufficiently sampled and large-scale samples might not be identified consistently or within the particular sampling period. Although the fit of those two opposing statistical scaling-laws, shown in Figure 6, to sampled data sets is sometimes ambiguous, they yet merge the limiting members of typical references of the scaling concepts in Earth sciences: the characterisation of representative objects contrary to fluctuating random change.

9 How is Change Organised?

Following the two distinguished types of statistical scale distribution, as denoted in Figure 6, we will finally consider the *characteristics of change*. For a normal distribution and the spatial aspect of scales, change is organised around a mean size with a drastic deficiency of both smaller and larger sizes. That would obviously imply that objects such as the water level of rivers, the amount of falling rain, the length of a fault and a river or the thickness of sediment layers definitely obey an average size statistically. Any deviation from the average is grouped around that mean size according a limiting variance with a rapidly decaying probability of smaller and larger dimensions. Focusing our attention onto the average, we are supposing a phenomenon being more or less on that average or just *normal*. Smaller changes will not question the rule by their assumed insignificance and larger changes are denoted exceptional, what we call *disaster*, yet they are believed to "prove the rule of averages". So far statistics is expressing our normal perspective, equivalent to the existence of representative scales, volumes or objects. Reviewing the statistical normal distribution in terms of the phenomenological context, we might become afflicted by doubts. We "made objects", first through the picking of a quality around a certain scale and second by adding an explanation in terms of a limited context through a restricted causation and function, as demonstrated in Figure 5. That means we perform this design procedure ourselves, through our restricting perception and our relational comprehension of space. On the other side our daily practical experience with the sustainability of objects raises doubts as well. Usually we fail in both the prediction and management of change, especially of the extremes which are exceeding our representative scales. Interested in a possible reason for this insufficient appreciation of change, we may find the limitations of our previous *experience* to be accountable. Our previous experience is consecutively limiting existing and new limitations, respectively objects guided and driven through specific motivations. Limitations of experience are thus driving mentally an ongoing loop of refining and reappearance of phenomena accompanied by the reluctance against change.

Our understanding of normality of objects might benefit from the resort to an even more advanced normality, that of change expressed by the power law. As extended scale statistics of many observations and models on change, such as earthquakes, river networks, landslides and many others follow the power law, we must accept the non-appearance of any reference scale, size or volume. From the viewpoint of phenomena, the characteristic scales are lost now in favour of an upper and lower spatial bound of qualities expressible by scales, as indicated in Figure 6. This range, respectively hierarchy of scales might more or less being related to the limits of a phenomenology. What remains is just change, random change within the spatial bounds which are defined through the respective power index of the distribution function. However, according to the power law distribution function change varies generally both ways up to infinity; between an infinitely increasing frequency approaching smaller scales and a decreasing frequency for scales approaching infinite size without any further preference. This means in a fractal sense, the number of smaller scales necessary to substitute larger ones and vice versa follows exactly the power law distribution. Because of this equivalence of scales among one another, their invariance, scales and in particular change find no other universally addressable allocation than a *random* variability. Generally one may conclude now that we experience just space by means of our limited phenomena. From the power law we find *space* tending to be infinite for both vanishing probabilities of increasing size as well as for a most probable increasing number of decreasing scales. This unconstrained equivalence provides the basis for the permanence of change of any composite properties in space. Phenomena are thus of no real significance, they appear non-specific and relational, they have neither a permanent existence nor an individual entity. Their design has been performed rather arbitrarily by means of a limiting experience through the mind. This perspective together with the infinite nature of space, allows the range of imaginable experience to become unrestricted or infinite as well. Change is thus universal. It appears to be assigned to any of our finite scale selections, it is occurring permanently and randomly. This is affecting the idea of an objective reality fundamentally. As discussed earlier, scale dependence corresponds to very accentuated states.

Surveying these qualities now from a more universal viewpoint by comparing power law and normal distribution function, as visible from Figure 6, the loss of properties through defining objects against their context becomes even evident statistically. Especially the share of smaller scales is drastically reduced, while that of larger ones is shrinking as well. Objects appear thus clearly effectuated through both the detachment from the universe of attributable scales and the additional accentuation of the mean reference scales. Sampling of a complex system with fluctuating change meets in fact the qualities of a power law, but remains still finite compared to infinity of space. The supposed *predictability* both based on established representative scales and of our complex systems remains in either case an illusion. Objective views fail as soon as the representative range of scales becomes lost. Let us, for instance, take a look on a global fracture system such as the split-up of lithosphere forming the entire Atlantic ocean. On this global scale we will not et all being able to pick the tremendous number of statistically equivalent small-scale fractures down to the order of crystal-scale fissures and dislocations. In between we will fall into nearly any trap animated through limited "representative" fracture formations and limited scale-consistent pattern. From a particular statistical analysis of complex phenomena, such as for example the observed size of fault-slip, water levels or land-slides, we can principally even not exclude extreme events trustworthy, as demonstrated in Figure 6. Those *catastrophic* samples, far right, might meet our distribution on a much lower probability level which we have never experienced before. As power law distributions are met via both samples and simulations, we might finally accept the limitations of our approach and try to adopt the valuable clues from modelling.

In terms of analysing self-organising systems statistically, the power law distribution function was associated with the favoured state of *self-organising criticality*. This critical condition has been found to be even a preferred state of earth systems compared with the sub-critical range of scale-dependent objects and the super-critical field of unrestricted random change, whenever the appropriate modelling rules for complexity are met. We might be interested in the specific circumstances for this kind of *universal preference* of scale-invariant changes. What makes it so attractive? Most likely, is seams now to be only a matter of diverging viewpoints between *objective reality* and *permanent change*. Therefore let us accomplish a little experiment together in our mind. Imagine, we like to move a huge sphere made of wet clay across a landscape, which is spangled with little crater-shaped pits and terraces of different depth and diameter. Only with the ultimate concentration of our strength we start to move the sphere toward the aimed straight direction. We push the sphere to overcome friction, the variation of its shape and the local roughness of the ground, passing step by step one obstacle after the other. Up and on, we even do need technical tools in order to assist our action of going properly straight. So far, the entire process is extremely exhausting and there is nothing attractive with this kind of slavery. Let us now, during a break, take notice of a rather moderately sweeping slope in the landscape. Forsaking the burden of going straight, we find the small share of support through the slope just sufficient to keep the sphere on the move. Only a tiny supporting push is required to surmount the most extreme obstacles. Thus the sphere starts immediately to find the ultimate solution for the move itself, while the entire network of conditions become integrated. Local and regional slopes, the ground friction, the internal mass distribution as well as the deforming surface of the sphere and finally our repeated slight pushing work smoothly together. Actually, the movement becomes *self-organising* in accomplishing the changing conditions instantaneously. The scaling distribution is expanding. The particular effort moving the sphere becomes remarkably low now, yet the process of motion through the landscape and the changing shape of the sphere appears no longer conceivable. The record of movements will follow a fractal distribution. It seems quite obvious, changes of the position are most attractive for a system whenever they are the identical expression of the entire complex relationship of conditions. From that point of view on change, the endeavour for a straight and clearly defined path, respectively the compulsive objective position, appears rather disproportionate. Therefore the object-centred view in conjunction with the statistics of scales appears again to be rather deliberate and exceptional.

10 A Dilemma of Change?

Obviously, sooner or later our effort in characterising and representing phenomena becomes overwhelmed from random change. The progression of change obeys not any preferred quality or assigned spatial scale-selection. But, how can we get accustomed to the experience of universal change? Unfortunately our own intention with sense based objects is rather conservative. Even the use of a model concept for complexity and fluctuating change keeps us depending on the denomination of specific phenomena. Denoted once, we experience them again and again in an unpredictable sequence of size and frequency. Our vital interests into prediction are in return motivated through specific concepts, such as disciplinary, scientifically, economical or even political aims. Change, however, is independent from any experience used for its investigation. It is like any of the fractal spatial subdivisions of the infinite space appear equivalent to another, there is nothing compulsory but rather equivalence. Statistics reveals that from both a simultaneous and a sequential sense of scale analysis. How can we benefit from the omnipresence of complexity scientifically? From a viewpoint of change the inconvenience with objects is that they are in general changeable, non-reliable and non-predictable. We will thus not escape from change through raising bounds, how narrow they might ever be drawn. However, we found the conditional relations being critical for change. When phenomena are thus not independent among each other and neither independent from our conceptual perspectives, we finally might start to question our own concepts. We may find, for example, the investigation of interaction and change more important than hitherto. Under the view of complex change, we may finally feel more confident that any of our intentions become reflected, unforeseen through a network of conditional relations. Our favour for management and sustainability may raise a complete change of our thinking against the background of random change. Amazing enough, we started out surveying nature and are now demanding our own responsibility. Complexity is thus primarily a question of perceiving nature.

Part II

Multi-Scale Representation of Data

Multi-Scale Representation of Data

Let us assume, that we have at our disposition, either numerical values of process variables on a four-dimensional (three space coordinates and time) grid or objects with adequate descriptions, context, and interrelation stored in a data bank. Analysis and understanding of the multi-scale geo-systems, we look at, now requires adequate mathematical tools to objectively quantify and/or deal with the scales in the observations, which might come either from measurements or from process model output. In this chapter we review and exemplify three approaches, namely wavelets, diffusion methods, and object-oriented data models.

Wavelet analysis is a modern tool to quantify scales in numerical data given on any regular grid. As an extension and generalisation of the classical Fourier approach, wavelets offer in addition the detection and localisation of transient (in space and time) scales so very characteristic for multi-scale systems. The contribution by Gerstner, Helfrich and Kunoth presents an easy to follow introduction and overview to wavelet analysis, explains the way how wavelets are generated in an effective way and finally delves into some very important aspects, when applying wavelet analysis to geo-data. Among these problems are the often high dimensionality of geo-data and the boundedness of the data available. An instructive example is given by analysing precipitation fields predicted by a complex meteorological weather forecast model. The possibility to impose particular constrains to the wavelet reconstruction (conservation of integrals, gradients etc.) make them especially valuable for representing complex data both accurately and efficiently, and with the application of the data already in mind.

A conceptionally somehow opposite method is represented by diffusion methods. While wavelet analysis define scales by convolution of the data with elementary functions (the wavelets) defined on all applicable length or time scales, Braunmandl, Canarius and Helfrich explain how scale representation with diffusion methods starts at the very small scale. Scales are explained with this concept in terms of generalised smoothing operators, which reproduce data efficiently on any desired scale. Starting with Gaussian scale space (Gaussian filter smoothing) as the most simple, both isotropic and homogeneous approach to scale representation, known characteristics of the data under consideration can be included by advanced methods e. g. in form of constraints, which allow important qualities of the data to be preserved over large scale intervals. A Digital Terrain Model (DEM) is used as an example to show, how drainage features can be preserved over ranges of scales using non-linear diffusion representations instead of the simple Gaussian filtering techniques.

Especially in case of geological records we are faced from the start on with a very different and in general much more complex type of data. How do we deal with the more semantic information, which specifies e. g. a 'stratigraphic surface'? In this concept measurements are conceived as objects of a class with certain attributes explaining the data structure, and possibly also operations, which specify their behaviour and interaction with other objects. Object-oriented modelling is the most appropriate concept to deal with this kind of data. Scales can be attributed to these object kind of data in a semantic way. One of the most challenging problems in this context, discussed by Breunig et al. in their contribution, is the treatment of geometries in different scales. They discuss in detail the problems they encounter when reconstructing a three- dimensional geological underground model of the Lower Rhine Basin. On a smaller scale they describe experiments with small-scale geometric-kinematic modelling of geological objects in an open pit mine.

Wavelet Analysis of Geoscientific Data

Thomas Gerstner[1], Hans-Peter Helfrich[2], and Angela Kunoth[1]*

[1] Institute for Applied Mathematics, University of Bonn, Germany
[2] Mathematical Seminar of the Agricultural Faculty, University Bonn, Germany

Abstract. Wavelets have shown to be an indispensable tool for a scale–sensitive representation and analysis of spatial or temporal data. In this chapter we illustrate the basic mathematical background for wavelets as well as fast algorithms for multivariate wavelet constructions on bounded domains. Finally, we will show an application of wavelets to analysis and compression of precipitation data.

1 Introduction

Ever since the first application of wavelet transforms in geophysics in Morlet et al. (1982) for the analysis of seismic signals, there has been a vast amount of wavelet-related research, both on the theoretical as well as on the practical level. An extensive description of different activities in geophysics can be found in Kumar and Foufoula-Georgiou (1997). The numerous applications cover not only the analysis and compression of all variants of signals and images but also the computation of solutions of partial differential equations for process simulation. These very different types of applications already indicate one of the special features of wavelets: the notion of *wavelets* does not describe a particular fixed object but rather stands for variants of a concept which is adapted according to the application at hand.

The underlying methodology may be sketched as follows. Many physical processes, signals or images, large amounts of data, or solutions of partial differential equations are in fact objects (such as functions) living on different *scales*. With these scales one may associate *meshes* of different *diameters* or mesh *sizes*. To describe the objects mathematically, one needs appropriate representations of the object. In order to obtain such representations, most applications ask for

(A) scalability;
(B) spatial locality.

Requirement (A) means that the representation is able to capture features on a coarse mesh as *efficiently* as on a fine mesh. Thus, the description of the object should in principle allow for a *multi-scale representation* of the object. Requirement (B) indicates that *local* features of an object should only be realized in a 'spatial neighbourhood'. (A) and (B) together would allow for an *efficient*

---
* kunoth@iam.uni-bonn.de

representation for the object. In addition, it ideally requires also only *minimal storage*.

Different mathematical descriptions of an object may satisfy one or the other of the two requirements. The concept of wavelets is that they provide in principle a methodology satisfying (A) *and* (B).

In order to specify these remarks, we mention two typical cases.

Case 1: Fourier Analysis of Data
Modern experimental methods and measuring devices deliver a huge amount of data. In order to handle the data properly, an abstract data model is required. One purpose of the model is to make the data be retrievable in an efficient manner.

One idea to do that is to assume a *hierarchy* and to develop a *multi-scale representation* of the data. The classical procedure is to make a *frequency analysis* of the data using the Fourier Theorem. In this way, one can decompose a (2π-periodic) function f of time t with limited energy into an integral (or a sum) of different signals, each of which gives a precisely defined frequency. That is, f is represented by a decomposition

$$f(t) = \frac{1}{\sqrt{2\pi}} \sum_{j=-\infty}^{\infty} c_j e^{ijt}, \qquad (1)$$

with $i = \sqrt{-1}$, and where the coefficients c_j are simply determined by

$$c_j = \frac{1}{\sqrt{2\pi}} \int_{-\pi}^{\pi} f(t) e^{-ijt} \, dt. \qquad (2)$$

For example, in the case of temperature data which is sufficiently dense, the Fourier model can very precisely exhibit seasonal and daily variations. All waves ω_j that generate the signal are obtained by scaling the base wave $\omega(t) = \exp(it)$, that is, one defines $\omega_j(t) := \omega(jt)$. However, in view of the above requirements, the Fourier model has very strong limitations. In the case of temperature data, the frequency space gives a clear picture of seasonal variations, but it does not show sudden climatic changes such as those occurring during the passage of a cold front. The reason is that the frequency synthesis is not localisable: each frequency spreads over all of the time axis. That is, the Fourier model satisfies requirement (A) but not (B). Apparently the reason is that the base wave ω is a function having *global support*, i.e., living on all of the (time) axis.

To overcome these restrictions, there have been various attempts to 'localise' the Fourier transform, leading to what is known as the *windowed Fourier transform*.

Case 2: Finite Elements
In the numerical simulation of physical processes, one often needs to compute an approximate solution of a partial differential equation. To this end, one can prescribe to the underlying domain where the process is to be modelled a mesh of a *uniform* mesh size. Solving the equation then corresponds to approximating the solution in terms of local piecewise polynomials, commonly called Finite

Wavelet Analysis of Geoscientific Data

Elements, which live on this mesh. Since these functions are usually *locally supported*, an expansion of the solution in terms of these functions would satisfy requirement (B). Of course, the finer the mesh is, the better one expects the solution to be. If one only works with one grid of the same mesh size, this would mean that requirement (A) is not met. Working on a very fine grid, one would then need a lot of time to solve the resulting systems of equations, and also much storage. This can be remedied by allowing for meshes of different sizes and Finite Elements living on them to represent the solution, thus, leading to a *hierarchical representation* of the solution.

Now the wavelet concept provides a methodology to overcome the prescribed limitations and offers a systematic way for both a practical as well as a powerful analytical tool. In the following description of the main features, we focus first on the general principles, independent of a certain application. We will later describe additional properties one can use for particular applications.

The concept is perhaps most conveniently described in terms of the multi-scale decomposition of a function $f : \mathbb{R} \to \mathbb{R}$ (which can, for instance, be a signal). The general idea is to split f into a sum of functions that are obtained by scaling and shifting one basis function ψ. That is, in the abbreviation

$$\psi_{j,k}(t) := 2^{j/2}\, \psi(2^j t - k), \qquad j, k \in \mathbb{Z}, \tag{3}$$

the index j stands for the *level of resolution* and k refers to the *location* of the function. In these terms, the function f is represented by a multi-resolution decomposition

$$f(t) = \sum_{j=-\infty}^{\infty} \sum_{k=-\infty}^{\infty} d_k^j \psi_{j,k}(t) \tag{4}$$

with *wavelet coefficients* d_k^j. In this decomposition the first sum is taken over the scales from the coarse to the finer scales and the second sum is taken over all locations. As in the case of a Fourier series all functions $\psi_{j,k}$ used in the representation are derived from one function, the *mother wavelet* ψ. In order to achieve localisation in time, besides the decomposition given by the outer sum in functions with varying oscillation speed, the inner sum gives a decomposition with different localisations. Thus, one can view j as the scaling parameter or the stage which replaces the frequency or the frequency band in the Fourier synthesis and k as the localisation parameter. The waves $\omega_j(t) = \exp(ijt)$ in the Fourier expansion (1) are here replaced by 'generalised' waves $\psi_{j,k}$ which have an *additional* index k representing the *location*. It is this analogy which gives the collection of all $\psi_{j,k}$ the terminology *wavelets*.

The important property of wavelets that is used in different contexts is that not only an expansion (4) exists but that in addition a norm involving the *expansion coefficients* controls the *function norm*. In mathematical terms, this means that the collection of $\psi_{j,k}$ is a *Riesz basis*, that is, each $f \in L_2(\mathbb{R})$ (the space of square Lebesgue integrable functions, or energy space) has a *unique* decomposition (4) and

$$\|f\|_{L_2(\mathbb{R})}^2 \sim \sum_{j=-\infty}^{\infty} \sum_{k=-\infty}^{\infty} |d_k^j|^2 \qquad (5)$$

holds independent of f. Here the relation $a \sim b$ means that a can be estimated from above and below by a constant multiple of b. Recall that the Fourier expansion of f also satisfies such a *stability estimate* with the equality sign. The stability estimate implies that small disturbances of the function f result in small deviations in the coefficients and vice versa.

The mother wavelet should have properties that lead to a meaningful interpretation of the decomposition. A proper choice makes it possible to split the signal recursively into a trend and a fluctuation. At each stage of the analysis, one may view the actual trend as the smooth part of the time series and the residual or fluctuations as the rough part. One obtains a decomposition ranging from very smooth to very rough components. The most important cases are the ones where the location effect can be most effective. In mathematical terms, this means that ψ has *compact support*, that is, ψ vanishes outside a fixed interval. Then each of the basis functions $\psi_{j,k}$ in the decomposition (3) has compact support which is centred around $2^j k$ with width proportional to 2^{-j}. That is, for high values of j corresponding to a high scale, the functions $\psi_{j,k}$ represent high-frequency parts of the functions f. The easiest and for many years known example of wavelets are based on choosing ψ as the Haar wavelet, consisting of piecewise constants,

$$\psi(t) = \begin{cases} 1, & t \in [0, 1/2), \\ -1, & t \in [1/2, 1). \end{cases} \qquad (6)$$

However, since the Haar wavelets lack smoothness it does not very well represent smooth functions. The triumphant march of wavelets through all sorts of applications then began in fact with the construction of smoother variants with compact support which are also L_2-orthogonal (see (10) below) in Daubechies (1987). In these cases, ψ is no longer a piecewise polynomial and even cannot be written explicitly as a function in closed form (which is actually never needed for analysis or computation purposes). For a compactly supported ψ, the above decomposition (4) provides not only a decomposition into different frequencies but in addition a decomposition into different local contributions of the same oscillation speed. This property of the set $\{\psi_{j,k}\}$ is commonly called *time-frequency localisation*.

The above concept is by no means restricted to time series. Similar considerations can be used to analyse spatial data. Consider elevation data that come from a measurement device with a very fine resolution. The data are influenced by the roughness of the ground and disturbances caused by the vegetation. It is not possible to derive form parameters such as slope and curvature without suitable generalisations that depend on the cartographic scale used in the representation of the data (compare also Chapter 1.3). Since it is not desirable to fix the scale once for all, we have to look for a suitable scaling procedure where according to the actual scale small disturbances are smoothed out.

Wavelet Analysis of Geoscientific Data

In this section we will show by several model problems how wavelets can be used to analyse geophysical data. We will explain some simple algorithms that can be efficiently used for different classes of problems. But first we have to collect a number of definitions and theoretical results on wavelets as an analysis tool.

2   More on the Theory of Wavelets

Like in the Fourier case, there exists both a discrete and a continuous wavelet transform, depending on the type of application.

In order to describe the continuous wavelet transform first, consider for a parameter $a \neq 0$, $a \in \mathbb{R}$, the family of functions

$$\{\psi_a : \ a \neq 0, \ a \in \mathbb{R}\}$$

which are generated by dilation of ψ as

$$\psi_a(x) := \frac{1}{\sqrt{|a|}} \psi\left(\frac{x}{a}\right).$$

For $|a| > 1$, the function is expanded, for $|a| < 1$ it is squeezed. Thus, a plays the role of the (inverse) frequency $1/j$ in (1) and can therefore be viewed as a *scale*. In addition, one introduces a second parameter b denoting the *location* by which ψ_a can be translated. Thus, one considers the family of functions

$$\{\psi_{a,b} : \ a \neq 0, \ a,b \in \mathbb{R}\}$$

where

$$\psi_{a,b}(x) := \frac{1}{\sqrt{|a|}} \psi\left(\frac{x-b}{a}\right).$$

The factor $|a|^{-1/2}$ assures that all functions have the same norm. Now the *Continuous Wavelet Transform* of a function f is an integral transformation of the form

$$(W_\psi f)(a,b) = \int_\mathbb{R} f(x)\, \psi_{a,b}(x)\, dx, \qquad (7)$$

in terms of the scale a and the location b. It is important to note that for $f \in L_2$ and appropriately chosen ψ, the continuous wavelet transform $(W_\psi f)$ is invertible. We later describe the requirements on ψ. The *inverse continuous wavelet transform* can be determined to be

$$f(x) = \int_\mathbb{R} \int_\mathbb{R} (W_\psi f)(a,b)\psi_{a,b}(x)\, a^{-2}\, da\, db. \qquad (8)$$

The terminology *continuous* comes from the fact that a and b are real numbers. Applying the continuous wavelet transform to a function f, one obtains for each pair of parameters a and b different information on f; the information on the scale being represented by a and different values of b measuring the signal at different locations at this scale.

By the wavelet transform (7) the function f is analysed, whereas by the inverse wavelet transform (8) the function is synthesised from its wavelet transform. Consider the case of the Haar wavelet (6). For fixed a and b, the right hand side of (7) gives the mean deviation of the function f from its mean value, that is the fluctuation of the function, in the interval $[b, b+a]$. The reconstruction by the inverse wavelet transformation may be done by taking the limit of a sum with the fluctuations as coefficients.

Of course, different choices of the mother wavelet ψ may well influence the outcome of an analysis. It is for this flexibility that we spoke at the beginning of this section of a methodology rather than a fixed method. In order to guarantee requirement (B), locality, a good choice of ψ is a function with compact support or at least sufficiently fast decay. In addition, ψ should have zero mean or, more generally,

$$\int_{\mathbb{R}} t^r \psi(t)\, dt = 0, \qquad r = 0, \ldots, d-1, \tag{9}$$

where d is a natural number. In this case, one says that ψ has *moment conditions up to order d*. The moment conditions in fact require a certain oscillatory behaviour of ψ, thus, the term 'wavelet' for describing a damped wave. The amount of moment conditions plays an important role in the analysis. Indeed, the conditions (9) imply that ψ (and also all its scaled and shifted versions $\psi_{a,b}$) are *orthogonal* to all polynomials of degree $d-1$. This implies that for sufficiently smooth f which is evaluated in a Taylor series, the continuous wavelet transform only depends on the d-th remainder of the Taylor series. The quantity $(W_\psi f)(a,b)$ gives a measure of the oscillatory behaviour of the function f in a vicinity of b whose size depends on a. Thus, for $a \to 0$, one has (up to a constant factor) $(W_\psi f)(a,b) \to f^{(d)}(b)$. Moreover, higher moment conditions of ψ imply a faster decay rate of $(W_\psi f)(a,b)$ as $|a| \to \infty$. This explains also why the Haar wavelet which only has moment conditions of order 1, i.e., only the mean value vanishes, is often not the best choice.

Like in the Fourier case, the continuous integral transform is a useful tool for theoretical purposes. For practically analysing signals, one resorts to the *Discrete Wavelet Transform* that corresponds to the transform (7) for discrete values of a and b. In fact, it can be viewed as choosing for $j \in \mathbb{Z}, k \in \mathbb{Z}$ the parameters as $a = 2^{-j}$, $b = ak$ and computing with the notation from (3) the expansion (4),

$$f(t) = \sum_{j=-\infty}^{\infty} \sum_{k=-\infty}^{\infty} d_k^j \psi_{j,k}(t).$$

For many applications the most interesting case is when the coefficients can be determined as

$$d_k^j = \int_{\mathbb{R}} f(t) \psi_{j,k}(t)\, dt. \tag{10}$$

Wavelet Analysis of Geoscientific Data

This corresponds to L_2-*orthogonal* or shortly *orthogonal wavelets* where the collection $\{\psi_{j,k}\}$ satisfies the orthogonality conditions

$$\int_{\mathbb{R}} \psi_{j',k'}(t)\,\psi_{j,k}(t)\,dt = \delta_{jj'}\delta_{kk'}, \qquad j,j',k,k' \in \mathbb{Z}, \tag{11}$$

where δ is the Kronecker symbol. Note that the coefficients determined by (10) have a form corresponding to that of the Fourier coefficients (2).

If f is known, one can theoretically determine its wavelet expansion (4) by computing all the coefficients d_k^j. But even if the $\psi_{j,k}$ are known explicitly, this would be much too expensive if one explicitly computed all integrals (10). Instead one uses the *Fast Wavelet Transform* which is usually introduced through a Multi-resolution Analysis of $L_2(\mathbb{R})$. This is indeed also a way to *construct* wavelets with different properties.

2.1 Multi-resolution Analysis

The concept of Multi-resolution Analysis (Mallat 1989) based on Haar wavelets may be introduced using a simple example as follows. Consider a function f which is piecewise constant on each of the intervals $[2^{-1}k,\,2^{-1}(k+1)]$ for integer k, see Figure 1. The function f can be additively decomposed into a function g which is piecewise constant on the intervals $[k, k+1]$ for integer k and a 'remainder' h. Thus, g would capture the 'coarse' part of f by averaging over two adjacent intervals and h the difference between f and g. Then $f = g+h$ can be considered as a two-scale decomposition of f, where g consists of piecewise constants with respect to the coarse scale $j = 0$ (that is, on intervals $[k, k+1]$) and f on a finer scale $j = 1$ (being piecewise constant on the intervals $[2^{-1}k,\,2^{-1}(k+1)]$). Both functions f and g can be represented in terms of (scaled) translates of the characteristic function $\phi(t)$ which equals 1 on $[0,1)$ and 0 elsewhere. That is, defining

$$\phi_{j,k}(t) := 2^{j/2}\,\phi(2^j t - k), \qquad j,k \in \mathbb{Z}, \tag{12}$$

we can write $f(t) = \sum_{k\in\mathbb{Z}} f_k \phi_{0,k}(t)$ and $g(t) = \sum_{k\in\mathbb{Z}} g_k \phi_{1,k}(t)$. In order to be able to use only one function ϕ to represent functions on different scales, it is important that $\phi(t)$ can be written as a linear combination of translates of scaled versions of ϕ on the *next finer* level. This means that there are some coefficients $\mathbf{a} := \{a_m\}_{m\in\mathbb{Z}}$ such that one has a relation of the form

$$\phi_{j,k}(t) = \sum_{m\in\mathbb{Z}} a_{m-2k}\,\phi_{j+1,m}(t), \qquad j,k \in \mathbb{Z}. \tag{13}$$

This relation is often denoted as *refinement equation* for ϕ. For ϕ being the characteristic function, the coefficients are just $a_0 = a_1 = \frac{1}{\sqrt{2}}$ and $a_\ell = 0$ for all other indices ℓ. The coefficient sequence $\mathbf{a}$ is called the *mask* of ϕ.

For each j, translates of $\phi_{j,k}$ generate a space V_j,

$$V_j = \left\{ v \in L_2(\mathbb{R}) :\ v(t) = \sum_{k\in\mathbb{Z}} c_k \phi_{j,k}(t) \right\}. \tag{14}$$

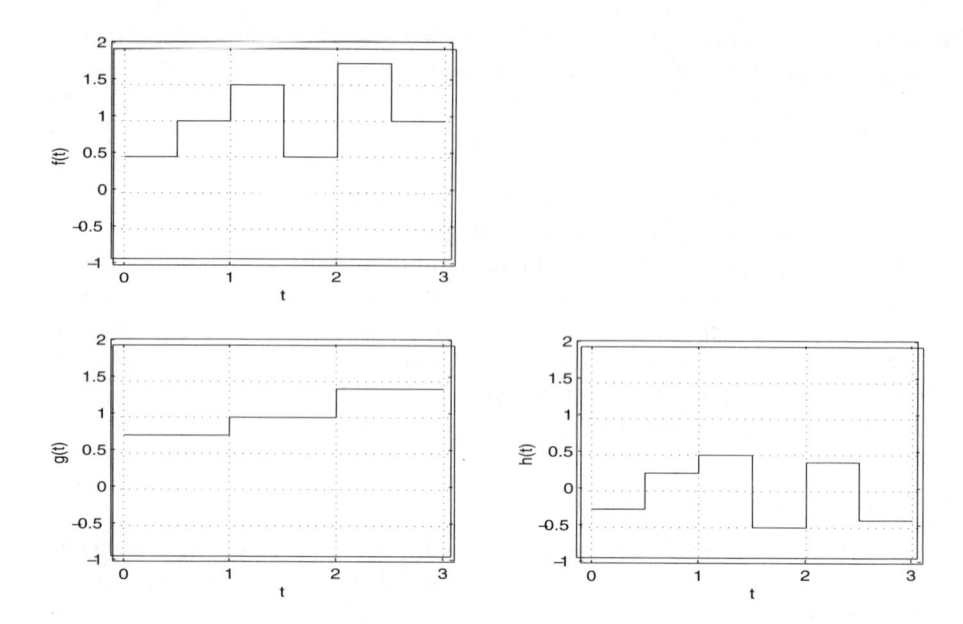

Fig. 1. $f \in V_1$ (top figure) is split into $g \in V_0$ (bottom left) and $h \in W_0$ (bottom right).

For ϕ chosen as the characteristic function, the space V_j consists of all functions which are piecewise constant on each interval $[2^{-j}k, 2^{-j}(k+1)]$, $k \in \mathbb{Z}$. Because of the refinement relation (13), the spaces V_j are *nested*, i.e., $V_j \subset V_{j+1}$. As the resolution level j increases, the V_j become larger and in the limit $j \to \infty$ approximate $L_2(\mathbb{R})$. The set of all spaces $\{V_j\}_{j \in \mathbb{Z}}$ which therefore satisfy

$$\cdots \subset V_{-2} \subset V_{-1} \subset V_0 \subset V_1 \subset V_2 \subset \cdots \subset L_2(\mathbb{R}) \tag{15}$$

is called a *Multi-resolution Analysis* of $L_2(\mathbb{R})$. Since the function ϕ plays a particular role in this concept, it is called the *generator* of the multi-resolution analysis. Because ϕ satisfies the refinement relation (13), it is sometimes also called *refinement* or *scaling function*.

So far the spaces V_j have been introduced from which f and g in Figure 1 come from. Now the question arises how the function h representing the 'detail' $f - g$ can be represented. In terms of spaces, this can be described as follows. Because of the inclusion $V_j \subset V_{j+1}$, there exists an orthogonal complement of V_j in V_{j+1} that we denote by W_j. That is, V_{j+1} can be decomposed into the direct sum

$$V_{j+1} = V_j \oplus W_j, \tag{16}$$

where V_j contains the 'low' frequencies and W_j the 'details' or differences from one resolution level j to the next finer resolution level $j + 1$. Repeating this process, one ends up with an orthogonal decomposition of $L_2(\mathbb{R})$,

$$L_2(\mathbb{R}) = \bigoplus_{j \in \mathbb{Z}} W_j. \tag{17}$$

Wavelet Analysis of Geoscientific Data

Finally, every function $f \in L_2(\mathbb{R})$ can be decomposed according to

$$f(t) = \sum_{j \in \mathbb{Z}} f_j(t), \quad f_j \in W_j, \tag{18}$$

where the part f_j corresponds to the frequency j. As it turns out, the spaces W_j are spanned by translates of the wavelets $\psi_{j,k}$ defined in (3), i.e.,

$$W_j = \left\{ w \in L_2(\mathbb{R}) : w(t) = \sum_{k \in \mathbb{Z}} d_k \psi_{j,k}(t) \right\}. \tag{19}$$

Since by (16) $W_j \subset V_{j+1}$, the wavelets $\psi_{j,k}$ can also be represented by linear combinations of the generator on level $j+1$, i.e., for some coefficients $\mathbf{b} := \{b_m\}_{m \in \mathbb{Z}}$ one has a relation of the form

$$\psi_{j,k}(t) = \sum_{m \in \mathbb{Z}} b_{m-2k}\, \phi_{j+1,m}(t), \quad j,k \in \mathbb{Z}. \tag{20}$$

The coefficients $\mathbf{b}$ are called the *mask* of the wavelet ψ.

If the scaling functions $\phi_{j,k}$ are orthogonal on each level j, that is,

$$\int_{\mathbb{R}} \phi_{j,k}(t)\, \phi_{j,\ell}(t)\, dt = \delta_{k,\ell}, \quad k,\ell \in \mathbb{Z}, \tag{21}$$

one can show that the coefficients b_k in (20) are indeed uniquely determined by

$$b_\ell = (-1)^{\ell-1} a_{1-\ell}, \tag{22}$$

where $\mathbf{a}$ is the sequence in the refinement relation (13). For ϕ being the characteristic function, one can thus see that the Haar wavelet given in (6) is the wavelet corresponding to the characteristic function, see Figure 2. Here $b_0 = \frac{1}{\sqrt{2}}$, $b_1 = -\frac{1}{\sqrt{2}}$, and all other mask coefficients vanish. Thus, given $\mathbf{a}$, one can immediately determine $\mathbf{b}$ and gets from this via (20) a representation of the wavelets $\psi_{j,k}$. Although this is no explicit representation, it is sufficient for computational purposes. In fact, the *Fast Wavelet Transform* which we describe next is only build from the mask coefficients $\mathbf{a}$ and $\mathbf{b}$ that determine $\phi_{j,k}$ and $\psi_{j,k}$.

2.2 The Fast Wavelet Transform

Consider a fixed parameter $J > 0$ which will stand for the highest 'desired' level of resolution and let $j = 0$ be the coarsest level. Recalling (16), the corresponding space V_J can be decomposed as

$$V_J = V_0 \oplus \bigoplus_{j=0}^{J-1} W_j. \tag{23}$$

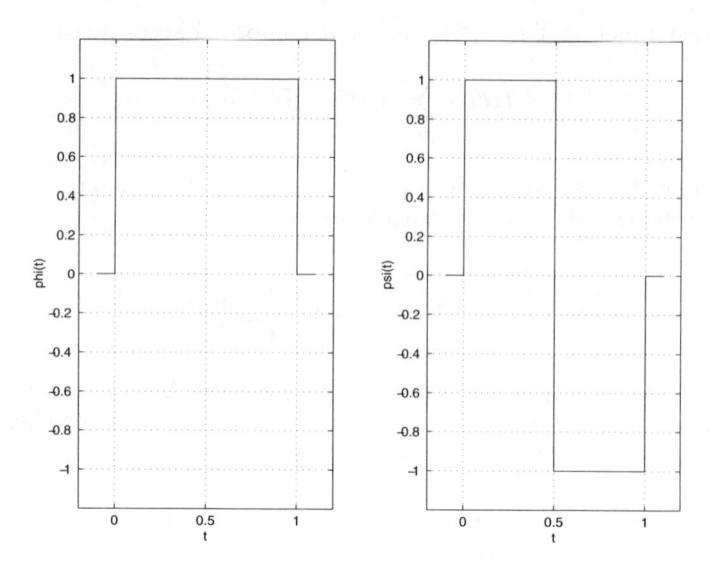

Fig. 2. ϕ as characteristic function and ψ Haar wavelet.

That means, each function $v \in V_J$ can be written either in its *single-scale expansion* involving only the generator functions on level J,

$$v(t) = \sum_{k \in \mathbb{Z}} c_k \, \phi_{J,k}(t), \qquad (24)$$

or in its *multi-scale expansion* involving the generators on level 0 and all wavelets on all intermediate resolution levels $0, \ldots, J-1$,

$$v(t) = \sum_{k \in \mathbb{Z}} c_k^0 \, \phi_{0,k}(t) + \sum_{j=0}^{J-1} \sum_{k \in \mathbb{Z}} d_k^j \, \psi_{j,k}(t). \qquad (25)$$

Thus, in either representation the function v is encoded in the expansion coefficients which we abbreviate as *vectors* $\mathbf{c}$ and $\mathbf{d} = (\mathbf{c}^0, \mathbf{d}^0, \mathbf{d}^1, \ldots, \mathbf{d}^{J-1})^T$ of the same length. Both representations convey different information. The coefficients c_k indicate a geometric location of v whereas the coefficients d_k^j have the character of differences. Usually *all* coefficients c_k are needed to represent v accurately, while many of the d_k^j may in fact be very small. This fact will be used later for data compression by thresholding, see Section 2.5. In order to benefit from both representations, one needs a mechanism to transform between them. That is, one looks for a linear transformation

$$\mathbf{T}_J : \mathbf{d} \to \mathbf{c} \qquad (26)$$

and its inverse $\mathbf{T}_J^{-1} : \mathbf{c} \to \mathbf{d}$. This transformation is indeed determined with the aid of the refinement relations (20) and (13). In order to explain the main mechanism, it is more convenient to write these relations in a matrix-vector form as follows. (This form will also be used later for describing wavelets on bounded intervals and domains.) Assemble in the (column) vector $\Phi_j(t)$ all functions $\phi_{j,k}(t)$ in their natural order,

$$\Phi_j(t) = (\ldots, \phi_{j,k-1}(t), \phi_{j,k}(t), \phi_{j,k+1}(t), \ldots)^T,$$

and correspondingly for $\Psi_j(t)$. Then the equations (13) and (20) can be written in abbreviated form as

$$\Phi_j(t) = \mathbf{M}_{j,0}^T \Phi_{j+1}(t) \tag{27}$$

and

$$\Psi_j(t) = \mathbf{M}_{j,1}^T \Phi_{j+1}(t). \tag{28}$$

Here the matrices $\mathbf{M}_{j,0}$ and $\mathbf{M}_{j,1}$ consist of the mask coefficients $\mathbf{a}$ and $\mathbf{b}$, that is,

$$(\mathbf{M}_{j,0}^T)_{k,m} = a_{m-2k}, \qquad (\mathbf{M}_{j,1}^T)_{k,m} = b_{m-2k}. \tag{29}$$

To derive $\mathbf{T}_J$, one starts from the multi-scale representation (25) which is abbreviated in matrix-vector form as

$$v = (\mathbf{c}^0)^T \Phi_0 + (\mathbf{d}^0)^T \Psi_0 + (\mathbf{d}^1)^T \Psi_1 + \cdots + (\mathbf{d}^{J-1})^T \Psi_{J-1}$$

and inserts the two relations (28) and (27) until the single-scale representation (24) with respect to level J,

$$v = \mathbf{c}^T \Phi_J,$$

is reached. Then one can derive that $\mathbf{T}_J$ has the representation

$$\mathbf{T}_J = \mathbf{T}_{J,J-1} \cdots \mathbf{T}_{J,j_0}, \tag{30}$$

where each factor has the form

$$\mathbf{T}_{J,j} := \begin{pmatrix} \mathbf{M}_j & \mathbf{0} \\ \mathbf{0} & \mathbf{I} \end{pmatrix}. \tag{31}$$

Here for each j $\mathbf{M}_j = (\mathbf{M}_{j,0}, \mathbf{M}_{j,1})$ is a 'square' matrix consisting of the two 'halves' from the relations (27) and (28) and $\mathbf{I}$ is the identity. Schematically $\mathbf{T}_J$ can be visualised as a pyramid scheme, see e.g. Dahmen (1997), Louis, Maass and Rieder (1997), and Meyer (1993). The transformation $\mathbf{T}_J$ is called the (discrete) *Wavelet Transform*. If each of the matrices $\mathbf{M}_{j,0}$, $\mathbf{M}_{j,1}$ has only a fixed finite number of entries in each row and column independent of j (which is the case when the masks $\mathbf{a}$ and $\mathbf{b}$ only involve a *finite* number of coefficients, like in the Haar basis case), it can be shown that applying the transform (30) in the above product form is an $\mathcal{O}(N)$ algorithm, where N is the number of unknowns on level J, i.e., the length of $\mathbf{c}$ or $\mathbf{d}$. Thus, applying the transform $\mathbf{T}_J$ needs even less operations than Fast Fourier Transform, justifying therefore the term *Fast Wavelet Transform*.

In the case of orthogonal wavelets, one has that $\mathbf{T}_J^{-1} = \mathbf{T}_J^T$. However, there are applications where one does not need orthogonality. In these cases, the inverse of $\mathbf{T}_J$ will also be derived as a pyramid scheme so that also the application of $\mathbf{T}_J^{-1}$ is a fast $\mathcal{O}(N)$ algorithm.

2.3 Different Types of Wavelets on $\mathbb{R}$ and $\mathbb{R}^n$

For many applications the orthogonal wavelets satisfying (11) with compact support constructed in Daubechies (1987) are the wavelets of choice. In these cases, the Riesz basis property (5) is satisfied with *equality* sign. One can show that wavelets with such properties cannot be made symmetric. Another reason to relax these conditions is that one sometimes asks for other properties of the wavelets. In principle, the wavelet methodology provides a mechanism to realize different species. In this subsection, a number of them which live on $\mathbb{R}$ (and can therefore be constructed by Fourier techniques) are listed.

From the application point of view, the property that is usually important is the requirement for the wavelets to have compact support, entailing that the masks **a** and **b** only have finitely many coefficients and that the correspondingly built Fast Wavelet Transform can be applied in $\mathcal{O}(N)$ operations, as described above.

Particular interest hold these types of wavelets that are constructed from generators which are *splines*, that is, piecewise polynomials. However, the different resulting *spline-wavelets* are then either globally supported or they are no longer orthogonal in the sense of (11). Relaxing this condition and only requiring orthogonality between different levels leads to *prewavelets*. These and other types have been intensively studied in Chui (1992). Another class that proves at times useful are the *interpolatory wavelets* which interpolate prescribed function values. In Bertoluzza (1997) these interpolating wavelets are used in an adaptive collocation method for the numerical solution of elliptic boundary value problems. The nice fact that these wavelets interpolate is paid for by the fact that they do not satisfy the Riesz basis property (5).

Another class that seems to be a very good compromise of different features are the *biorthogonal spline-wavelets* derived in Carnicer, Dahmen and Peña (1992). In this setting, one works in fact with *two* multi-resolution analyses $\{V_j\}$, $\{\tilde{V}_j\}$ of $L_2(\mathbb{R})$ (called the *primal* and *dual multi-resolution analysis*, respectively) and has therefore also the flexibility of two generators ϕ for $\{V_j\}$ and $\tilde{\phi}$ for $\{\tilde{V}_j\}$. Consequently, one also obtains two mother wavelets ψ and $\tilde{\psi}$ whose dilates and translates satisfy the *biorthogonality conditions*

$$\int_{\mathbb{R}} \psi_{j',k'}(t) \, \tilde{\psi}_{j,k}(t) \, dt = \delta_{jj'} \delta_{kk'}, \qquad j, j', k, k' \in \mathbb{Z}. \tag{32}$$

The interesting point is that one can take as ϕ piecewise polynomials (specifically B-Splines or Finite Elements). The corresponding wavelets $\psi_{j,k}$ are then just linear combinations of these piecewise polynomials and are often more useful to compute with, for instance, in the numerical analysis for partial differential equations, see e.g. Dahmen, Kunoth and Urban (1996). The dual multi-resolution analysis is in fact only used for analysis purposes, e.g., to determine whether the Riesz basis property (5) holds. (This in turn is important for questions like preconditioning when it comes to the efficient solution of linear systems.)

Another generalisation are many variations of *multiwavelets*. There one works with instead of just one generator ϕ with a finite number of generators $\phi_1, \ldots, \phi_m$ to generate multi-resolution analyses.

There are by now also a number of techniques to transform wavelets with specific properties into other ones with other properties. A special case of the general principle called *stable completion* in Carnicer, Wahmen and Peña (1996) is referred to as *lifting scheme* in Sweldens (1996). This principle is used in Section 3.

So far all these investigations are for univariate functions on all of $\mathbb{R}$. For more than one dimension, it is more difficult to construct wavelets, see e.g. Louis, Maass and Rieder (1997). For this reason, one often just works with *tensor products* of the univariate ones to generate *multivariate wavelets*.

2.4 Wavelets on Bounded Domains and Manifolds

Essentially all the previous constructions of wavelets with different properties are based on the massive use of Fourier techniques. However, as soon as one needs for any application (like the numerical solution of partial differential equations) wavelets on a finite interval, a bounded domain or a manifold in $\mathbb{R}^n$, this is no longer possible. On the other hand, there are a number of very favourable properties of wavelets (among them the Riesz basis expansion into functions with compact support) that one would like to use for the discretisation of partial differential or integral equations that typically live on bounded domains. A first idea to use periodisation is, of course, only applicable for periodic problems.

When constructing wavelets on bounded subsets Ω of $\mathbb{R}^n$, one cannot simply restrict the expansion (4) to Ω since one would loose the Riesz basis property (5). All the different constructions of orthogonal and biorthogonal wavelets on the interval therefore essentially use the same idea. Namely, to leave as many as possible of the functions living in the interior of the interval or the domain untouched, and then to build *fixed* linear combinations of all generators overlapping the ends of the interval in such a way that the moment conditions (9) are still satisfied. For constructing the corresponding wavelets from the generators, one then uses the stable completion techniques (Carnicer, Dahmen and Peña 1996) or the lifting scheme (Sweldens 1996), see Dahmen, Kunoth and Urban (1999) for a general construction of biorthogonal interval-adapted wavelets. Such generators and functions also satisfy refinement relations of the form (27) and (28) where Φ_j and Ψ_j are then vectors of finite length and the matrices $\mathbf{M}_{j,0}^T$ and $\mathbf{M}_{j,1}^T$ are up to the boundary effects of the form (29). Examples of interval-adapted primal wavelets consisting of piecewise linear and piecewise quadratic polynomials are displayed in Figure 3. For domains or manifolds in $\mathbb{R}^n$ which can be represented as the union of disjoint smooth parametric images of a cube, these constructions can then be used to generate wavelets on such domains or manifolds. Two techniques have been used there, either to *compose* the functions by glueing them together, or by using a sophisticated mathematical tool, the characterisation of function spaces by Ciesielski and Figiel, see Dahmen (1997) for the main ideas. An example for such a piecewise linear continuous wavelet at the corner of a box (taken from a private communication by Dahmen, Harbrecht, Schneider) is displayed in Figure 4.

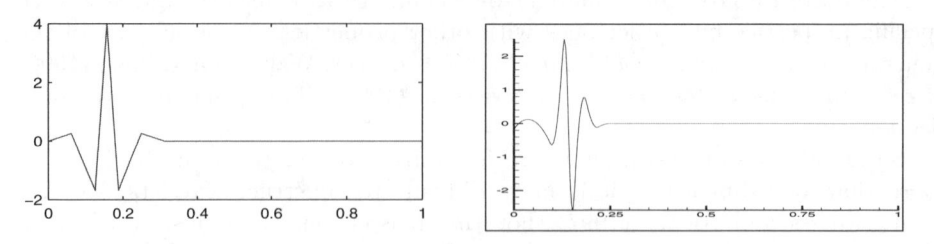

Fig. 3. Examples of primal wavelets for $d = 2, 3$ adapted to the interval $[0, 1]$.

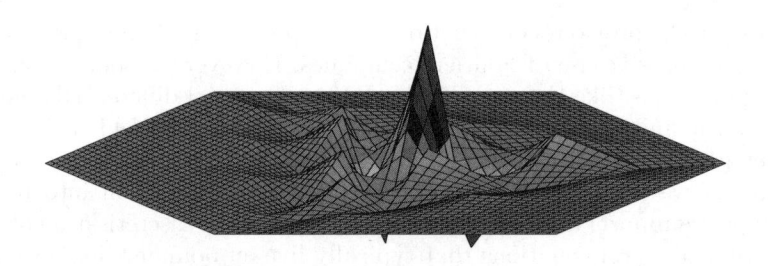

Fig. 4. Piecewise linear continuous wavelet at the corner of a box.

2.5 Methods for Analysing Data by Wavelets

In this section we describe some common techniques for analysing data by wavelets, cf. Ogden (1997). Consider discrete values $\mathbf{c} = (c_1, \ldots, c_N)^T$ with $N = 2^J$ of a function that is transformed by a discrete orthogonal wavelet transform of the form

$$\mathbf{d} = \frac{1}{\sqrt{N}} \mathbf{T}_J^T \mathbf{c} \tag{33}$$

into the vector of wavelet coefficients $\mathbf{d}$ representing the signal, see (26). For simplicity, we assume that by a suitable preprocessing of the data the first sum on the right hand side of (25) is not needed in the representation, i.e., $\mathbf{c}^0 \equiv 0$. Furthermore, it has been assumed that the matrix $\mathbf{T}_J$ is orthogonal. Thus, the reconstruction of the function values is achieved by the transformation

$$\mathbf{c} = \sqrt{N} \, \mathbf{T}_J \, \mathbf{d}.$$

Now the scaling of the matrix in (33) is chosen such that

$$\mathbf{d}^T \mathbf{d} = \frac{1}{N} \sum_{i=1}^{N} c_i^2 \tag{34}$$

Wavelet Analysis of Geoscientific Data

holds, i.e., the right hand side of (34) should be of the same order as the L_2-norm of the function.

There are several methods for analysing the signal **c** depending on the purpose. As pointed out earlier, many of the coefficients d_k^j are very small and can be neglected without altering the original signal too much.

For the purpose of *compact storage* and *fast transmission* of data, one can suppress all coefficients that are smaller than a given threshold λ. The value of λ controls the amount of compression and the level of detail visible when transforming back the distorted signal. By this way, we get a *filtering* of the data. An application to precipitation data is given in the next section.

One purpose of filtering is the *recovery* of a function from *noisy*, sampled data. Assume that the data comes from discrete values $f(t_i)$ of an unknown function disturbed by white noise, i.e.,

$$c_i = f(t_i) + \epsilon_i,$$

where $\epsilon_i \sim \mathcal{N}(0, \sigma)$ are statistically independent random variables that are normally distributed with variance σ^2. By (33) the wavelet coefficients are also normally distributed and independent. We have for $\mathbf{d} = (d_k^j)$

$$d_k^j = \theta_k^j + \frac{1}{\sqrt{N}} z_k^j,$$

where the unknown wavelet coefficients θ_k^j come from the function values $f(t_i)$ and $z_k^j = \mathcal{N}(0, \sigma)$ is white noise. In Donoho and Johnstone (1994) two ways of thresholding are proposed:

1. *Hard thresholding:* All coefficients with $|d_k^j| \leq \lambda$ are set to zero and the other coefficients remain unchanged.
2. *Soft thresholding:* Set

$$\hat{\theta}_k^j = \begin{cases} d_k^j - \lambda, & d_k^j > \lambda, \\ 0, & |d_k^j| < \lambda, \\ d_k^j + \lambda, & d_k^j < -\lambda, \end{cases}$$

as an estimate $\hat{\theta}_k^j$ for θ_k^j.

A common choice of the threshold is

$$\lambda = \frac{\sigma}{\sqrt{N}},$$

which ensures that the dropped coefficients are of the same order of magnitude as the noise of those coefficients.

The advantage of recovery methods of functions by wavelets is that the estimation procedure can easily be adapted to handle cases where the noise level depends on time or space. This is a consequence of the good localisation properties of wavelets. The disadvantage of the method is that the variance σ^2 or a good estimate is needed in order to carry out a reconstruction of optimal order. But it is possible to carry over cross-validation methods to function estimation by wavelets (Hurvich and Tsai 1998).

Finally, we present an application of the wavelet analysis technique.

3 Application: Wavelet Analysis of Precipitation Data

In order to improve the prediction of the weather forecast for central Europe, inside the global weather forecast model GME the non-hydrostatic local model LM is nested (Theis and Hense 2000). The current resolution of the local model LM is 7 km and provides a weather prediction for the next two days. Since the atmosphere is a chaotic and turbulent system, depending on the current weather situation the size of the prediction uncertainty may vary strongly in space and time with varying amplitudes in different scales. Therefore it is important to include into a prediction also an estimate for its probability. One possibility is to adequately *filter* the data, see (Theis and Hense 2000) for details. This is done here in such a way that first the given data (which can be viewed as data living on the smallest resolution level) is transformed into multi-scale data, living on different levels, by means of the Fast Wavelet Transform defined in (26), (30). Recalling that the new data has the character of difference information from one scale to the next, discarding the coefficients that live below a certain threshold provides then the desired filtering. For the wavelet transform, the

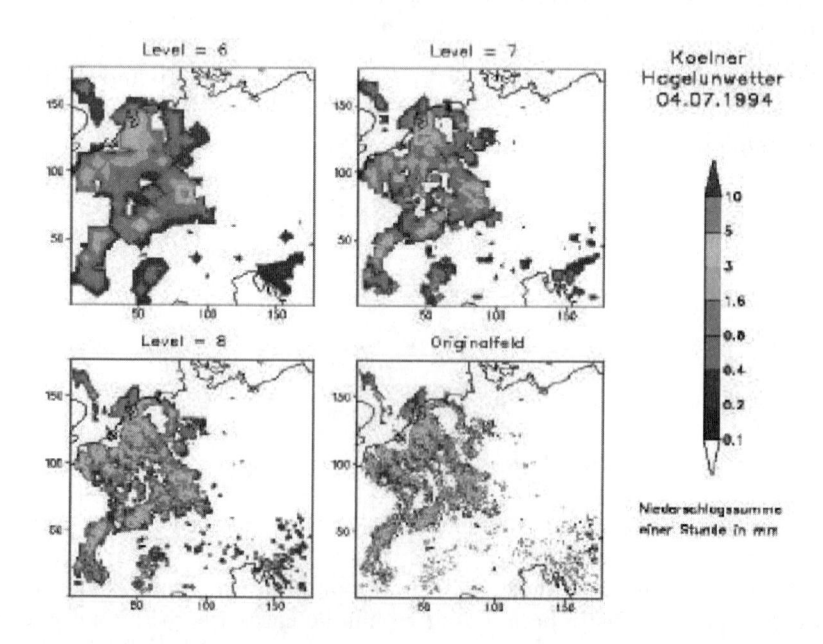

Fig. 5. Wavelet analysis of modelled precipitation data "Kölner Hagelunwetter".

Wavelet Analysis of Geoscientific Data

wavelets constructed in terms of the lifting scheme (Sweldens 1996) are used (Gerstner 1999).

Figures 5 and 6 show two situations where this filtering has been applied. The original data is visualised in the lower right graphic. The other three graphics each show the filtered data, filtered at different levels. Level=6 means that wavelet coefficients belonging to resolution level 7 or higher are discarded. Accordingly, as Level becomes higher, more and more detail of the data is seen.

Uncertainty estimates may be derived from the discarded levels, but this is a problem not yet solved. One in fact obtains different results when thresholding of the wavelet coefficients in different norms is compared. For another data set (courtesy of P. Gross, Meteorological Institute, University of Bonn) we have compared wavelet representations with respect to the L_1-norm (Figure 7) and with respect to the L_∞-norm (Figure 8). While structures at higher levels are similar, at lower levels (top right, bottom right in 3D) quite different patterns begin to exhibit which would allow different precipitation forecasts. This important aspect of wavelet analyses will be addressed in further studies.

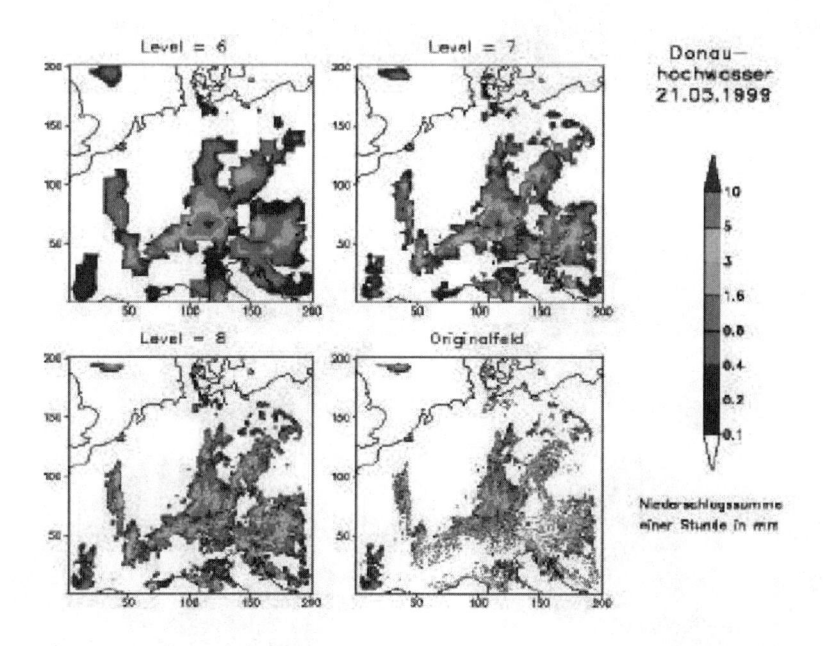

Fig. 6. Wavelet analysis of modelled precipitation data "Donauhochwasser".

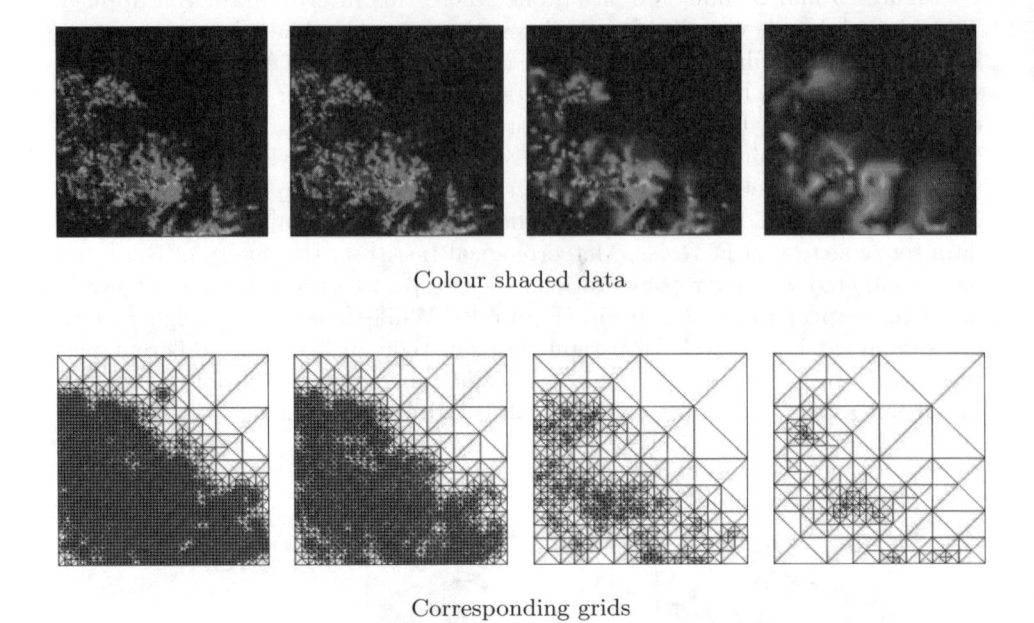

Colour shaded data

Corresponding grids

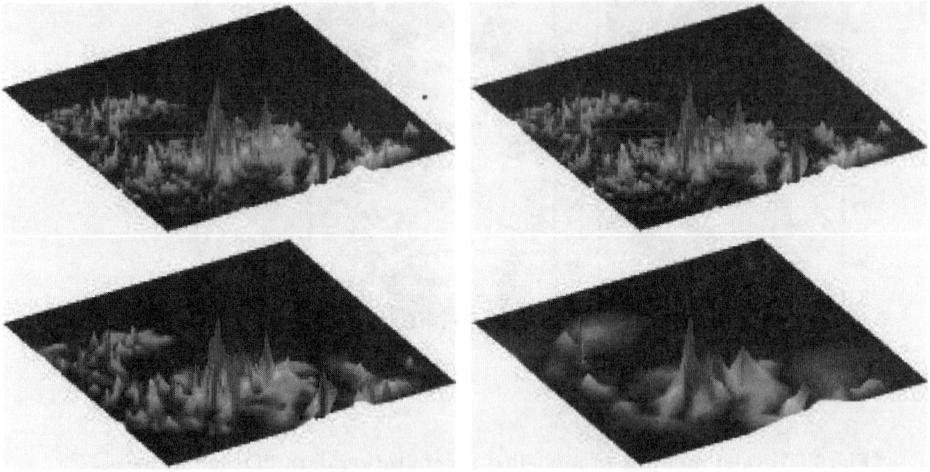

Colour shaded 3D view

Fig. 7. Wavelet analysis of modelled precipitation data in L_1-norm.

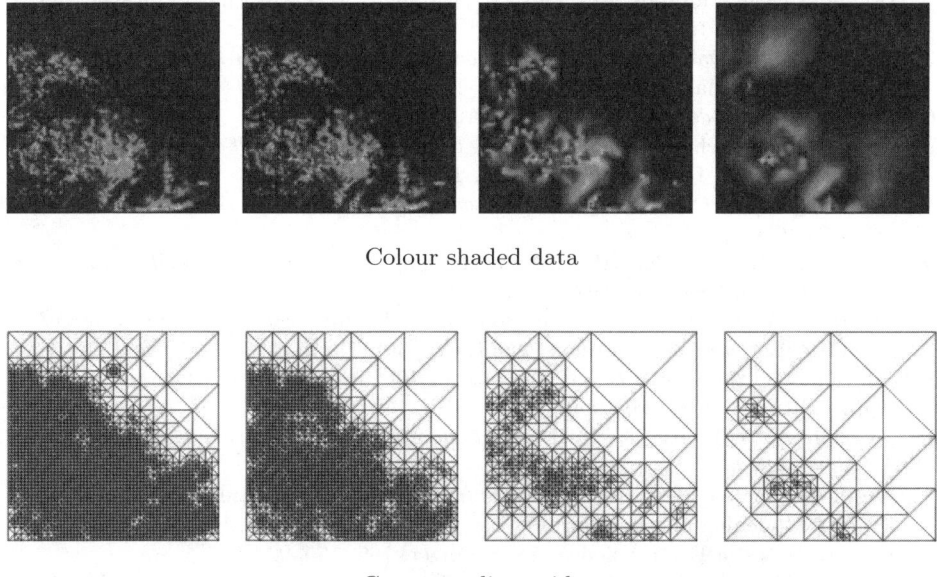

Wavelet Analysis of Geoscientific Data

Colour shaded data

Corresponding grids

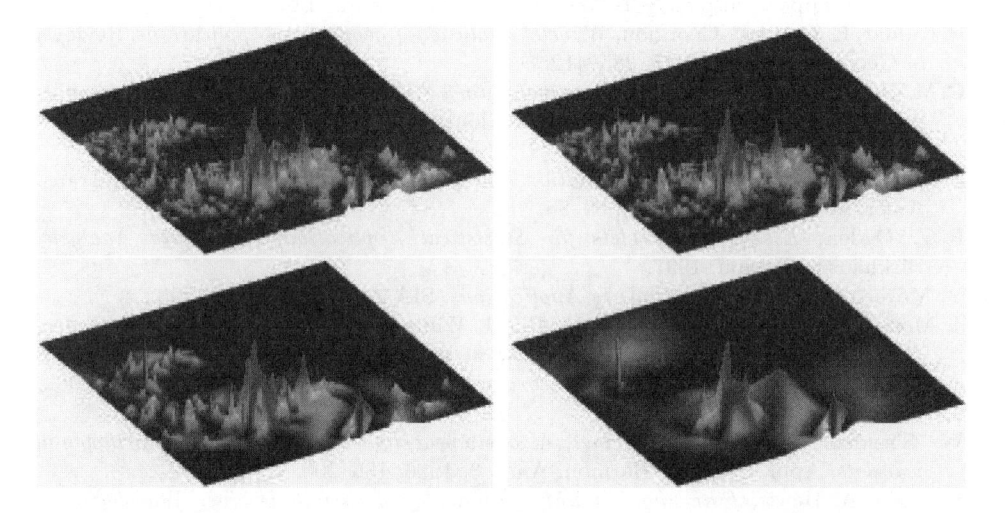

Colour shaded 3D view

Fig. 8. Wavelet analysis of modelled precipitation data in L_∞-norm.

References

S. Bertoluzza, *An adaptive collocation method based on interpolating wavelets*, in: Multiscale Wavelet Methods for PDEs, W. Dahmen, A. J. Kurdila, P. Oswald (eds.), Academic Press, 1997, 109–136.

J. M. Carnicer, W. Dahmen, J. M. Peña, *Local decomposition of refinable spaces*, Appl. Comp. Harm. Anal. 3, 1996, 127–153.

C. K. Chui, *An Introduction to Wavelets*, Academic Press, 1992.

A. Cohen, I. Daubechies, J.-C. Feauveau, *Biorthogonal bases of compactly supported wavelets*, Comm. Pure Appl. Math. 45, 1992, 485–560.

W. Dahmen, *Wavelet and Multiscale Methods for Operator Equations*, Acta Numerica 1997, 55–228.

W. Dahmen, A. Kunoth, K. Urban, *A wavelet Galerkin method for the Stokes problem*, Computing 56, 1996, 259–302.

W. Dahmen, A. Kunoth, K. Urban, *Biorthogonal spline-wavelets on the interval — Stability and moment conditions*, Appl. Comput. Harm. Anal. 6, 1999, 132–196.

I. Daubechies, *Orthonormal bases of wavelets with compact support*, Comm. Pure and Appl. Math. 41, 1987, 909-996.

D. L. Donoho, I. M. Johnstone, *Ideal spatial adaption via wavelet shrinkage*, Biometrika, 81, 1994, 425–455.

T. Gerstner, *Adaptive hierarchical methods for landscape representation and analysis*, in Process Modelling and Landform Evolution, S. Hergarten, H. J. Neugebauer (eds.), Lecture Notes in Earth Sciences 78, Springer, 1999, 75–92.

T. Gerstner, *Multiresolution visualization and compression of global topographic data*, in revision for GeoInformatica, 2000 (shortened version in Proc. Spatial Data Handling 2000, to appear, also as SFB 256 report 29, Universität Bonn, 1999).

P. Kumar, E. Foufoula-Georgiou, *Wavelet analysis for geophysical applications*, Reviews of Geophysics 35, 4, 1997, 385–412.

C. M. Hurvich and C.-L. Tsai, *A crossvalidatory AIC for hard wavelet thresholding in spatially adaptive function estimation*, Biometrika, 85, 701–710, 1998.

A. Louis, P. Maass, A. Rieder, *Wavelets. Theory and Applications*, Wiley, 1997.

S. Mallat, *Multiresolution approximation and wavelet orthonormal bases of L^2*, Trans. Amer. Math. Soc. 315, 1989, 69–88.

R. T. Ogden, *Essential Wavelets for Statistical Applications and Data Analysis*, Birkhäuser, Boston, 1997.

Y. Meyer, *Wavelets: Algorithms & Applications*, SIAM, Philadelphia, 1993.

J. Morlet, G. Arens, E. Fourgeau, D. Giard, *Wave propagation and sampling theory, Part 1. Complex signal and scattering in multilayered media*, Geophysics 47, 2, 1982, 203–221; *Part 2, Sampling theory and complex waves*, Geophysics 47, 2, 1982, 222–236.

W. Sweldens, *The lifting scheme: A custom-design construction of biorthogonal wavelets*, Appl. Comput. Harmon. Anal. 3, 1996, 186–200.

S. Theis, A. Hense, *Filterung von LM-Vorhersagefeldern mittels einer Wavelettransformation* (in German), Manuscript, Meteorologisches Institut, Universität Bonn, April 2000.

Diffusion Methods for Form Generalisation

Andre Braunmandl[1], Thomas Canarius[2], and Hans-Peter Helfrich[3]*

[1] Institute of Photogrammetry, University of Bonn, Germany
[2] Institute for Applied Mathematics, University of Bonn, Germany
[3] Mathematical Seminar of the Agricultural Faculty, University of Bonn, Germany

Abstract. In this section we focus our interest on techniques originating in computer vision. There we have the problem of distinguishing fine scale structures from coarse scale structures in digital images. A major step in solving this problem is related to the concept of a *scale space*. The classical example is the Gaussian scale space, which gives a *linear* method for the transition from fine to coarse scales. We demonstrate relations of this scale space to smoothing and averaging procedures.

We show that the concept of a scale space is also applicable to form generalisations of digital elevation models (DEM). Anisotropy and inhomogeneity of DEM demand an application of nonlinear diffusion methods that have to be customised by steering parameters, which have to be interpretable in the context of DEM form generalisation. An example is given for elevation data from the Broeltal near Bonn.

1 Introduction

The interpretation of objects is based on the scale of perception. Their mere existence as semantic objects is restricted to a certain range of scales. As an example take a pencil. If we look at it from some distance we see only the desk it is lying on. Getting closer we will notice it as a small round — maybe coloured — stick. Getting still a little bit closer reveals that this stick is not perfectly round, but has a hexagonal profile. It has a printing on it and a sharp tip on one end. And with a more thorough investigation of the tip, we will probably notice that it is not really sharp, because someone has already written with it. In normal cases, our interest in this pencil will cease at this point. We do not have to look any closer, because for our purpose — writing — we know enough about it. But it is quite obvious that there are many more details on this pencil, such as the structure of its wood or the roughness of its surface. It is possible to find details down to atomic dimension and more, but in this context it would be senseless to speak of a pencil.

The question is, how much detail do we actually need for our purposes? If this is clear before the mensuration process, it is possible to acquire exactly the data that we need. But there are cases — especially in computer vision — where data is already available and the task is to identify structures belonging to certain objects. Two kinds of problems can arise in this case. First, there is not

---
* helfrich@uni-bonn.de

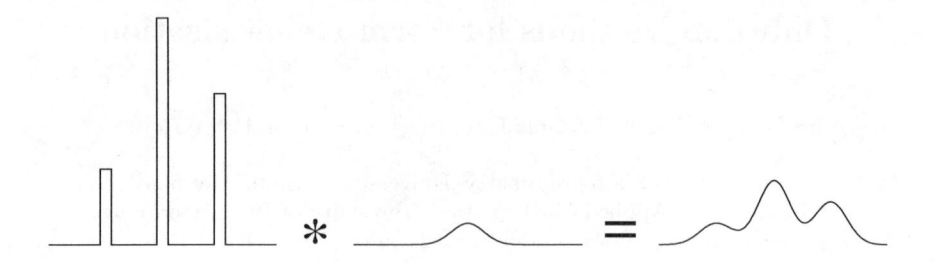

Fig. 1. A qualitative example for convolution of a function with a Gaussian.

enough detail in the data to identify an object; that is, relevant information is missing. This problem can only be solved by using more detailed data. Second, there is too much detail in the data. The overall object structure is hidden by insignificant details. One solution is smoothing the data. This way small details are omitted and larger patterns become visible.

In these cases, where we need to switch between different levels of detail or just have to evaluate what is the desired level of detail, we can use the scale space framework. In the following, we will explain how scale space is defined mathematically and show examples of how to apply it in geoscientific context.

As an application, we mention the computation of geomorphological parameters such as slope and curvature from data of a digital elevation model. High resolution data is influenced by the roughness of the ground and small disturbances. In order to get slope and curvature, first and second partial derivatives of the height functions are needed. Numerical procedures for computing the derivatives are very sensitive to small perturbations in the data. By an appropriate presmoothing of the data, the numerical differentiation can be stabilised.

2 Gaussian Scale Space

The notion of *scale space* has been introduced by Witkin (1983). In fact the basic theory was developed by Iijima (1962) and refined by Otsu (1981), but their work did not become popular because it was published in Japanese.

The classical example of a scale space is the Gaussian scale space, which can be interpreted as an idealised erosion process with a rough surface as an initial condition. To be more specific, let the surface be given by $z = u_0(x,y)$, where z denotes the height at a point with Cartesian coordinates (x,y). The Gaussian filter can be obtained by the solution of the heat equation

$$\frac{\partial u}{\partial t} = \Delta u, \qquad (1)$$

$$u(0,x,y) = u_0(x,y), \qquad (2)$$

where Δ denotes the Laplacian operator. The parameter t delivers the scale in the functions $u(t, x, y)$, which by the initial condition approximate the rough function $u_0(x, y)$. We get

$$u(t, x, y) = u_0(x, y) * g(t, x, y). \tag{3}$$

The $*$ is the sign for the convolution operation and

$$g(t, x, y) = \frac{1}{4\pi t} \exp\left(-\frac{x^2 + y^2}{4t}\right) \tag{4}$$

is the well known Gaussian kernel function. notation to write

In order to apply equation (3), we need to know each detail of the rough surface $z = u_0(x, y)$. In practice, we have to assume that the observation is given by a local smoothing of the "real" surface. As we see later under weak hypotheses, the asymptotic state of iterative linear smoothing results in convolution with the Gaussian kernel. It may be justified to assume that the observation process is given by equation (3) for some fixed $t > 0$. In order to get $u(t + h, x, y)$ with $h > 0$, which represents an observation at a coarser scale, by exploiting the associative property of the convolution with Gaussians, we obtain

$$u(t + h, x, y) = u(t, x, y) * g(h, x, y). \tag{5}$$

That is, we have access to all observations with $t > t_0$ just by convoluting our observation data with the Gaussian kernel (4), where t is replaced by h. Property (5) is called *recursivity*, because we can get the scale for parameters of an equally spaced grid with mesh size h by applying iteratively the convolution with a fixed kernel function.

The observation process model, presented above, obviously describes a homogeneous (translational invariant) and isotropic (rotational invariant) observation. It is not surprising that this model does not fit all requirements of every "real" observation process occurring in practice. As an example, sharp lines could not be observed in this model. But the fundamental requirement of *causality* is fulfilled. This criterion states that no spurious details are introduced by the generalisation process leading from fine to coarse scales, that is, every local extrema in coarse scale representations can be traced back to the original observation data. This has already been shown in the pioneering papers of Witkin (1983) and Koenderink (1984).

Since differentiation commutes with convolution,

$$\frac{\partial^n}{\partial x^n} \frac{\partial^m}{\partial y^m} u(t, x, y) = \frac{\partial^n}{\partial x^n} \frac{\partial^m}{\partial y^m} u_0(x, y) * g(t, x, y)$$

$$= u_0(x, y) * \frac{\partial^n}{\partial x^n} \frac{\partial^m}{\partial y^m} g(t, x, y), \tag{6}$$

the Gaussian scale space offers a direct way to uniquely determine all spatial derivatives. Obviously the causality criterion also applies to all spatial derivatives.

An Example: Geometric Relief Classification In context of a *Digital Elevation Model (DEM)* the direction of gravity plays a central role and — in the scale range of some meter to some kilometre — curvature is an attribute which triggers many environmental processes. The local shape of a surface in the vincity of a point can be described as the deviation from the tangent plane (Taylor's formula)

$$z = \frac{1}{2}(rx^2 + 2sxy + ty^2),$$

where the origin $(0,0,0)$ of the coordinate system (x,y,z) is shifted to the tangent point. Here we use the notations

$$p = \frac{\partial u}{\partial x}, \quad q = \frac{\partial u}{\partial y}, \quad r = \frac{\partial^2 u}{\partial x^2}, \quad s = \frac{\partial^2 u}{\partial x \partial y}, \quad t = \frac{\partial^2 u}{\partial y^2}.$$

By an rotation of the coordinate system, we can achieve that the middle term on the right hand side vanishes. Moreover, we can choose the coordinate system $(\bar{x}, \bar{y}, \bar{z})$ such that $\bar{x}$-axis and $\bar{y}$-axis are lying in the tangent plane and the $\bar{z}$-axis is orthogonal to that plane, and

$$\bar{z} = \frac{1}{2}(\kappa_1 \bar{x}^2 + \kappa_2 \bar{y}^2) \tag{7}$$

holds. In that case, the transformed coefficients κ_1 and κ_2 are the *principle curvatures*, i.e., the maximum and the minimum of curvatures of plane curves generated by intersection of the surface with planes perpendicular to the tangent plane. From a differential geometric point of view, equation (7) describes precisely the local behaviour of the surface up to terms of order two. Unfortunately, the principle curvatures do not depend on the direction of the gravity vector. From this point of view, these quantities are not sufficient for a geomorphometric description.

Therefore, it seems to be preferable to introduce new coordinates (X, Y, z) by rotation around the z-axis, such that the axis opposite to the direction of the gravity vector remains unchanged. By choosing the rotation angle such that the X-axis is directed along the direction of the steepest descent and the Y-axis along the level curves, where both new axes are again in the horizontal plane, we obtain (Koendering and van Doorn, 1997)

$$z = \frac{1}{2}(RX^2 + 2SXY + TY^2),$$

where the transformed quantities R, S, T are given by

$$R = \frac{p^2 r + 2pqs + q^2 t}{p^2 + q^2}$$

$$S = \frac{(p^2 - q^2)s - pq(r - t)}{p^2 + q^2}$$

$$T = \frac{q^2 r - 2pqs + p^2 t}{p^2 + q^2}.$$

These quantities have now a geometrical meaning, which makes more sense in a geomorphometric context. For example, the classification of relief units given in Dikau (1988) relies on the signs of R and S. In a two dimensional setting, the curvature of the level curve is equals to $-T\cos\phi$, and $S/\tan\phi$ is the curvature of the slope line projected to the horizontal plane, where ϕ denote the slope angle determined by $\tan\phi = \sqrt{p^2+q^2}$. In the three-dimensional setting, $-R\cos^3\phi$ is the curvature of the slope line viewed as space curve.

It is not reasonable to compute the partial derivatives from rough observations. For example, if we approximate the first derivative u_x by a finite difference quotient

$$u_x \approx \frac{u(x+h,y) - u(x,y)}{h},$$

then for small values of the mesh size h this expression is very sensitive to small disturbances of the data, whereas the approximation to the derivative is better for small values of h. By applying equation (6), we can calculate derivatives depending on the scale parameter t in a stable way. Clearly, in order to get the derivatives we have to discretise equation (6), but for a suitable discretisation process the result should not get worser for small mesh sizes.

3 Diffusion and Smoothing

Consider elevation data that comes from a measurement device with a very fine resolution. The data is influenced by the roughness of the ground and disturbances caused by the vegetation. It is not possible to derive form parameters such as slope and curvature without suitable generalisations that depend on the cartographical scale used in the representation of the data. In order to achieve this goal, we have to look for a suitable procedure where according to the actual scale small disturbances are smoothed out. In this section, we show that two common methods of smoothing lead in the limit to Gaussian filtering.

3.1 Smoothing by Averaging

By the law of large number, one may conclude that a conveniently rescaled iterated convolution of a linear positive kernel converges to a convolution with a Gaussian kernel. It follows that all iterative *linear* smoothing methods converge to the application of the heat equation to the signal. Starting with a fine-structured image with various inhomogeneities and local characteristics it proceeds by coarsening the information typically using local information. The application of the heat equation uniformly and unconditionally to the whole signal or image results in a smoothing of the whole image by a convolution with the fundamental solution of the heat equation which is the Gaussian kernel. This process is by no means selective nor in any sense structure driven. The main effect that one observes is a smoothing of coarse grain structure which is on the other hand sensitive to noise. To stress the relation between diffusion and

a) scale ≈ 50 m   b) scale ≈ 200 m

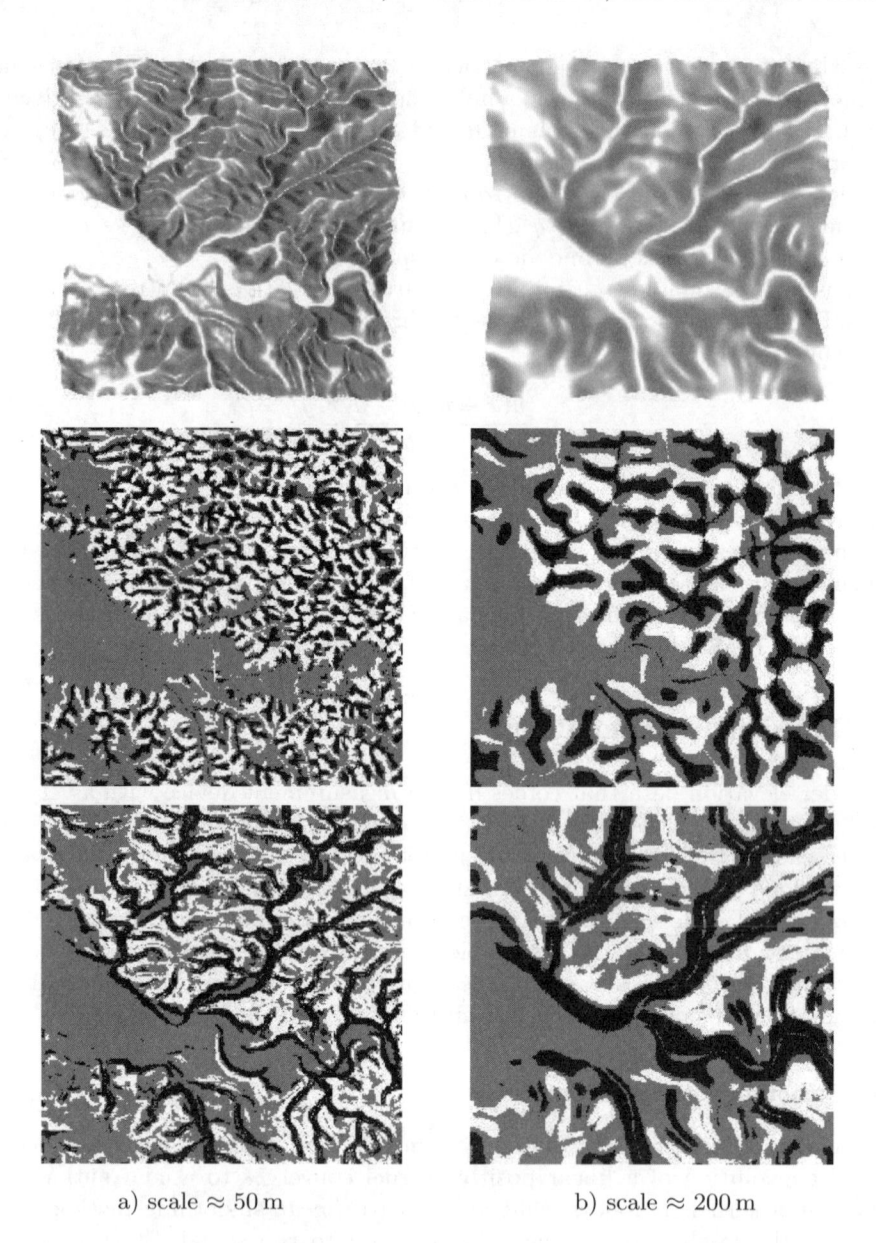

Fig. 2. Relief visualisation and curvature classification of a DEM (mesh size 50 m) based on the *Gaussian scale space*. The middle row shows a classification using the horizontal curvature component and the bottom row shows the same using the vertical curvature component for both selected scales. Regions of concave, respectively convex curvature, are displayed in black, respectively in white. Regions of curvature below a threshold are displayed in gray.

Diffusion Methods for Form Generalisation

smoothing, we define a discrete evolution operator by taking mean values over a circle with centre (x,y) with radius h:

$$M_h u(x,y) := \frac{1}{\pi h^2} \int_{B_h(x,y)} u(x',y')\, dx'dy'.$$

Using Taylor's formula, $u(x,y)$ can be represented around $(0,0)$ as

$$u(x,y) = u(0,0) + u_x(0,0)x + u_y(0,0)y$$
$$+ \frac{1}{2}\left(u_{xx}(0,0)x^2 + u_{yy}(0,0)y^2 + 2u_{xy}(0,0)xy\right) + o(x^2+y^2)$$

Taking the mean value over $B_h(x,y)$ we get

$$\frac{M_h u(x,y) - u(x,y)}{h^2/8} = \Delta u(x,y) + o(1)\ .$$

The same holds for convolution with a Gaussian with $t = h^2/8$ and we get

$$M_h u(x,y) = u(x,y) * g(t,x,y) + o(t). \tag{8}$$

If we replace the image (signal, data set) iteratively by the mean value over a small neighbourhood we get from equation (8) in the limit

$$\lim_{n\to\infty} \left(M_{h/n}\right)^n u(x,y) = u(x,y) * g(t,x,y).$$

As a conclusion, we want to state that the heat equation (or the convolution by a Gaussian kernel at different scales) is the asymptotic state of the class of iterative linear smoothing filters with radial symmetric, positive weights.

3.2 Smoothing as a Compromise Between Data and Model Fitting

In order to get a reasonable smoothing behaviour for data from a Digital Elevation Model, we have to consider methods that give a compromise between the degree of approximation to the rough data and the degree of smoothness. We will show that a more direct solution of this problem may again lead to Gaussian filtering.

By introducing norms in suitable function spaces, we can give a formalisation of the smoothing problem :

Find an approximation $v(x,y)$ to the data $u_0(x,y)$ such that the two quantities
1. $\|u_0 - v\|_0$
2. $\|v\|_1$
are small.

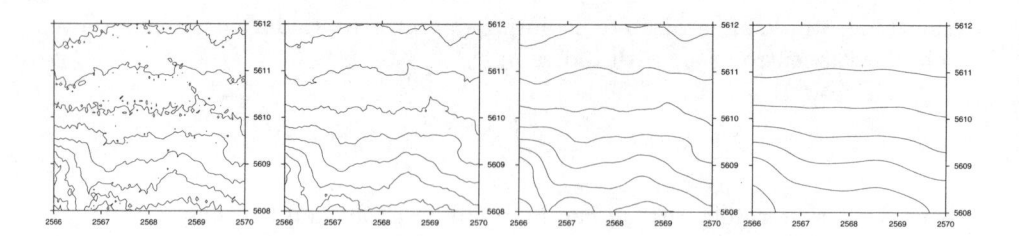

Fig. 3. Isolines from a digital elevation model and smoothing with three parameters

The first term $\|u_0 - v\|_0$ stands for a measure for the deviation $u_0 - v$ of the smoothed data v from the original data u_0. Such a measure is called *norm*, if some simple properties are fulfilled. The term $\|v\|_1$ should give a measure for the smoothness of the data. Of course, the problem to be solved is now to choose suitable norms. In general, it is not possible to make both terms small simultaneously. The behaviour of the two terms should again be controlled by a *scaling parameter t*.

For the approximation norm $\|\cdot\|_0$, we would like to choose some weak norm that does not depend on the smoothness of the data. A common choice is the L_2-norm that gives the root of mean squared deviation between the original and the smoothed data.

For the control of the smoothness, we can use a Sobolev norm, i.e., the root of the integral over the sum of squares of all derivatives up to the order m. In this case, the appropriate function space is the Sobolev space $H^m(\Omega)$, which consists of all functions with square integrable derivatives (in the distributional sense) up to the order m.

The compromise between approximation and smoothness is achieved by minimising a weighted sum of the two quantities. We have now to solve the variational problem:

Find $u(t,\cdot) \in H^m(\Omega)$ such that

$$\int_\Omega (u_0(x,y) - u(t,x,y))^2 + t \int_\Omega \sum_{k=0}^m \binom{m}{k} \left(\frac{\partial^m u(t,x,y)}{\partial x^{m-k} \partial y^k} \right)^2 dx dy \quad (9)$$

is minimal.

For convenience, we have taken squares of the norms that simplifies the solution of (9). In the case $m = 1$, we obtain by a standard variational principle (Euler equations) a linear elliptic boundary-value-problem

$$u(t,x,y) + t\Delta u(t,x,y) = u_0(x,y) \quad \text{in } \Omega, \tag{10}$$

$$\frac{\partial u(t,x,y)}{\partial \nu} = 0 \quad \text{on } \partial\Omega. \tag{11}$$

The condition (11) states that the derivative of the function u in direction of the outer normal ν on the boundary $\partial\Omega$ of the domain Ω vanishes. In the case

$\Omega = \mathbb{R}^2$, the requirement $u(t,x,y) \in L_2(\Omega)$ replaces the boundary condition. The solution of (10,11) may be formally written as

$$u(t,x,y) = (\text{Id} - t\Delta)^{-1} u_0(x,y), \qquad (12)$$

where Id denotes the identity operator, which maps each function $u \in L_2(\Omega)$ to itself. It may be interesting to note that equation (12) gives for small t an approximate solution of the initial value problem (1, 2) for the heat equation. By this way, we see that the regularisation process in the case $m = 1$ leads to an approximation of the Gaussian scale space. From semi-group theory for elliptic linear operators, it can be concluded that

$$u_0(x,y) * g(t,x,y) = \lim_{n \to \infty} (\text{Id} - \frac{t}{n}\Delta)^n u_0(x,y)$$

holds. That means that iterative smoothing by this method leads to the Gaussian scale space.

For $m > 1$, a partial differential equation of order $2m$ is obtained. In this case the regularisation method has better smoothing properties.

The mentioned smoothing methods have the disadvantage that sharp structures are smoothened out. In contrast to this, the Mumford-Shah procedure (Mumford and Shah, 1989) that minimises

$$E_f(u,K) = \int_\Omega (u_0(x,y) - u(t,x,y))^2 \, dxdy + t \int_{\Omega \setminus K} |\nabla u|^2 \, dxdy + |\alpha||K|$$

with nonnegative parameters α and β preserves edges. Here K denotes the set of discontinuities with one dimensional Hausdorff measure $|K|$. In the simple case, where the set K of discontinuities consists of smooth curves, the *Hausdorff measure* $|K|$ denotes the total length of the curves.

This free discontinuity problem is studied in detail in (Morel and Solimini 1995). The Mumford Shah functional can be regarded as a continuous version of the Markov random field method of Geman and Geman (1984), which can be found in some books on stochastic optimisation and uses simulated annealing to find global minimisers.

4 Nonlinear Diffusion Scale-Space

Gaussian scale space is not the optimal choice for every application. Its smoothing properties are very strong and it tends to remove all sharp structures in the data. Large differences in the data are leveled out very effectively. In recognition of objects, this property is obviously not very useful. Even if coarse scale objects are in the centre of interest, if their border is sharp in the original data it should remain sharp in scale space until the specific scale of the object is reached. Only structures belonging to finer scale objects should be eliminated. Obviously knowledge about the objects of interest (domain specific knowledge) is necessary to achieve this goal.

4.1 The Perona-Malik Model

In 1987 Perona and Malik presented an approach for edge detection in digital images using an adapted scale space method. One problem in edge detection is noise. Edges are defined by their gray level gradient, but noise also leads to gray level gradients. The only way to distinguish both effects is their relevant scale. Noise is an effect that occurs on the scale of pixel width, which is the finest scale representable by a digital gray level image. Edges occur on nearly all representable scales. As we see, edge detection is a typical scale space problem and Gaussian scale space is recognised to be a major help for this task. But edges with a low gray level gradient are very difficult to detect and the use of Gaussian smoothing reduces the detectability of this low edges in the same degree as the noise.

Perona and Malik pointed out that it is useful to reduce smoothing on the edges and to have normal Gaussian smoothing in the rest of the image. In their context of application, images consist of regions of constant gray level divided by edges with gray level gradient, which we denote here by $p = \nabla u$. Their idea was to find an indicator function for edges and use this function to stop smoothing where edge probability is high. The most obvious choice is based on the gray level gradient, so they constructed the following indicator function

$$d(p) = \frac{1}{1 + \frac{p^2}{\lambda^2}}, \qquad \lambda > 0, \tag{13}$$

which determines the *diffusivity*, and the corresponding diffusion equation

$$\frac{\partial u}{\partial t} = \nabla \cdot (d(|\nabla u|)\nabla u) = \nabla \cdot \left(\frac{\nabla u}{1 + \frac{|\nabla u|^2}{\lambda^2}} \right). \tag{14}$$

This nonlinear diffusion method combines domain specific knowledge — the information about edges — with properties of the Gaussian scale space approach. This way the two processes of edge detection and smoothing interact, instead of being applied subsequently and — as experiments have shown — performs better than two step methods, where in the first step edges are detected, and in the second step a modified Gauss filter is applied.

4.2 Geometric Generalisation of Digital Elevation Models

We will now focus our interest on the generalisation of digital elevation models (DEM). Weibel (1989) proposed an automatic method for cartographic generalisation with visually very impressive results. This concept does not imply the generalisation of slope or curvature information, which would be necessary for a morphometric DEM analysis. At this point, we will utilise methods from the Gaussian scale space approach that have the potential to handle slope and curvature information.

DEM generalisation is a selective process that describes local surface form structures depending on their spatial context, where non relevant form structures are omitted. Only structures that are relevant with respect to the chosen spatial context remain. The larger the spatial context, the more structures are omitted, and the simpler the DEM description becomes. That is, two contradicting objectives steer a generalisation process:

1. Simplicity of description.
2. Preservation of (relevant) information, which splits up in two major aspects:
   (a) Exactness of height approximation.
   (b) Preservation of form character.

The degree of generalisation is given by the weighting between both objectives. A maximum generalisation leads to the simplest case of a flat plain where all information of the original DEM is lost but the mean height value. No generalisation is done if all information is preserved, that is, the original DEM remains unchanged and no simplification has been performed. In between these two extrema, a compromise between both objectives takes place.

Gaussian smoothing without modification is able to fulfil the objective of simplicity. Spatially small structures are smoothed out while spatially large structures remain nearly unchanged. The problem is that spatially small structures may be relevant in a spatially large context and in this case they should remain, e.g., the crest of a ridge should remain sharp until the whole ridge is omitted. We have to find quantitative local measures describing the DEM surface form with respect to a given extend of spatial context, to determine relevance of surface structures. Measures of this kind have been extensively investigated by Förstner (1991) in the context of the interpretation of digital images.

One approach is to measure relevance of surface structures by its curvature properties. Following the Perona-Malik model and ideas of Weickert (1998) we propose the diffusion equation

$$\frac{\partial u(t)}{\partial t} = \left(\mathbf{D}(\mathbf{H}^2(u(t)) * g(t))\nabla u(t)\right), \qquad (15)$$

as a tool for generalisation of DEM, where the diffusity is tensorial and depends on the Hessian matrix $\mathbf{H}(u(t))$. The smoothing of the Hessian has been introduced to refer the curvature information to the correct scale (see Förstner 1991, Weickert 1998).

In Braunmandl (2002), it is shown how the form of the diffusitivity tensor $\mathbf{D}$ has to be designed such that the resulting diffusion equation fulfils the objective of information preservation better than Gaussian smoothing, while leading to the same amount of simplicity. As an example of the results of this work, we show in Figure 4 the results of the geometric relief classification given in Section 2 based on the modified diffusion equation (15).

Acknowledgement The authors are indebted to the *Landesvermessungsamt Nordrheinwestfalen, Germany,* for supplying the 50 m Digital Elevation Model, from which the data for the examples was taken.

a) scale ≈ 50 m   b) scale ≈ 200 m

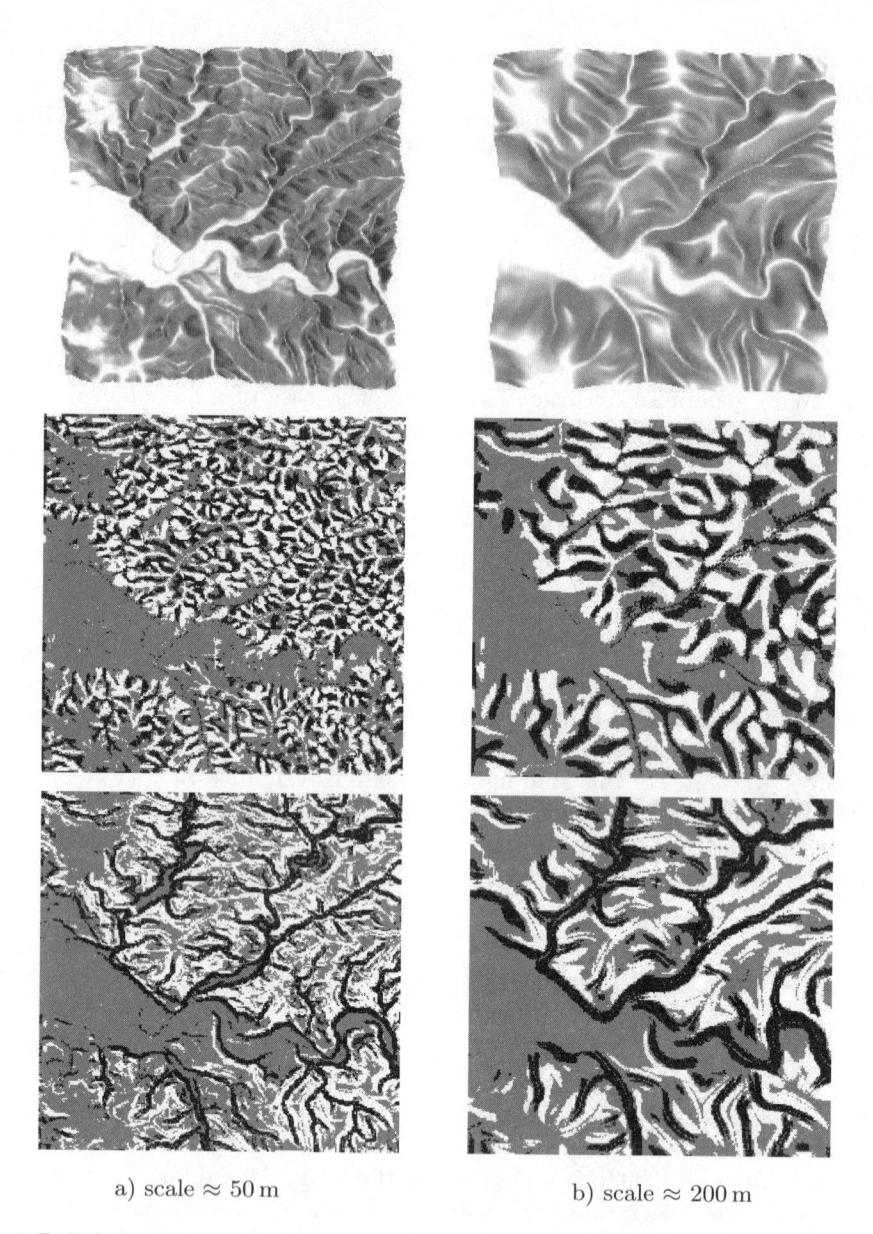

Fig. 4. Relief visualisation and curvature classification of a DEM (mesh size 50 m) based on the *nonlinear scale space* (Braunmandl 2002). The middle row shows a classification using the horizontal curvature component and the bottom row shows the same using the vertical curvature component for both selected scales. Regions of concave, respectively convex curvature, are displayed in black, respectively in white. Regions of curvature below a threshold are displayed in gray.

These results should be compared with the ones in figure 2. The nonlinear scale space preserves the sharpness of large scale structures and removes small scale structures, whereas the smoothing properties correspond to the Gaussian scale space.

References

Luis Alvarez, Freédéric Guichard, Pierre-Louis Lions, and Jean-Michel Morel. Axioms and Fundamental Equations of Image Processing. Archive Rational Mechanics and Analysis, 123:199–257, 1993.

Andre Braunmandl. Geometrische Generalisierung von Digitalen Höhenmodellen, Dissertation, Bonn, 2002

Wolfgang Förstner. Statistische Verfahren für die automatische Bildanalyse und ihre Bewertung bei der Objekterkennung und -vermessung, volume 370 of C. Deutsche Geodätische Kommission, München, 1991.

R. Dikau Entwurf einer geomorphographisch-analytischen Systematik von Reliefeinheiten. In: Heidelberger Geographische Bausteine, H. 5, Heidelberg, 2002

Stuart Geman and Donald Geman. Stochastic Relaxation, Gibbs Distributions, and the Bayesian Restoration of Images. IEEE Transactions on Pattern Analysis and Machine Intelligence, 6(6):721–741, 1984.

Taizo Iijima. Basic Theory on Normalization of a Pattern (in Case of Typical One-Dimensional Pattern). Bulletin of Electrical Laboratory, 26:368–388, 1962. (in Japanese).

Jan J. Koenderink. The Structure of Images. Biological Cybernetics, 50:363–370, 1984.

Jan J. Koenderink and A. J. van Doorn, Image Structure, In E. Paulus and F. Wahl, editors, Mustererkennung 1997, 3–35, 19. DAGM-Symposium, Braunschweig, Springer Verlag, 1997

Jean-Michel Morel and Sergio Solimini. Variational Methods in Image Segmentation. Birkhäuser, 1995.

D. Mumford and J. Shah. Optimal Approximation by Piecewise Smooth Functions and Associated Variational Problems. Communications on Pure and Applied Mathematics, XLII:577–685, 1989.

Nobuyuki Otsu. Mathematical Studies on Feature Extraction in Pattern Recognition. PhD thesis, Researches of the Electrical Laboratory, Ibaraki, Japan, 1981. (in Japanese).

P. Perona and J. Malik. Scale-Space and Edge Detection Using Anisotropic Diffusion. In IEEE Computer Society Workshop on Computer Vision (Proceedings), 16–22, 1987.

R. Weibel. Konzepte und Experimente zur Automatisierung der Reliefgeneralisierung, volume 15 of Geo-Processing. Geographisches Institut, Universität Zürich, 1989.

J. Weickert Anisotropic diffusion image processing. Teubner, 1998.

Andrew P. Witkin. Scale-space filtering. In Eighth International Joint Conference on Artificial Intelligence (IJCAI '83 Proceedings), volume 2, 1019–1022, 1983.

Multi-Scale Aspects in the Management of Geologically Defined Geometries

Martin Breunig[*], Armin B. Cremers, Serge Shumilov, and Jörg Siebeck

Institute of Computer Science, University of Bonn, Germany

Abstract. The interactive geological modelling and the geometric-kinetic modelling of geological objects need an efficient database support. In this chapter we present multi-scale aspects in the management of geologically defined geometries. From the computer science side, object-oriented data modelling provides a framework for the embedding of multi-scale information by enabling geoscientists to model different "semantic scales" of the data. In particular, the concepts of inheritance and aggregation implicitly deal with multi-scale aspects. According to these concepts, the database-backed software system GeoStore has been developed. It handles geological data in two different scales: the large-scale data of the Lower Rhine Basin and the small-scale data of the Bergheim open pit mine. We describe the database models for both scales as well as the steps undertaken to combine these models into one view. Furthermore, there are interesting connections between interactive multi-scale modelling and geo-database query processing techniques. In this context, we have developed the "multi-level R*-tree", an index structure allowing "perspective region queries." Multi-scale landform data modelling and management imposes manifold requirements. We discuss some of them and conclude that they can be met by a hierarchical data type: a directed acyclic graph structure which, if implemented in an object-oriented database system, is readily extensible. Another important aspect is the interoperability between modelling and data management tools. We describe the tasks to be carried through for accessing and recycling data in different scales contained in heterogeneous databases. In particular, we present the eXtensible Database Adapter (XDA) which performs network-transparent integration of client-side GIS tools and object-oriented spatial database systems.

Introduction

The interactive geological modelling and the geometric-kinetic modelling of geological objects (Siehl 1993, Alms et al. 1998) need an efficient database support to handle large sets of spatial and spatio-temporal data. Important issues to be considered are the development of graphical database query languages, the support of an efficient spatial (3D) and spatio-temporal (4D) database access and the access to heterogeneous and distributed geological data. The problems to be

---

[*] mbreunig@iuw.uni-vechta.de

solved in the field — on the geological as well as on the computer science side — are manifold. However, first success has been achieved in the development of geological 3D/4D information systems. An example of such a development is GeoStore (Bode et al. 1994). In a close cooperation between geologists and computer scientists, this information system for geologically defined geometries has been developed on top of GeoToolkit (Balovnev et al. 1997), an object-oriented 3D/4D database kernel system. GeoStore has been developed to handle geological data of the *Lower Rhine basin* in two different scales: the large-scale data of the basin with focus on the so-called *Erft block* and the small-scale data of the *Bergheim* open pit mine. Whereas the first mentioned area covers about 42 × 42 km, the last spans an area of about two square kilometres, which allows a detailed study of the geological structures.

Seen from a data handling perspective, the mentioned large-scale geological application focuses on spatial scales. The small-scale application, however, aims at the construction of a kinematic model. That is why this application also involves temporal scales. In this paper we present first approaches for these two applications to the modelling and management of multi-scale geological data of the Lower Rhine Basin.

1 General data modelling aspects: generalisation and specialisation used for object-oriented geological data modelling

Object-oriented data modelling implicitly provides a framework for the embedding of multi-scale information. The generalisation of sub-classes and the specialisation of a super-class into several sub-classes, respectively, describe two different directions of the same abstraction modelled for specific and general objects inclusively their methods. A typical example from geology are specific stratigraphic surfaces of the Lower Rhine Basin that can be grouped into a class "Stratigraphic Surfaces" for convenient data processing. Objects with the same data structure (i.e. their attributes) and the same behaviour (i.e. their operations) usually are grouped into a common class. The interface of this class abstracts from implementation details like the internal data structure of a geological fault or the method used to interpolate between two geological sections for the construction of stratigraphic boundary surfaces. This framework allows the geoscientist to model different "semantic scales" of the data, i.e. different abstraction levels of the same objects. Stratigraphic surfaces, for example, have the same attributes and operations as pure geometric surfaces. However, they add their own attributes and operations like their lithologies or their interpolation method used in geometric-kinetic modelling of geological objects. Furthermore, "sub-strata" can be modelled to indicate different parts of a single stratum between different geological faults. According to the terminology of object-oriented modelling the stratigraphic surface "inherits" the structure and behaviour of the (more general) pure geometric surface class. In this way inheritance, expressing generalisation and specialisation in the so-called "is-a relationship", and aggregation, expressing the so-called "part-of relationship" implicitly deal with multi-scale aspects of object-oriented data modelling. Inheritance enables the sharing of attributes and operations between geological classes which are typically based on hierarchical relationships. The use of inheritance also leads to a less redundant database design. This makes the modelling and the management of geological objects clearer and more comfortable.

1.1 Modelling of multi-resolution geometries in geology

One of the most challenging problems in geological data modelling is the treatment of geometries being represented in different scales. This multi-resolution problem of geological data modelling is so manifold that we cannot treat all of its facets in this contribution. In the following we will try to emphasise specific aspects that are closely connected to the construction of a 3D geological underground model of the Lower Rhine Basin. Furthermore, we focus on first experiments dedicated to small-scale geometric-kinetic modelling of geological objects in the Bergheim open pit mine (see also Siehl and Thomsen, in this volume).

The central point is to ensure the geometric and topological consistency of the geological underground model in space and time. This means that the surfaces and bodies have to "fit" against each other in a plausible way so that the geologist finds the most likely scenery of geological states and processes in the model. Beginning with the recent state the balanced backward reconstruction has to consider a spatially consistent time series of the geological past.

Large-scale model In the first step of the large-scale geological modelling process, sections of the Lower Rhine Basin have been digitised (Klesper 1994). In this step, point data has been transferred to line data. In the second step, the line data on the sections have been interpolated by triangulated stratigraphic and fault surfaces. Figure 1 shows the large-scale object model of GeoStore.

The main classes are sections, strata and faults, which can be geometrically represented as lines (on the sections) and as surfaces (between the sections), respectively. There are also spatial relationships explicitly stored in the model like "left and right fault" or "hanging stratum". The surfaces are geometrically represented as triangle networks, i.e. lists of triangles in 3D space. For each triangle, its three neighboured triangles are explicitly stored in GeoStore. With this spatial representation, the topology of the triangle networks can be efficiently scanned in geometric algorithms like intersections.

Small-scale model Concerning the small-scale examinations, the detailed structural study of the Bergheim open pit mine aims at the construction of a kinematic model of a small faulted domain (Siehl and Thomsen, in this volume). The balanced volume-controlled 3D-restoration of the geological blocks is not possible without developing a suitable geometric and topological data structure for the small-scale surfaces and bodies. The small-scale 3D-model (Figure 2) has been

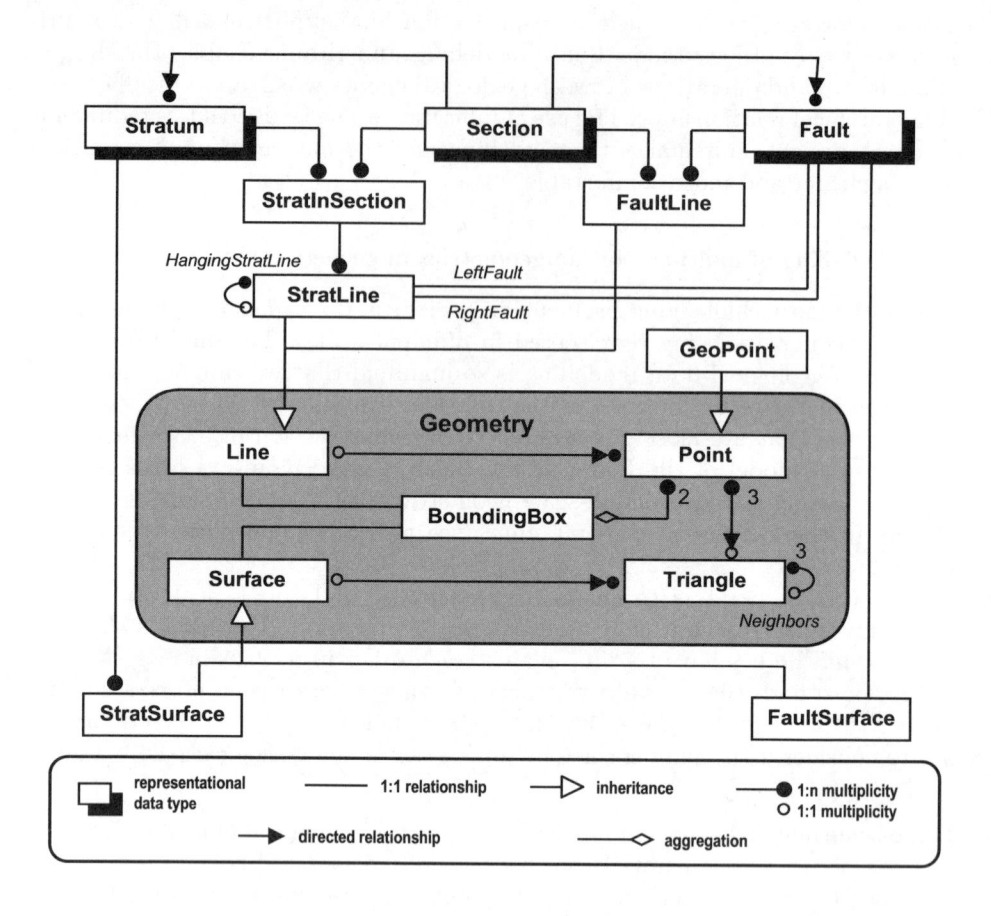

Fig. 1. Large-scale object model of GeoStore

designed as extension of GeoToolKit. Model3D and Region are the main classes. Every region can be represented as a closed box (Figure 3) made from connected parts of stratigraphic, fault and section surfaces — TFaces. It is a boundary box representation for volumes. Model3D consists of several regions and stores topology relations between them. It allows effective search and extraction of neighboured regions during spatial queries. For each region, its neighboured regions are explicitly stored in GeoStore. With this spatial representation, the topology of the triangle networks can be efficiently scanned in geometric algorithms like intersections.

In future it will also be interesting to examine different versions of objects to represent time series of the 4D model. This means that the time units for the backward restoration could be varied to test different variants of time series. The temporal scale will become an important aspect to investigate for geological data handling.

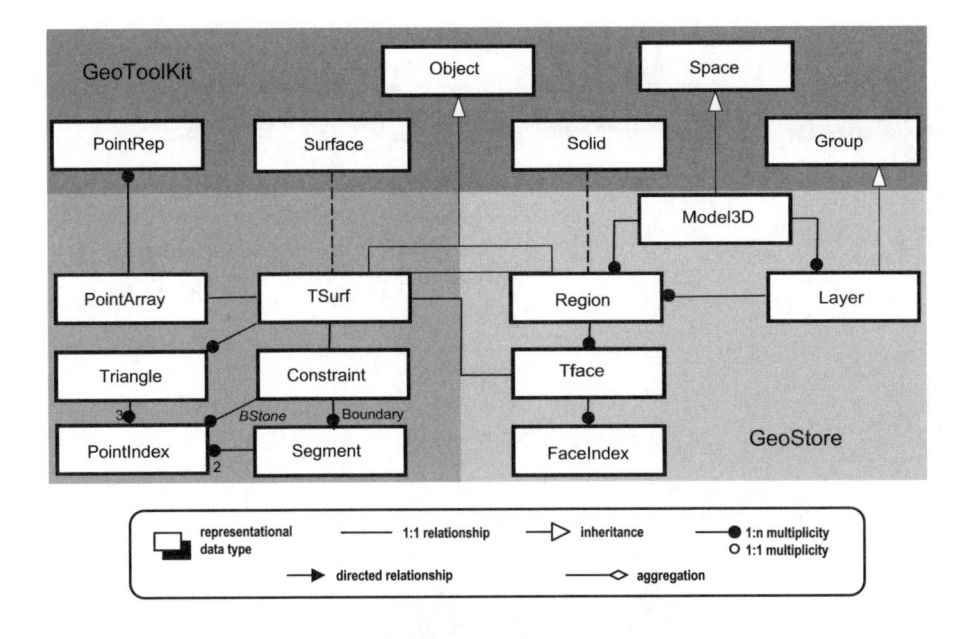

Fig. 2. Small-scale object model of GeoStore

Combination of large-scale and small-scale model Finally, it is important "to switch" between the large-scale and the small-scale geometric model in GeoStore. This means that the 3D model has to be considered at two levels of abstraction: to get a general overview, the geologist should use the large-scale model. For a detailed analysis of small regions, however, it should be possible "to zoom" into the detailed geological structures of the small-scale model. Furthermore, the well data can help to synchronise the time of the small-scale and the large-scale geological model. Geometrically, this data are additional points which can be extrapolated to line- and surface-data, spatially located between different sections and faults.

1.2 Management of multi-resolution geometries in a database

Object-oriented databases Object-oriented database management systems (ODBMS) directly support the object-oriented modelling techniques described above. They map the user-defined objects directly, i.e. in a 1:1-mapping into the database objects. We will give examples for the management of geological objects below. However, until today most databases are relational database management systems (RDBMS) that store the data as "relations", i.e. as tables. Unfortunately, they do not allow the direct management of objects. During the execution of database queries, the "objects" can be constructed out of the values of the data sets in different tables. However, such so-called join-operations are expensive in their run time, because they compare each element of each data

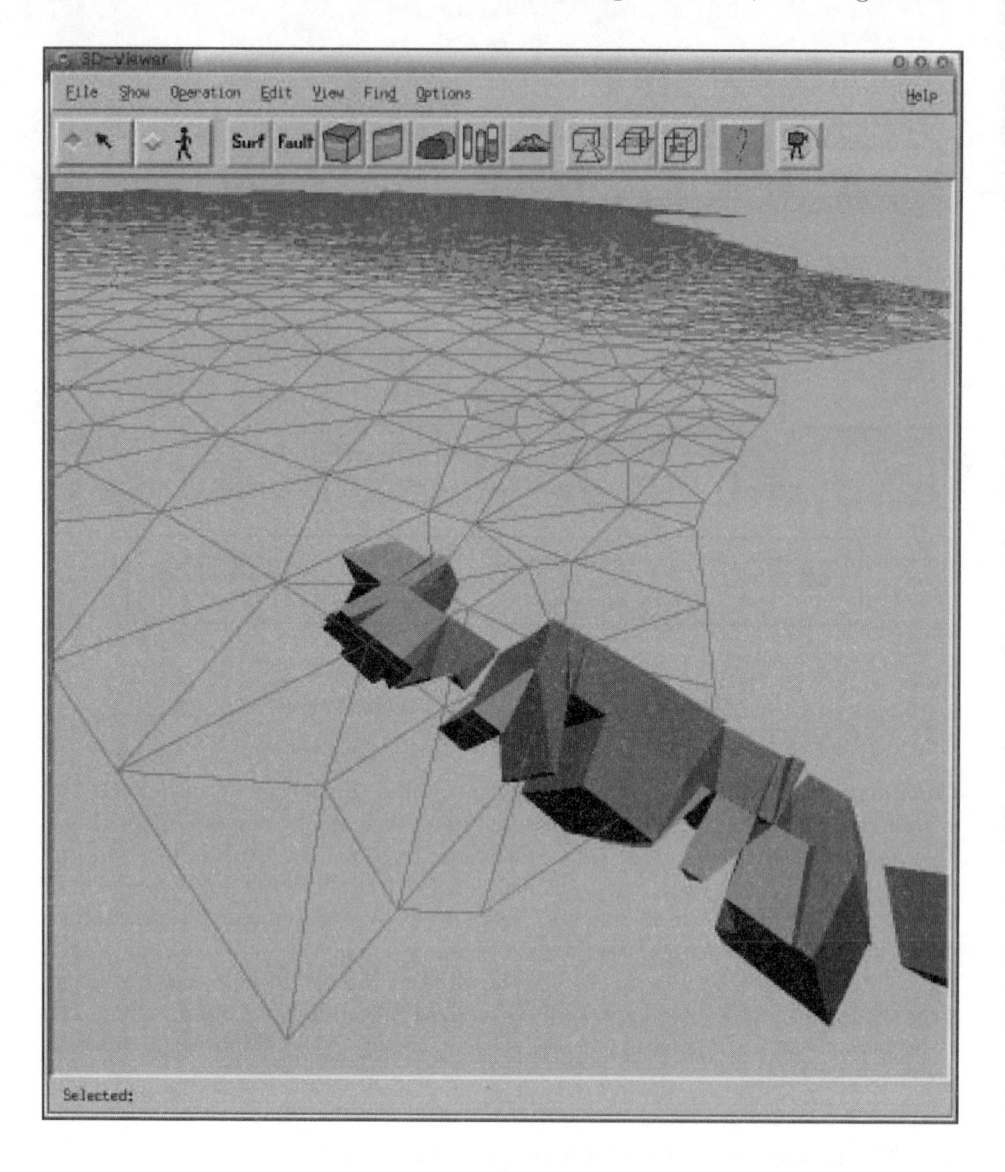

Fig. 3. Detailed geometries of the Bergheim open pit mine

set against each other. To support data abstractions, RDBMS provide aggregate functions like COUNT, AVG, SUM, MIN, MAX and others, which can be used in the expressions of the query languages. These functions count the number of data sets or give the average, sum, minimum or maximum value for the values of a certain attribute in a given data set. They allow the abstraction from concrete data values to aggregated values in a different "semantic scale". A typical example is the database query returning the average value of density in a set of given geological blocks.

Geo-databases Let us investigate the aspect of multi-scale data modelling and management with geo-databases. The "heart" of these software systems is responsible for the processing of spatial database queries. Selecting objects that contain or intersect a given 2D- or 3D-box, spatial database queries are well suited to support geological analysis and interpretations. The spatial object retrieval can be divided into the two steps of the coarse filter and the fine filter selection (Nievergelt & Widmayer 1997). During the coarse filter selection the coordinates of a box are compared with the "minimal bounding boxes", i.e. the approximated geometries of the objects stored in the database. The comparisons are executed at different object approximation hierarchies with the *intersect* or the *contains* predicate, respectively. The coarse filter reduces the search space of the objects stored on secondary memory significantly. It is also often used in geometric algorithms as a pre-step for computations to reduce the average time complexity. Against that, the fine filter selection is responsible for the comparison of the exact geometries with the geometric predicate, i.e. the query box, in main memory. A typical example is the intersection of geological surfaces with a given box in 3D space.

Examples of spatial access methods that directly use the spatial object retrieval described above are the R-tree (Guttman 1994) and the R*-tree (Beckmann et al. 1990). Their balancing operations automatically provide the optimal height and the balance of the tree. R-trees subdivide space into overlapping rectangles (2D) or boxes (3D), respectively.

Database support for the interactive geological modelling To support the interactive geological modelling in an appropriate way, an efficient spatial database access for multi-resolution geometries in space and time has to be provided. However, even for the 'more easy case' of multi-resolution space there are still many open problems like up- and downscaling between different geometric resolutions for geological modelling data, the efficient spatial access for multi-resolution objects and the consistent spatial data management for objects in different scales.

Our experiments started from two different directions: from the application side, i.e. from geology, and from the computer science side by developing special data structures.

As we described before for the data modelling, the geological data of the Lower Rhine Basin stored in GeoStore (Bode et al. 1994, Balovnev et al. 1997) correspond to two different scales: in the first step digitised large-scale data like sections, stratigraphic boundary surfaces and faults of the Erft block have been inserted into the database. The user can access this data by analysing the results of spatial 2D and 3D database queries. The resulting data sets can be visualised in the 2D map and in the 3D viewer of GeoStore, respectively.

Then the well data of the examination area have been inserted into GeoStore. In the map, the wells are represented as point data. In the 3D viewer, however, the wells can be viewed as lines. The lithologies in the well can be compared with the lithologies of near sections. A typical spatial database query in this context is: "Select all wells that are nearer than 3 km to section 117C".

In a separated object-oriented well management tool, which was developed in close cooperation between the computer scientists and the geologists at Bonn University, the measured values of electro logs like natural radioactivity can be visualised and selected from the well database. In this sense, the fine scale values of the electro logs can be compared with the large-scale values of the lithology. The analysis of this data leads to a comparison of geological data with different scales. The similarity of patterns between different electro log curves also allows interpretations for facies analysis (Klett & Schäfer 1996).

GeoStore was developed on top of GeoToolKit (Balovnev et al. 1997), an object-oriented 3D/4D database kernel. GeoToolKit is closely connected with the object storage system ObjectStore[1]. The generic spatial access methods of GeoToolKit allow the efficient storage of geometric 3D objects. The corresponding GeoToolKit interface provides a "space" in which objects can be inserted in, deleted and retrieved from. In first tests we also investigated the multi-scale management of data in spatial access methods. However, our experiments were restricted to fixed scales, i.e. no up- and downscaling was performed. The objective was to support the visualisation of large 3D data sets selected by the database. For this, a "multi-level R*-tree" has been developed (Wagner 1998), which manages geologically defined triangle networks used for geological surfaces with three different resolutions. Figure 4 shows geologically defined surfaces of the Lower Rhine Basin in three resolutions, retrieved by a perspective region query of GeoToolKit's R*-tree. The corresponding space is divided into three distance zones.

We have carried out some experiments with objects of a fixed number concerning the resolution of their triangle meshes. For each zone one view cone is provided. In comparison to "usual region queries" the perspective region query in the R*-tree additionally considers a reference point with a given view direction and a view opening angle. The objects being in the respective distance zone, have to be loaded from the database as soon as the position of the observer is changing. A copy of the visualised data is given to a rendering machine, which transforms the world coordinates into the two-dimensional plane of the picture. Hidden surfaces are determined, the light sources and shadows are computed and the colour values are transferred.

This approach is particularly useful to support the zooming of object parts during the loading of large data sets. It allows "perspective region queries", which consider different views of the geological observer in a scene. The approach has been tested with our large-scale data. It could also be used for advanced spatio-temporal geological applications. Such applications also require an efficient approach for temporal data modelling like the one proposed in (Polthier & Rumpf 1995).

---

[1] ObjectStore is registered trademark of eXceleon Corporation

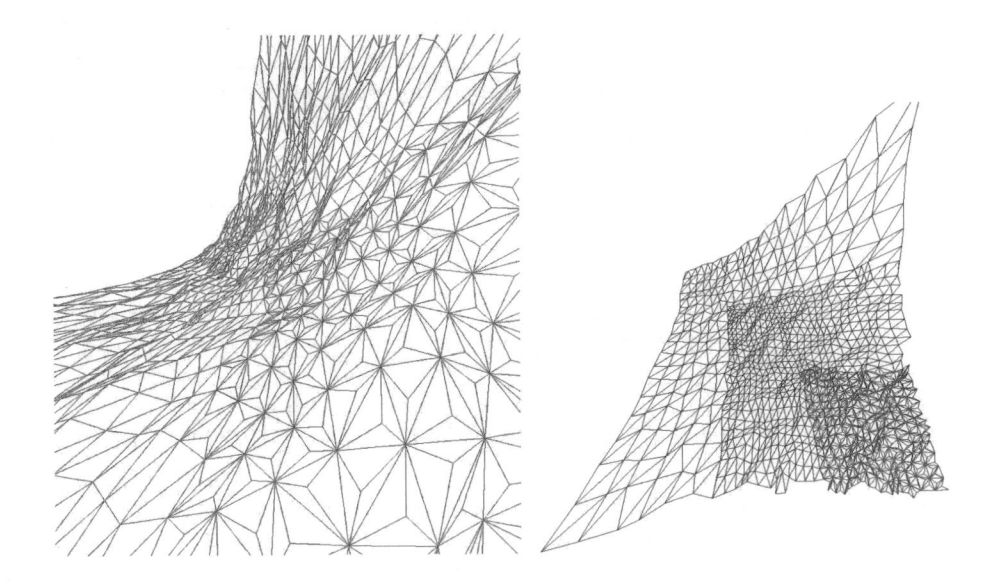

Fig. 4. Result of a spatial query retrieved by the multi-level R*-tree with large data geological data sets of the Erft block in the Lower Rhine Basin

2 Modelling and management of multi-scale landform data

GeoStore can also handle digital elevation models (DEMs) like that of the Lower Rhine Basin. The spatial and temporal management of DEMs is another interesting point in our investigations, which aim at the support for the analysis of landform phenomena by spatio-temporal database queries. In this paragraph we give a brief sketch of some of the aspects that are involved within multi-scale modelling of landform phenomena in a database system's context.

Landform phenomena are scale-dependent. A typical classification for 'scale' in this context is the classification from pico- to mega-scale (see table 1). The landform objects under investigation may roughly be grouped into these classes. One common approach to multi-scale landform modelling is to decompose complex forms into simpler units and thus building a *hierarchy* of landforms. On each scale so-called form facets and form elements can be defined (Kugler 1974), which are composed to landforms and aggregated to landform associations. The geometry of form facets and form elements are of type *surface*. Dikau (Dikau 1988) gives a system for the meso-scale, in which the classification of landform units is based on geometric DEM-parameters like slope, exposition and curvature. In the following we will focus on aspects of the database support for this approach of multi-scale landform modelling.

Although the object-oriented approach of data-modelling adapts very well to object hierarchies, much more than e.g. the pure relational model (see above), we claim that the multi-scale aspects of landform modelling should not rest

Table 1. Typical classification of geomorphological scale (from Dikau 1988, altered).

	Main type of size order (width of unit in m.)	representative example
Mega	10^6 or more	continental shields
Macro ^B_A	10^4–10^6	mountain area
Meso ^B_A	10^2–10^4	valley, hill
Micro ^B_A	10^0–10^2	dune, terrace
Nano	10^{-2}–10^0	erosion rills
Pico	10^{-2} or less	glacial striations

completely on the application's layer of a geo-information-system. Rather, some of the responsibilities should be pushed towards the database kernel — thus adopting the strategy of GeoStore (Balovnev et al. 1997, see above). Among these responsibilities are:

- Offering data types which reflect the spatial hierarchy and that admit to insert, delete, update and query landform objects (including landform associations) on different scales.
- Possibility to store a generalised spatial extent along with a landform. If a higher scale form is missing a spatial extent, then its spatial extent may be the result of a function of the spatial extents of all its aggregated forms on the next smaller scale. The most obvious function is the set-based union of spatial extents.
- Ensuring spatial integrity constraints, e. g. the topological constraint that an aggregated form may not overlap with its aggregation form.
- Built-in efficient data access: low-level clustering of the objects on secondary storage (such as disks), use of access methods, topological support and query evaluation techniques.

These responsibilities can be met by a hierarchical data type. Through the object-oriented modelling technique (see above) the elements of the hierarchy may be internally implemented as a base class containing a minimal set of attributes and operations, e. g. ancestors and predecessors of a given element or the operation of returning a spatial extent. On an application level one adds inherited classes which can then also be part of a hierarchy and are intended to be representatives for instance of the landforms listed in Table 1. Note that it is even possible to seamlessly replace operations (as *virtual functions*), e. g. the one for returning a spatial extent which may be an implementation of an application-specific generalisation algorithm.

Hierarchies can very often be represented adequately by directed-tree-like data-structures[2]. Note, that this is not necessarily the case in hierarchical landform modelling. As Kugler (Kugler 1974) points out, a form element may be regarded as belonging to the adjacent valley *or* the mountain. Thus, two different landforms may 'share' a common form element. Consequently, the tree-like structure should be replaced by a directed acyclic graph (DAG) which gives support for node sharing. On the other hand there may be application-driven unique assignments of a form to a higher scale landform: the geoscientist who investigates concavities would ascribe the aforementioned form element to the valley, but not to the mountain. At a first glance, these are contrary positions. In fact they correspond to so called *views* on the data. In the example above the DAG would be stored internally, with the form element having *two* ancestors, but the concavity-investigator only sees one. Put together, the embedded data type should be able to handle these DAG-structures and should furthermore support application-driven views on them.

Another point to note about the hierarchical view concerns the typical use cases which lead to the operations to be performed on sets of hierarchies. There seem to be two different overall approaches: (a) Starting from basic landform units and assembling them to higher scale objects and (b) identifying objects on a higher scale and disassembling them to smaller ones. Obviously, both approaches are needed and must be supported. For short, given two or more hierarchies: Establishing a new root (i.e. object on a higher scale) with the input hierarchies as descendants is an operation on hierarchies that corresponds to approach (a). Given a node of a hierarchy: Appending another hierarchy to the list of descendants of the given node corresponds to approach (b). Those operations belong to an algebra to be defined on these hierarchies. Such an algebra formalises the operations and establishes a clear semantic to them. It can be used within a multi scale query language. The methods which *derive* or *group* landform units can

- be incorporated within the database kernel itself,
- rest outside the system in different tools or applications that deliver their output through a well-defined interface to the database system, or finally
- can be realized as a module linked at runtime to the database system and that contains routines which are being called-back by the database system.

The first alternative works fastest during runtime, but is obviously not well extensible, because the addition of another computation method implies recompiling parts of the database kernel. The second one offers the most flexibility including not only computational methods but also — very important — manual work of a geoscientist. Ideally, the tools in use are part of an integrated geo-informational infrastructure (see below). Yet, the last alternative is best, since it can simulate the second one and is also extensible and fast.

---

[2] The tree must in fact be directed because only those trees admit a unique assignment of roots and subtrees.

3 Interoperability between modelling and data management tools

Interoperability between geological modelling tools and a database guarantees the consistent storage and multi-dimensional retrieval of shared geological objects. In the last years it could be shown that object-oriented software architectures are well suited to support interoperability between geoscientific software tools (Shumilov & Breunig 2000). They also serve as a basis for the integration of heterogeneous data sources. Furthermore, the object-oriented modelling technique provides a graphical presentation for the design of databases and software tools, independent of programming languages.

The access and the recycling of data contained in different projects or databases is an important aspect of multi-scale data management. Providing a uniform access to multiple independent data sources, we have to deal with multiple scales of objects on the following three levels: in a concrete database, in the common integrated data repository and in the client side. Specifically, the tasks required to carry out these functions are comprised of:

- Locating, accessing and retrieving relevant data from multiple heterogeneous resources;
- Abstracting and transforming retrieved data to a common model and level so that they can be integrated;
- Integrating the transformed data according to matching keys;
- Reducing the integrated data by abstraction to increase the information density in the result to be transmitted.

The problem of access to heterogeneous data sources usually is solved by the integration of the database system with a middleware platform like the Common Object Request Broker Architecture (CORBA) (OMG 1999). In contrast to database systems, middleware provides a flexible transparent distribution model with a larger-scale set of services across heterogeneous networks. In this environment the database objects will be made accessible to remote clients through corresponding transient mediators (Balovnev et al. 1999). Mediators act as usual CORBA objects as well as database clients. In other words, they mediate between CORBA client and database converting data parameters and delegating all function calls in both directions.

The transience of mediators has several advantages, each with their own price. Since the mediators are transient, it is necessary to control their lifetime — the creation and deletion. Care should be taken to avoid the creation of multiple mediators tied with the same database object. In GeoStore these questions, as good as other related problems such as distributed garbage collection of unused mediators are solved by the development of an eXtensible Database Adapter XDA (Shumilov & Cremers 2000).

CORBA clients work with remote server objects through the CORBA Interoperable Object References (IORs). A reference identifies an object in the distributed CORBA environment. References can be used inside the database, but they are especially beneficial when applied to external databases, allowing to build local data configurations from large-scaled distributed data repositories. With the persistence of object references, it makes perfect sense for the client to store an object reference for later use or exchange it with other clients. For example, references to the persistent CORBA objects implemented by a server X can be stored by a server Y (a client of server X) and vice-versa, thereby enabling the construction of ORB-connected multi-scalable-databases.

Not every object stored in the database should be exposed to the distributed environment. Geological applications are characterised by a high granularity. They usually consist of an extremely large number of small-scale objects which do not have unique names. The lightweight objects should be hidden from the remote clients view to keep remote method invocation's overhead to the minimum. Traditionally, those issues will be addressed by adopting a convenient way of database views. The database view defines a subset of all persistent-capable classes (i.e. all classes, of which instantiations — objects — are stored in the databases). And further, for each individual class from that subset, the view can define a subgroup of all member variables/functions. In GeoStore, the database view is implemented by CORBA transient mediator objects.

The CORBA/ODBMS integration allows us to speed up all operations which need to process large amounts of data by performing them locally at the server site. In general, we can conclude that for successful use of a CORBA/ODBMS integration layer, the amount of data transfer between client and server is critical. The basic questions here are: which amount of data should be transferred to the client and how this data should be represented for the client. For example, not all of the triangles stored in the database for one geological boundary surface will be necessary for it's visualisation in different scales. It is useful to dynamically scale the surface representation and to minimise the network traffic between the server and the client.

Future developments of CORBA/ODBMS integration process will be certainly influenced by the modern rapidly evolving technologies such as XML and Java. In many real-world application domains CORBA/ODBMS integration simplifies the database access and improves the database navigation facilities. This is especially interesting for web browser access to databases. With the integration of the Java language into the ORB environment, Java applets can interact with the persistent CORBA objects through domain-specific interfaces, without any knowledge of how the objects are actually stored. The other question is how to describe data that will be transported to the client. What meta-information about the objects does the client need and how should this meta-information be represented? Therefore in the future we have to consider the application of XML for the development of generic definition of the data types transmitted, as XML is a standard and will be employed to enable data transfer between different applications across heterogeneous platforms.

References

Alms A., Balovnev O., Breunig M., Cremers A. B., Jentzsch T., Siehl A. (1998): Space-Time Modelling of the Lower Rhine Basin Supported by an Object-Oriented Database, in: Physics and Chemistry of the Earth, Vol. 23, No. 3, pp. 251–260.

Balovnev O., Bergmann A., Breunig M., Shumilov S. (1999): Remote Access to Active Spatial Data Repositories. In: Proceedings of the First International Workshop on Telegeoprocessing (TeleGeo'99), Lyon-Villeurbanne, France, pp. 125–130.

Balovnev O., Bergmann A., Breunig M., Cremers A. B. & Shumilov S. (1998): A CORBA-based approach to Data and Systems Integration for 3D Geoscientific Applications. In: Proceedings of the 8th Intern. Symposium on Spatial Data Handling (SDH'98), Vancouver, Canada, pp. 396–407.

Balovnev O., Breunig M., Cremers A. B. (1997): From GeoStore to GeoToolKit: The second step. In: Proceedings of the 5th Intern. Symposium on Spatial Databases. LNCS, Vol. 1262, Springer, Berlin, pp. 223–237.

Beckmann N., Kriegel H.-P., Schneider R., Seeger B. (1990): The R*-tree: an efficient and robust access method for points and rectangles. In: Proceedings of ACM SIGMOD, Atlantic City, N.Y., pp. 322–331.

Bode T., Breunig M., Cremers A. B. (1994): First experiences with GeoStore, an information system for geologically defined geometries. In: Proceedings IGIS '94, Ascona, LNCS, Vol. 884, Springer, Berlin, pp. 35–44.

Dikau, R. (1987): Entwurf einer geomorphographisch-analytischen Systematik von Reliefeinheiten. Heidelberger Geographische Bausteine 5.

Guttman A. (1984): R-trees: a dynamic index structure for spatial searching. In: Proceedings of ACM SIGMOD, Boston, MA, pp. 47–57.

Klesper C. (1994): Die rechnergestützte Modellierung eines 3D-Flächenverbandes der Erftscholle (Niederrheinische Bucht). PhD thesis, Geological Institute, University of Bonn. In: Berliner Geowissenschaftliche Abhandlungen (B) 22, 117 p.

Klett M., Schäfer A. (1996): Interpretation des Tertiärs der Niederrheinischen Bucht anhand von Bohrungen. — abstract; 148. Jahrestagung der Deutschen Geologischen Gesellschaft, 1.–3. Okt., Bonn.

Kugler, H. (1974): Das Georelief und seine kartographische Modellierung. Dissertation B. Martin-Luther-Universität Halle, Wittenberg.

Polthier K., Rumpf M. (1995): A concept for time-dependent processes. In: Goebel et al. (eds.), Visualization in Scientific Computing, Springer, Vienna, pp. 137–153.

Nievergelt J., Widmayer P. (1997): Spatial data structures: concepts and design choices. In: M. van Krefeld et al. (eds.), Algorithmic Foundations of Geographic Information Systems, LNCS, Tutorial No. 1340, Springer, Berlin, pp. 153–198.

Object Management Group (1999): CORBA/IIOP 2.3.1 Specification. OMG formal document. October 1999.
http://www.omg.org/corba/c2indx.htm.

Siehl A. (1993): Interaktive geometrische Modellierung geologischer Flächen und Körper, in: Die Geowissenschaften, Berlin, 11. Jahrg., Heft 10–11, pp. 342–346.

Shumilov S., Breunig M. (2000): Integration of 3D Geoscientific Visualisation Tools. In: Proceedings of the 6th EC-GI & GIS Workshop, Lyon, France.

Shumilov S., Cremers A. B. (2000): "eXtensible Database Adapter — a framework for CORBA/ODBMS integration", In: Proceedings of the 2nd International Workshop on Computer Science and Information Technologies (CSIT'00), Ufa, Russia.

W3C (1998): World Wide Web Consortium, Extensible Markup Language (XML) 1.0. http://www.w3.org/TR/1998/REC-xml-19980210.

Wagner M. (1998): Effiziente Unterstützung von 3D-Datenbankanfragen in einem GeoToolKit unter Verwendung von R-Bäumen, Diploma Thesis, Institute of Computer Science III, University of Bonn, 90 p.

Part III

Scale Problems in Physical Process Models

Scale Problems in Physical Process Models

Scale problems in physical process models always occur, when processes cannot be resolved anymore by the integrations of underlying first principles equations. Reasons for unresolved processes in models are manifold. Computer power constraints, which allow no higher-resolution grids, can be the reason. The contributions by Gerstner and Kunoth, and by Canarius, Helfrich and Brümmer, introduce and demonstrate novel mathematical analytical concepts to deal efficiently with sub-scale processes, when appropriate process models are known explicitly. The prior contribution explains the basic features of wavelets and multi-grid methods in the context of partial differential equations, which always form the backbone of most of our physical process models. It is shown with examples how the efficiency of numerical process models can be optimised using these concepts. The latter contribution dwells on analytical coupling of scales by mathematical homogenisation of the governing equations. With the processes and media qualities on the very fine scale known it is possible under certain conditions to derive equations for a larger scale, which include the processes on the finer scale. An example is shown by deriving a physical process model, which simulates water and solute transport in a porous medium including the uptake by roots.

Hydrology has other problems with scales in addition. Even when the processes are assumed to be known accurately enough, it is the variability of the medium which renders the determination of its transport characteristics on larger scales (interesting for applications, e. g. river catchments modelling) by measurements impossible. Further, the results of the physical processes on the large scales are dependent on the variability of the medium on the smaller scales. This implies, that averaging soil characteristics over larger scales and applying the model equations on this scale does not lead to a correct solution. Upscaling soil parameters from the small measurement scale to so-called effective ones on the model scale is thus a tricky business, which is discussed in the contribution by Diekkrüger in depth.

Often, however, the nature of the processes changes and the process complexity increases dramatically when approaching smaller scales from the larger scales. This is the typical situation in dealing with fluid motion in meteorology and geophysics. The problems are confounded when phase transitions are involved or when we come close to the boundaries of media. Here we have no hope to improve the situation by increasing resolution, either explicitly or analytically in the equations. In addition, only the effects of the processes on the small scales are known to some degree, like the turbulent fluxes from profile measurements, cloud processes from the rain at the surface, or sediment transport from ripples on the surface. The solution is often a sub-model based on the assumption of an analogy process with some constants to be determined from experiment — a process often termed parameterisation. The contributions by Raschendorfer, Simmer, and Gross introduce the reader in depth into the general parameterisation concept applied in meteorology and illustrate the methodology for turbulent transports in the boundary layer to solve these problems. Heinemann and Reudenbach explain the intricate problems, scale concepts and interaction modelling used for cloud and precipitation processes and satellite remote sensing. How the transport of particulate matter close to the fluid boundary can be adequately dealt with and ripple production quantitatively explained by continuum models of sediment transport is demonstrated by Hergarten, Hinterkausen and Küpper in their contribution. Finally, Mendel, Hergarten and Neugebauer develop an explicitely scale-separated, multi-scale model, which encompasses the fluid processes in an arbitrarily structured soil, both far away and infinitely close to the fine roots, water uptake by physiological processes at the root surface, and the transport of fluid within the carrying vessels of a tree, thus crossing the frontier between physics and life science.

Wavelet and Multigrid Methods for Convection-Diffusion Equations

Thomas Gerstner, Frank Kiefer, and Angela Kunoth*

Institute for Applied Mathematics, University of Bonn, Germany

Abstract. Multi-scale methods such as wavelet and multigrid schemes are efficient algorithms for the numerical solution of large classes of partial differential equations (PDEs). In this chapter we illustrate the basic ideas behind iterative methods for the solution of large linear systems which arise from the discretisation of PDEs. Besides describing the concepts of wavelet preconditioning, we summarise the main ingredients of multigrid methods. Examples from computational geosciences will show the importance of multi-scale methods for the simulation of large and complex systems.

The simulation of physical processes typically requires the numerical solution of various types of partial differential equations that model the processes (see e. g. Griebel, Dornseifer and Neunhoeffer (1998), Hattendorf, Hergarten and Neugebauer (1999)). Discretising the equations on realistic domains in three dimensions, one finally faces the problem to solve very large but sparsely populated systems of linear equations. While the amount of unknowns no longer allows for direct solution techniques to decompose the system matrices, classical iterative methods converge only very slowly. This is due to the fact that standard (single–scale) discretisations lead to system matrices with condition numbers depending on the discretisation in such a way that they grow very fast as the discretisation step size becomes finer. These large condition numbers in turn are the reason for the slow convergence of iterative methods.

In contrast, wavelet and multigrid methods provide the solution of very large and sparse systems up to prescribed accuracy quickly in a *fixed* number of iterations. The key to their success is the fact that the differential equations are discretised with respect to *different* discretisation step sizes or *grids* providing a hierarchical decomposition of the domain in order to capture the multi-scale behaviour of the solution. The solutions on the different scales are then appropriately catenated to build the solution of the initial system. This can be viewed as *preconditioning* the linear system of equations: by applying these multilevel methods one can achieve what is asymptotically the best possible, namely, that the convergence speed of the methods does not depend any more on the number of unknowns.

The necessary details for describing wavelet techniques are collected in Chapter 1.2. We briefly explain here how wavelets are used for preconditioning. The

---
* kunoth@iam.uni-bonn.de

basic mathematical ideas for multigrid methods are described in terms of *intergrid operators* called restriction and prolongation operators that map matrices and vectors living on one grid into structures on a grid of another length scale. Below we describe the construction and usage of both classical and wavelet-based multigrid methods for process simulation. However, before we do so, we derive a standard discretisation for a (partial) differential equation and discuss classical iterative methods in order to be able to precisely describe the effects of applying the multilevel techniques.

Discretisation of Partial Differential Equations There is indeed a variety of multilevel methods that can be viewed rather as a methodology than a single method. However, they all rely on a few basic principles that can be sufficiently described in terms of the simplest example of process modelling, a stationary (not time-dependent) one-dimensional partial differential equation. Namely, consider the boundary value problem

$$-u''(x) = f(x), \qquad 0 < x < 1,$$
$$u(0) = u(1) = 0. \qquad (1)$$

In order to solve this problem by a numerical method, one introduces a grid on the underlying domain $\Omega := (0,1)$ (which is an open interval here) and then computes the solution u of (1) approximately with respect to the grid points. Namely, decompose Ω into $N+1$ subintervals by defining (equidistant) grid points $x_j := jh$ where $h = (N+1)^{-1}$ is the grid size or resolution. We use the notation $\Omega_h := \{x_j\}_{j=0,\ldots,N+1}$ to refer to the grid.

In order to derive a discretisation of the differential equation (1) relative to the grid Ω_h, there are different methods available. Modern methods are often based on deriving a weak formulation of the differential equation which then allows for using for instance finite elements or wavelets. Since we want to describe the main ideas, here we apply a classical difference method. That is, one approximates the second partial derivative by a (second order) difference quotient, evaluated at the (interior) grid points,

$$-u''(x_j) \approx \frac{u(x_{j+1}) - 2u(x_j) + u(x_{j-1})}{h^2}, \qquad j = 1,\ldots,N. \qquad (2)$$

By setting $u_j := u(x_j)$ and observing that the boundary values in (1) give $u_0 = u_{N+1} = 0$, one obtains a linear system of N equations for the N unknowns $v := (u_1,\ldots,u_N)^T$ that approximate u at the grid points x_j,

$$Av = b. \qquad (3)$$

Here the vector $b \in \mathbb{R}^N$ is just the right hand side $f(x)$ of (1) evaluated on Ω_h, and A is a $N \times N$-matrix with entries determined by the differences in (2), that is,

$$A = \frac{1}{h^2} \begin{pmatrix} 2 & -1 & & & \\ -1 & 2 & -1 & & \\ & -1 & 2 & -1 & \\ & & \ddots & \ddots & \ddots \\ & & & -1 & 2 \end{pmatrix}. \quad (4)$$

Since A is nonsingular, the system (3) has for any given b a unique solution v. Note that A is a tridiagonal matrix because it has nonzero entries only on the diagonal and its side diagonals.

In order to describe different methods for solving the system (3), it is more instructive at this point to consider the two-dimensional version of the differential equation (1) and its discretisations. To this end, consider the Poisson problem on the unit square, involving a second order partial differential equation,

$$-\Delta u(x,y) := -\left(\frac{\partial^2}{\partial x^2} + \frac{\partial^2}{\partial y^2}\right) u(x,y) = f(x,y), \quad (x,y) \in \Omega := (0,1)^2,$$
$$u|_{\partial \Omega} = 0, \quad (5)$$

where $\partial \Omega$ denotes the boundary of Ω. For discretisation, associate with Ω again a grid Ω_h that consists of subdividing the x and the y axis into $N+1$ subintervals, yielding $\Omega_h := \{(x_i, y_k), \ 0 \leq i, k \leq N+1\}$. The discretisation of (5) is then performed analogously to the one-dimensional case, approximating now the second order partial derivatives by second order differences in the x and y direction each. To set up the corresponding linear system of equations, the grid points (x_i, y_k) are ordered (for instance) lexicographically, yielding the problem to determine $n := N \times N$ unknowns $v_j \approx u(x_i, y_k)$ that solve

$$Mv = b \quad (6)$$

where

$$M = \begin{pmatrix} A & -\tilde{I} & & & \\ -\tilde{I} & A & -\tilde{I} & & \\ & -\tilde{I} & A & -\tilde{I} & \\ & & \ddots & \ddots & \ddots \\ & & & -\tilde{I} & A \end{pmatrix}. \quad (7)$$

Here A is the matrix from the univariate problem (4), $\tilde{I} := h^{-2} I$ where I is the identity matrix of size N and b contains the values of the right hand side f evaluated on Ω_h. Like A above, the matrix M is nonsingular, admitting for any given right hand side a unique solution v^*. Note that M has an amount of entries which is proportional to the number of unknowns $n = N^2$. Such matrices are called *sparse*. However, in contrast to A, the matrix M has in addition to the tridiagonal pattern two side diagonals which are 'far away' from the main diagonal, its distance being determined by the number of unknowns in the respective spatial direction. These always occur in some form or another (using any ordering of the unknowns) when the spatial dimension is larger than one. Of course, if one increases the grid refinement and enlarges N, the size of the matrix grows quadratically. Sparse matrices are usually stored in a sparse format, that is, one stores only the nonzero entries and the necessary information on their position, therefore requiring $\mathcal{O}(n)$ amount of storage. If now the linear system $Mv = b$ is solved by any direct method like Gaussian elimination or QR-decomposition, this creates *fill-in*, that is, all the diagonals consisting of zeros between the main diagonals and the diagonal 'far away' fill up with non-zero entries, requiring a much higher amount of storage. In three spatial dimensions, this effect is even more dramatic.

Iterative Methods For that reason, one usually solves sparse systems of the form (6) by iterative methods. These require in principle only asymptotically the same amount of storage. Also by the discretisation of a boundary value problem (5) one already introduces a *discretisation error*. Thus, also the linear system needs to be solved only up to an accuracy determined by the discretisation error. Using a direct method instead which would solve $Mv = b$ exactly (up to machine precision) would be a waste of time.

Next we describe some basic iterative methods for the solution of linear systems $Av = b$, see e.g. Hackbusch (1994). For simplicity, for the remainder of this chapter, we consider the system stemming from the one-dimensional equation (1). The basic iterative methods are derived from writing the (scaled) equations (3) as a fixed point equation $v = v + C(b - Av)$ from which the iteration

$$v^{(i+1)} = (I - CA)v^{(i)} + Cb, \qquad i = 0, 1, 2, \ldots \tag{8}$$

follows. Then, $C \in \mathbb{R}^{n \times n}$ is some nonsingular matrix determined later and $I \in \mathbb{R}^{n \times n}$ is the identity. That is, starting from some initial guess $v^{(0)}$, one obtains by the assignment (8) the next iterate $v^{(1)}$, uses this to determine $v^{(2)}$, and so on, until a prescribed tolerance is met. Any iteration is called convergent if the iterates $\{v^{(i)}\}$ converge for an arbitrary initial guess $v^{(0)}$ to the (unique) solution v^* of (6). The iteration (8) converges if and only if for the spectral radius ρ,

$$\rho(I - CA) < 1,$$

holds, that is, the modulus of all eigenvalues of $I - CA$ is strictly smaller than 1. For convergent iterations, the value of $\rho(I - CA)$, the *convergence factor*, determines the *speed* of convergence. Roughly speaking, it is the factor by which the error $e^{(i)} := v^{(i)} - v^*$ is reduced in each iteration sweep. From this, one can also determine the amount of iterations that are required to reduce an initial error $e^{(0)} = v^{(0)} - v^*$ by a factor of 10^{-c} for a given constant c as follows. For some vector norm $\|\cdot\|$ let m be the smallest integer such that the error in the i-th iterate satisfies

$$\|e^{(i)}\| \leq 10^{-c} \|e^{(0)}\|,$$

which is approximately the case if

$$(\rho(I - CA))^m \leq 10^{-c}.$$

Wavelet and Multigrid Methods for Convection-Diffusion Equations

This is satisfied if
$$m \geq -\frac{c}{\log_{10}(\rho(I-CA))}.$$

The quantity $\log_{10}(\rho(I-CA))$ is called the *convergence rate*. It is the amount of iterations which are required to reduce the error by one decimal digit. Note that the closer $\rho(I-CA)$ is to 1, the smaller the convergence rate and thus the slower the convergence is.

Up to this point, we have not yet specified the role of the matrix C in (8). If $C = A^{-1}$, the inverse of A, then $I - CA = 0$ and the iteration (8) terminates after one step. But determining A^{-1} explicitly is just the problem that is to be avoided. Nevertheless, it gives an idea how one should choose C, namely, to approximate A^{-1} as good as possible at ideally low cost. Different possibilities stem from splitting A additively into

$$A = D + L + U, \tag{9}$$

where D is the diagonal, L the strict lower triangular and U the strict upper triangular part of A. Choosing now $C = D^{-1}$ is the approximation to A^{-1} which is easiest to compute, but also has mostly only very limited effects. This method (possibly with an appropriate weighting term) is called *Jacobi method*. Observing that triangular matrices can also be easily inverted (by forward or backward substitution), a more sophisticated choice which is still quickly computable is to take $C = (D+U)^{-1}$. This variant is commonly called *Gauss-Seidel method*. Both Jacobi and Gauss-Seidel methods are nowadays seldom used by themselves since their convergence (provided that they converge) is usually too slow. However, they are used as one of the basic ingredients in multigrid methods where these basic iterative methods are usually referred to as *smoothing* iterations. This is due to the fact that one can observe e.g. by Fourier mode analysis that small oscillations in the errors are quickly 'smoothed' out by such iterations.

Up to this point, the description of the iterative methods and the remarks on convergence apply to any linear system of equations. For the particular case (6), the matrix stemming from the discretisation of the partial differential equation (1) or (5) has particular properties. Namely, A (and also M defined in (7)) is symmetric and positive definite, thus A has only positive real eigenvalues. For such matrices, both the Jacobi and the Gauss-Seidel iterations converge. Their convergence speed depends on the (spectral) *condition number* of A which is given in terms of the largest and smallest eigenvalues of A,

$$\kappa_2(A) := \frac{\lambda_{\max}(A)}{\lambda_{\min}(A)}.$$

Essentially the spectral radius ρ for each of these methods satisfies

$$\rho(I-CA) \approx \frac{\kappa_2(A)-1}{\kappa_2(A)+1} = 1 - \frac{2}{\kappa_2(A)+1}$$

which is strictly smaller than one. However, for matrices A stemming from the discretisation of the partial differential equation (5), the condition number is

proportional to h^{-2},
$$\kappa_2(A) = \mathcal{O}(h^{-2}) \tag{10}$$
resulting in a spectral radius which is very close to 1. Thus, if one wants to decrease the grid size h in order to obtain a better approximation to the solution of the partial differential equation, this results in growing condition numbers and exceedingly slow convergence of such methods. Here one speaks of an *ill-conditioned* matrix A.

Preconditioning The other extreme is when the matrix C in the iteration (8) can be found in such a way that the spectral radius of $I - CA$ does *not* depend on the grid spacing h at all. There are essentially two strategies which can achieve that. They can by summarised as *preconditioning techniques* since they massively improve the condition number of the system matrix. The first strategy is denoted by *additive preconditioning technique*. There are basically two techniques that make up this class, namely, the *Fast Wavelet Transform* and the *BPX Preconditioner* (Dahmen and Kunoth 1992). The first technique consists of applying an appropriately scaled version of the Fast Wavelet Transform described in Chapter 1.2, yielding a transformed matrix $\tilde{A}$ which satisfies $\kappa_2(\tilde{A}) = \mathcal{O}(1)$ *independent* of $h \sim 2^{-J}$.

Table 1. Condition numbers for $-u'' + u = 1$ using periodised biorthogonal spline-wavelets.

Level J	$d = 2$, $j_0 = 0$		
	no prec.	D^{-1}	$A_{\Psi J}^{-1}$
3	2.73e+02	5.17e+00	1.62e+00
4	1.11e+03	5.59e+00	1.95e+00
5	4.43e+03	6.13e+00	2.27e+00
6	1.77e+04	6.50e+00	2.51e+00
7	7.08e+04	6.83e+00	2.72e+00
8	2.83e+05	7.09e+00	2.88e+00
9	x	7.32e+00	3.01e+00

Here we present some numerical results from Kunoth, Piquemal and Vorloeper (2000). In Table 1 we have discretised the one-dimensional problem $-u'' + u = 1$ with periodic boundary conditions $u(0) = u(1)$ by means of periodised biorthogonal spline-wavelets which are based on the piecewise linear hat function (indicated by $d = 2$) and coarsest level 0. The second column labelled no prec. shows the effect (10) on the condition number with growing finest level J. ('x' indicates that the condition number has not been computed.) Using the wavelet preconditioner indicated by D^{-1}, one sees in the third column that the condition number stays uniformly bounded independent of J. A further reduction in the absolute values of the condition numbers is achieved by taking into

Table 2. Condition numbers for $-u''+u=1$, $u(0)=u(1)=0$, using boundary adapted wavelets.

Level J	$d=2$, $j_0=3$		
	no prec.	D^{-1}	$A_{\Psi J}^{-1}$
4	1.81e+02	4.59e+01	2.34e+01
5	7.40e+02	5.51e+01	2.48e+01
6	2.96e+03	6.12e+01	2.58e+01
7	1.19e+04	6.66e+01	2.61e+01
8	4.75e+04	7.07e+01	2.64e+01
9	1.90e+05	7.41e+01	2.67e+01

account a scaling by the inverse of the diagonal of A in terms of the wavelet representation instead of the diagonal scaling, listed in the last table and denoted by $A_{\Psi J}^{-1}$.

For many applications, periodic boundary conditions are not realistic. Therefore we have also tested the effect of using wavelets which are adapted to bounded domains, which is an interval here, described in Chapter 1.2, (see also Dahmen, Kunoth and Urban 1999). The effect of the boundary adaption on the absolute values of the corresponding condition numbers can be seen in Table 2. Asymptotically, they still remain uniformly bounded. It is known that these numbers can be further reduced by more sophisticated strategies to construct wavelets adapted to intervals, see Barsch (2001) and Dahmen, Kunoth and Urban (1997). Another influence comes here from the fact that the coarsest level is 3 (for technical reasons in the construction due to the fact that then the functions living at each end of the boundary do not overlap).

The other strategy for preconditioning is the class of *multigrid methods* that can also be viewed as *multiplicative preconditioning techniques*. Both strategies are *optimal* in the sense that they provide the solution in a fixed amount of steps *independent* of the grid spacing h.

Multigrid Methods In the following, we will describe the basic ingredients of multigrid methods for the efficient solution of sparse linear systems which arise from the discretisation of elliptic partial differential equations (see Briggs, Henson and McCormick 2000, for an introduction as well as Braess 2001, and Hackbusch 1985). Multigrid methods consist of two complementary components,

- the *smoothing iteration*,
- and the *coarse grid correction*.

If one applies classical iterative methods of the form (8) (e.g. a Gauss-Seidel or Jacobi method) for the solution of such linear systems, these traditional solvers strongly damp out high frequency (*rough*) parts of the error between the discrete solution and the iterates with only a few iterations. However, low frequency (*smooth*) error components will be reduced only slowly and the error reduction is minimal as soon as the error essentially consists of only smooth components. In this sense traditional iterative methods are called *smoothers*. Now, low frequency error components may be represented on a coarser scale without loss of accuracy but with considerably less degrees of freedom. Thus, the central idea is to transport the smoothed error onto a coarser grid to reduce it there with the help of an appropriate coarse grid problem, implying lower computational costs. Since ideally the problem on the coarse grid has the same structure as the original problem on the fine grid, this procedure may be continued recursively, resulting in what is called *multigrid methods*.

In order to describe the coarse grid correction in somewhat more detail, it will be sufficient to consider only *two* different grids. The previously introduced grid Ω_h will play the role of the *fine grid*. In addition, we assume that there is a second grid Ω_H with respect to a coarser discretisation $H > h$ and such that the grid points of Ω_H are also grid points of Ω_h. Here we consider for simplicity

$$H = 2h.$$

We denote by $V_h = \mathbb{R}^{N_h}$ and $V_H = \mathbb{R}^{N_H}$ the spaces of corresponding grid functions, respectively. For transporting vectors and matrices from one grid to the other, one needs two linear *intergrid operators*. The one that takes vectors from the fine to the coarse grid is called *restriction operator*,

$$R_{h \to H} : V_h \to V_H; \tag{11}$$

its counterpart *prolongation operator*,

$$P_{H \to h} : V_H \to V_h. \tag{12}$$

For instance, one chooses $R_{h \to H}$ and $P_{H \to h}$ of the following form

$$R_{h \to H} = \frac{1}{2} \begin{pmatrix} \frac{1}{2} & 1 & \frac{1}{2} & & & & \\ & & \frac{1}{2} & 1 & \frac{1}{2} & & \\ & & & & \frac{1}{2} & 1 & \frac{1}{2} \\ & & & & & \ddots & \ddots & \ddots \end{pmatrix} \in \mathbb{R}^{N_h \times N_H}, \tag{13}$$

and

$$P_{H \to h} = \begin{pmatrix} \frac{1}{2} & & & \\ 1 & & & \\ \frac{1}{2} & \frac{1}{2} & & \\ & 1 & & \\ & \frac{1}{2} & \frac{1}{2} & \\ & & 1 & \\ & & \frac{1}{2} & \frac{1}{2} \\ & & & \ddots & \ddots & \ddots \end{pmatrix} \in \mathbb{R}^{N_H \times N_h}. \tag{14}$$

Wavelet and Multigrid Methods for Convection-Diffusion Equations 131

The prolongation and restriction operators are usually chosen such that they satisfy the *consistency conditions* $P_{H\to h} = c\,R_{h\to H}^T$ with some constant c which is 2 here.

Now the coarse grid correction can be described as follows. Let $v_h \in V_h$ be some approximation of the solution of the linear system $A_h v = b_h$ discretised with respect to the fine grid Ω_h. Denote by $r_h := b_h - A_h v_h$ the *residual*. The residual is transported onto the coarse grid Ω_H by applying the restriction operator $R_{h\to H}$ to r_h, yielding the vector $r := R_{h\to H} r_h$. Then the equation $A_H e = r$ is solved up to some accuracy by a basic iterative method, giving an approximation e_H to e. Finally one transports e_H back to the fine grid by applying the prolongation operator $P_{H\to h}$ and adds this to the previous approximation v_h, giving a new approximation

$$\tilde{v}_h := v_h + P_{H\to h} e_H. \tag{15}$$

The terminology 'coarse grid correction' stems from the fact that $\tilde{v}_h$ can formally also be expressed as

$$\tilde{v}_h := v_h + P_{H\to h} A_H^{-1} R_{h\to H}(b_h - A_h v_h) \tag{16}$$

where, of course, instead of the explicit computation of A_H^{-1} on the coarse grid the linear system is solved iteratively.

From the viewpoint of the wavelet terminology of Chapter 1.2 and subdivision schemes in Computer Aided Geometric Design, the intergrid operators $R_{h\to H}$ and $P_{H\to h}$ can be interpreted also as *mask matrices* (Dahlke and Kunoth 1994).

If one restricts the smoothing procedure to those degrees of freedom which are represented by grid points that do not belong to the coarser grid, one obtains the *Hierarchical basis Two-Grid Method* (Bank 1996, Bank, Dupont and Yserentant 1988). From the point of view of subspace correction methods (Xu 1992), the corresponding smoothing iteration can be considered as a correction step which is performed only with respect to a hierarchical complement space. This complement space is spanned by fine grid functions that are located in grid points which do not belong to the coarser grid anymore. Note that the Hierarchical Basis Two-Grid Method may also be interpreted as a block Gauss-Seidel method for the two scale hierarchically transformed linear system. In this sense it can be seen as an approximate block LU-factorisation of the hierarchically transformed discrete operator.

Working with more than two grids and applying the above procedures recursively, one can immediately generate (*Hierarchical Basis*) *Multigrid Methods*. There are many different variants when working with several grids. Some of the possibilities are starting from the finest and descending through all grids to the coarsest ('V-cycle') or moving up and down between intermediate grids ('W-cycle').

If one chooses wavelet-like basis functions to span the complement spaces within the hierarchical basis multigrid method, one obtains *Multi-scale Methods* which can be viewed as wavelet-like multigrid or generalised hierarchical basis methods (Griebel and Kiefer 2001, Kiefer 2001, Pflaum 2000, Vassilevski and

Wang 1998). They can be interpreted as ordinary multigrid methods which use a special kind of *multi-scale smoother* (Kiefer 2001). Furthermore, one can show an optimal convergence behaviour similar to classical multigrid (Vassilevski and Wang 1997) for 'nicely' elliptic problems. From a theoretical point of view this is due to the enhanced stability properties of the corresponding subspace decompositions (Carnicer, Dahmen and Peña 1996, Oswald 1994). From the multigrid interpretation the reason for this is the stronger effect of the respective smoothing iteration compared to the hierarchical one. However, since wavelet-based multi-scale smoothers still compute corrections only for parts of the unknowns, it is rather difficult to construct *robust* multi-scale solvers which do not depend on the mesh size used for discretisation *and* the coefficients of the partial differential operator involved. One way to construct robust solvers based on direct splittings of the underlying function spaces is a smart choice of the multiscale decompositions itself. In Kiefer (2001) a rather general Petrov-Galerkin multiscale concept is developed. Here, problem-dependent coarsening strategies known from robust multigrid techniques (matrix-dependent prolongations, algebraic multigrid (AMG)), (see Alcouffe et al. 1981, Briggs, Henson and McCormick 2000, Dendy 1982, Dendy 1983, Griebel 1990, Hemker et al. 1983, Oeltz 2001, Ruge and Stüben 1987, Stüben 1999, de Zeeuw 1990, de Zeeuw 1997) are applied together with specific wavelet-like and hierarchical multi-scale decompositions of the trial and test spaces on the finest grid. The main idea is to choose one complement space hierarchically (to implicitly generate physically meaningful coarse-grid operators) and the other one wavelet-like (to stabilise the multi-scale setting with respect to the mesh size, and to strengthen the smoother). The numerical results in Kiefer (2001) show that by this choice wavelet-type multi-scale solvers can be constructed which result in robust and efficient solvers. There also tables of multigrid convergence rates can be found for the following example. On the

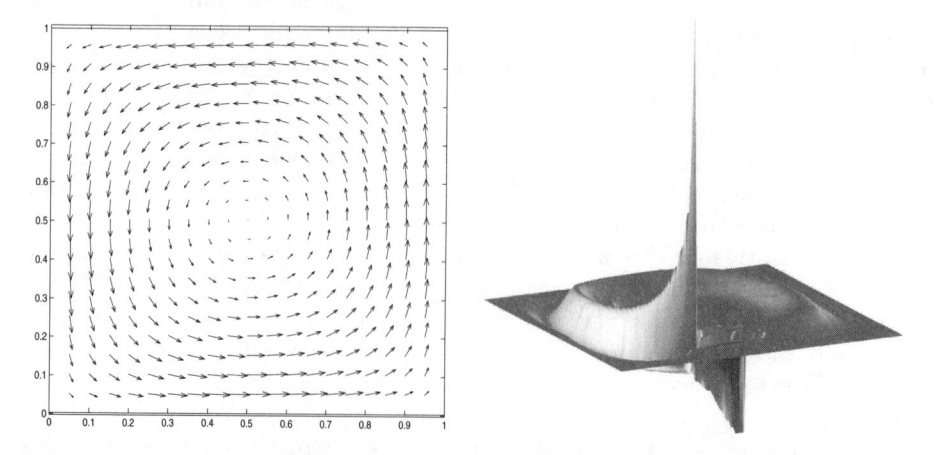

Fig. 1. Left picture: circular convective vector field; right picture: AMGlet generated by a wavelet-like stabilisation of the corresponding AMG-based hierarchical basis decomposition of the trial space.

left hand side of Figure 1 a circular convective vector field is shown. It belongs to the singularly perturbed convection-diffusion operatorq

$$A = -\Delta - 4 \cdot 10^6 \left(x(x-1)(1-2y)\, \partial_x - y(y-1)(1-2x)\, \partial_y \right)$$

subject to homogeneous Dirichlet boundary conditions on the unit square. Since the wavelet-type basis functions are implicitly generated by the AMG-approach, we call them *AMGlets*. On the right of Figure 1, one can see that apparently the AMGlet follows the convective flow and takes positive *and* negative values as a wavelet should.

References

R. E. Alcouffe, A. Brandt, J. E. Dendy, J. W. Painter. *The multi-grid method for the diffusion equation with strongly discontinuous coefficients*, SIAM J. Sci. Comput. 2, 1981, 430–454.

R. E. Bank, *Hierarchical bases and the finite element method*, Acta Numerica 5, 1996, 1–43.

R. E. Bank, T. F. Dupont, H. Yserentant, *The hierarchical basis multigrid method*, Numer. Math. 52, 1988, 427–458.

T. Barsch, *Adaptive Multiskalenverfahren für elliptische partielle Differentialgleichungen — Realisierung, Umsetzung und numerische Ergebnisse* (in German), IGPM, RWTH Aachen, Dissertation, 2001.

W. L. Briggs, V. E. Henson, S. F. McCormick, *A Multigrid Tutorial*, SIAM, Phiadelphia, 2000.

D. Braess, *Finite Elements: Theory, Fast Solvers and Applications in Solid Mechanics*, 2nd. ed., Cambridge University Press, Cambridge, 2001.

J. M. Carnicer, W. Dahmen, J. M. Peña, *Local decomposition of refinable spaces*, Appl. Comp. Harm. Anal. 3, 1996, 127–153.

S. Dahlke, A. Kunoth, *Biorthogonal wavelets and multigrid*, in: Adaptive Methods — Algorithms, Theory and Applications, Proceedings of the 9th GAMM Seminar, W. Hackbusch, G. Wittum (eds.), NNFM Series Vol. 46, Vieweg, Braunschweig, 1994, 99–119.

W. Dahmen, A. Kunoth, *Multilevel preconditioning*, Numer. Math. 63, 1992, 315–344.

W. Dahmen, A. Kunoth, K. Urban, *Biorthogonal spline-wavelets on the interval — Stability and moment conditions*, Appl. Comp. Harm. Anal. 6, 1999, 132–196.

W. Dahmen, A. Kunoth, K. Urban, *Wavelets in numerical analysis and their quantitative properties*, in: Surface Fitting and Multiresolution Methods, A. Le Méhauté, C. Rabut, L. L. Schumaker (eds.), Vanderbilt University Press, Nashville, TN, 1997, 93–130.

J. E. Dendy, *Black box multigrid*, J. Comput. Phys. 48, 1982, 366–386.

J. E. Dendy, *Black box multigrid for nonsymmetric problems*, Appl. Math. Comput. 13, 1983, 261–283.

M. Griebel, *Zur Lösung von Finite-Differenzen- und Finite-Element-Gleichungen mittels der Hierarchischen Transformations-Mehrgitter-Methode*, Technical Report 342/4/90 A, TU München, 1990.

M. Griebel, T. Dornseifer, T. Neunhoeffer, *Numerical Simulation in Fluid Dynamics, a Practical Introduction*, SIAM, Philadelphia, 1998.

M. Griebel, F. Kiefer, *Generalized hierarchical basis multigrid methods for convection-diffusion problems*, Manuskript, IAM, Universität Bonn, 2001.

W. Hackbusch, *Multigrid Methods and Applications*, Springer, New York, 1985.

W. Hackbusch, *Iterative Solution of Large Sparse Systems of Equations*, Springer, New York, 1994.

I. Hattendorf, S. Hergarten, H. J. Neugebauer, *Local slope stability analysis*, in: Process Modelling and Landform Evolution, S. Hergarten, H. J. Neugebauer (eds.), Lecture Notes in Earth Sciences, Springer No. 78, 1999, 169–185.

P. W. Hemker, R. Kettler, P. Wesseling, P. M. de Zeeuw, *Multigrid methods: Development of fast solvers*, Appl. Math. Comput. 13, 1983, 311–326.

F. Kiefer, *Multiskalen-Verfahren für Konvektions-Diffusions Probleme* (in German), Dissertation, IAM, Universität Bonn, 2001.

S. Knapek, *Matrix-dependent multigrid homogenization for diffusion problems*, SIAM J. Sci. Comput. 20, 1999, 515–533.

A. Kunoth, A.-S. Piquemal, J. Vorloeper, *Multilevel preconditioners for discretizations of elliptic boundary value problems — Experiences with different software packages*, IGPM-Preprint #194, RWTH Aachen, July 2000.

D. Oeltz, *Algebraische Mehrgitter-Methoden für Systeme partieller Differentialgleichungen* (in German), Diplomarbeit, IAM, Universität Bonn, 2001.

P. Oswald, *Multilevel Finite Element Approximations*, Teubner Skripten zur Numerik, Teubner, Stuttgart, 1994.

C. Pflaum, *Robust convergence of multilevel algorithms for convection-diffusion equations*, SIAM J. Numer. Anal. 37, 2000, 443–469.

J. W. Ruge, K. Stüben, *Algebraic multigrid*, in: Multigrid Methods, S. McCormick (ed.), SIAM, Philadelphia, 1987.

K. Stüben, *Algebraic multigrid (AMG): An introduction with applications*, Technical Report 53, GMD-Forschungszentrum Informationstechnik GmbH, St. Augustin, 1999.

P. S. Vassilevski, J. Wang, *Stabilizing the hierachical basis by approximate wavelets. I: Theory*, Numer. Lin. Algebra Appl. 4, 1997, 103–126.

P. S. Vassilevski, J. Wang, *Stabilizing the hierachical basis by approximate wavelets, II: Implementation and Numerical Results*. SIAM J. Sci. Comput. 20, 1998, 490–514.

J. Xu, *Iterative methods by space decomposition and subspace correction*, SIAM Rev. 34, 1992, 581–613.

P. M. de Zeeuw, *Matrix-dependent prolongations and restrictions in a black-box multigrid solver*, J. Comput. Appl. Math. 33, 1990, 1–27.

P. M. de Zeeuw, *Acceleration of Iterative Methods by Coarse Grid Corrections*, PhD thesis, University of Amsterdam, 1997.

Analytical Coupling of Scales — Transport of Water and Solutes in Porous Media

Thomas Canarius[1], Hans-Peter Helfrich[2]*, and G. W. Brümmer[3]

[1] Institute for Applied Mathematics, University of Bonn, Germany
[2] Mathematical Seminar of the Agricultural Faculty, University of Bonn, Germany
[3] Department of Soil Science, University of Bonn, Germany

Abstract. In almost all phenomena in the natural sciences many scales are present. Each of these scales contains specific information about the underlying physical process. *Upscaling* techniques allow the transfer of information, i. e., laws which are given on a micro-scale to laws valid on a *higher scale*. This means that the laws valid on the micro-scale are generalised to a higher scale, but does not mean that the laws are of the same type. By some typical examples, we describe the results of mathematical homogenisation. As an application, we consider the problem of fluid flow in porous media involving strong capillary forces as well as water uptake by roots.

1 Introduction

There are many applications for the technique of *mathematical homogenisation*, for example in material sciences, quantum physics and chemistry, continuum mechanics, hydrology, and many more.

Homogenisation was first developed for periodic structures because in many fields of science one has to solve boundary value problems in periodic media. Quite often the size of the period is small compared to the size of the total system and we look for an asymptotic description when the size of the period tends to zero. This results in a coarse scale description of processes that take place on different scales (in space, in time), which is coupled in an intricate way to a lower scale description. Furthermore, it can yield *effective operators* and *effective parameters*. On the micro- and meso-scale, the kind and rate of water and solute transport processes essentially depend on the aggregate and grain size distribution, on the porosity, and on the chemical and mineralogical composition of the porous media (soils, sediments, rocks). Already on the micro-scale, aggregate or particle sizes and pore sizes comprise more than ten orders of magnitude in porous media and vary from coarse aggregates/particles or macro-pores (diameter: cm, mm) to colloids or micro-pores (μm, nm) down to ions and molecules or nano- and subnano-pores (nm, pm). Combined with the different size ranges of aggregates or particles and pores, different kinds and rates of water and solute movement take place. In the range of coarse aggregates or particles and

---

* helfrich@uni-bonn.de

macro-pores, water exists in mobile form, and water flow can be described by Darcy's Law. Combined with the water flow, a transport of solutes takes place, strongly influenced by the affinity of the solutes to the solid and liquid phases and exchange processes between both phases of the porous media. In the range of colloids, molecules, and ions combined with micro- and nano-pores, water exists in immobile forms (Selim et al. 1995) and serves as a medium for diffusion of ions and molecules. The diffusion processes, which can be described by Fick's Laws, lead to an exchange of dissolved substances between immobile and mobile water phases and between immobile water and surfaces of solid particles, especially of colloid surfaces. In particular, the movement of substances with a high affinity to colloid surfaces (e. g., phosphorus, potassium, heavy metals) is influenced in this way by ad-, desorption and cation/anion exchange processes on colloid surfaces of humic substances, clay minerals, and iron and manganese oxides. In many cases, minerals of colloid sizes contain micro- and nano-pores or lattice defects. After the adsorption on external colloid surfaces, therefore, a diffusion into the interior of colloids (solid-state-diffusion) can take place (e. g., Brümmer et al. 1988). These processes on the molecular level, combined with the binding capacity of the soil colloids and their degree of saturation, determine the kind and rate of transport processes, especially of substances with a high matrix affinity. Therefore, deterministic models for the description of water and solute transport processes in porous media have to consider these processes on the micro-scale, which can be described by partial differential equations.

Mathematical homogenisation allows for the *upscaling* of models given by (partial) differential equations and provides a *homogenised* model. This incorporates interaction between the scales. The number of scales under consideration is not limited. There could be a high number of them or even infinitely many (see Allaire 1992, for further reading and references). In some cases, it is possible in the homogenised description to decouple the different scales by introducing a memory term τ. In this context, integro-differential equations (Vogt 1982, Peszynska 1995) or (nonlinear) storage terms (Bourgeat et al. 1996, Hornung et al. 1994) occur. For the case of a model for a chromatograph, the homogenisation limit was rigorously obtained (Vogt 1982) and the validity of the model was experimentally tested (Sattel-Schwind 1988).

Transport of solutes in porous media, especially in soil, in connection with diffusion processes can be modelled on mathematical homogenisation by partial differential equations with *memory terms* (Hornung and Showalter 1990, Hornung et al. 1994). For numerous percolation experiments, such models incorporating sorption processes were tested and applied to parameter estimation of relevant soil parameters (Niemeyer 2000, Spang 2000).

To describe the fluid flow in porous media, which has been done in the engineering literature and which has been justified in particular cases with rigorous mathematical methods, one can deploy *double porosity models*. Such models will be obtained by averaging local properties of the system over a representative elementary volume that contains both pore space and matrix space. A double porosity model of a porous media is a coarse grain media in which the fissures

Analytical Coupling of Scales

play the role of *pores* with high conductivity and the blocks of porous media play the role of *grains* with less conductivity, but which exist in a large number. In Hornung et al. 1994 a problem of diffusion, convection, and nonlinear chemical reactions in a periodic array of cells is studied. There it is assumed that in the cells there are porous bodies which are surrounded by semi-permeable membranes, i.e., fluxes and concentrations are coupled nonlinearly at the interfaces between the bodies and the surrounding fluid. Substances are diffusing and reacting in the fluid and in the solid part. On the interface between the fluid and the solid part continuity of the flux and nonlinear transmission conditions are considered and a homogenisation limit is rigorously derived.

2 Examples for Mathematical Homogenisation

In this section, we will give two examples for mathematical homogenisation. The characteristics of typical problems in this field are:

- The process is described by an initial boundary value problem for a partial differential equation.
- The medium to be considered is periodical.

Alternatively, one can consider in particular settings a medium with randomly placed heterogeneities, i.e., a porous medium where the permeability and the porosity are statistically homogeneous random fields oscillating at a dimensionless scale ε. Furthermore, the following assumptions are usually made:

- Essential parameters vary in a periodic way.
- The period ε is small in relation to the size of the domain.
- The scale parameter ε usually corresponds to the relation between the scale of the heterogeneities and the scale of the reservoir.

To illustrate the main ideas of the above, we will now give two simple, but instructive, examples. Later on, we will discuss one particular problem in more detail.

2.1 A Diffusion Problem

As a first example, we consider a stationary diffusion problem, where the diffusivity tensor A varies very rapidly. In order to give a mathematical description, we associate with the problem a small parameter ε that gives the period of the oscillation. We assume that the diffusivity tensor is given by $A = A(x, \frac{x}{\varepsilon})$, where $A(x,y) = (A_{i,j})(x,y)$ is a uniformly elliptic, essentially bounded matrix. The first variable $x \in \mathbb{R}^3$ is utilised for the slow variations of the matrix, whereas the second variable y gives, after scaling to $y = x/\varepsilon$ with a small *scale parameter* ε, the rapid oscillations. It is assumed that $A(x,y)$ is periodic in each component

y_i, $i = 1, 2, 3$ with period 1. Having this in mind, we are looking for a solution u^ε (e.g., concentration of a solute) of the boundary value problem

$$-\text{div}(A(x, \frac{x}{\varepsilon})\nabla u^\varepsilon) = f(x) \text{ in } \Omega,$$
$$u^\varepsilon = 0 \text{ on } \partial\Omega.$$

It is well known that the problem has a unique solution u^ε for each fixed ε. Using straightforward homogenisation (Allaire 1992), the homogenised problem

$$-\text{div}(A^*(x)\nabla u) = f(x) \text{ in } \Omega, \qquad (1)$$
$$u = 0 \text{ on } \partial\Omega \qquad (2)$$

is obtained, with

$$A^*_{i,j}(x) = \int_Q A(x,y)(\vec{e_i} + \nabla_y w_i(x,y)) \cdot (\vec{e_j} + \nabla_y w_j(x,y)) dy,$$

where $\vec{e_i}$ denotes the ith unit vector and $Q = [0,1]^3$ denotes the unit cube in $\mathbb{R}^3$. The functions $w_i(x, \cdot)$ are the periodic solutions of

$$-\text{div}_y(A(x,y)(\vec{e_i} + \nabla_y w_i(x,y))) = 0$$

with period 1 in the variable y.

Thus, we notice that the *effective equations (1),(2)* have the same type as the original equation, but the *effective operator* $A^*(x)$ is *not* obtained by simple averaging. In order to illustrate this, consider the one-dimensional case, where the matrix $A(x,y)$ is simply a scalar, positive coefficient $a(y)$ with period 1. Then the homogenised coefficient is given by (Jikov et al. 1994)

$$a^* = \frac{1}{\int \frac{1}{a(y)} dy}.$$

From a numerical point of view, it would require much effort to solve the equation (1),(2) for small ε. This is due to the fact that the number of elements for a fixed level of accuracy grows like $1/\varepsilon^N$.

2.2 Flow in a Porous Medium

The next example concerns the flow of a fluid in an unsaturated porous medium. Because it is only intended to show the occurrence of storage terms, we consider an oversimplified linear model. Here the porous medium is considered to be the soil or aggregated soils. We assume that the porous medium has a *double porosity* structure, i.e., it consists of two pore systems with different pore conductivity. The porous medium has a periodic structure as shown in Figure 1, where the bright region is a region of high conductivity, the dark one is of low conductivity.

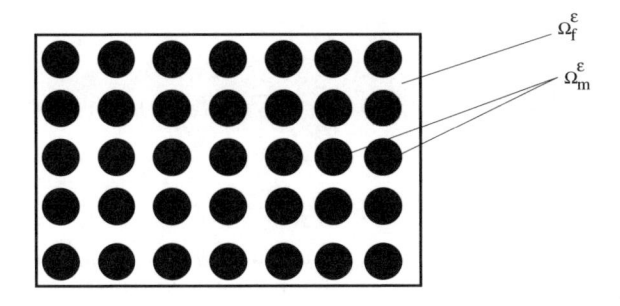

Fig. 1. Sketch of the geometry

A simple linear model for the fluid flow is given by

$$\frac{\partial \sigma_f^\varepsilon}{\partial t} - \text{div}(k(x/\varepsilon)\nabla \sigma_f^\varepsilon) = 0, \ x \in \Omega_f^\varepsilon, \ t > 0,$$

$$\frac{\partial \sigma_m^\varepsilon}{\partial t} - \varepsilon^2 \text{div}(k(x/\varepsilon)\nabla \sigma_m^\varepsilon) = 0, \ x \in \Omega_m^\varepsilon, \ t > 0.$$

Here σ_f^ε, σ_m^ε denote the flow in the region of high conductivity, resp. low conductivity. The scale factor ε^2 causes the process in Ω_m^ε proceeding on a different time scale than that in Ω_f^ε, i.e., we have t as time scale in Ω_f^ε and $\tau = \varepsilon^2 t$ as time scale in Ω_m^ε. In order to get a unique solution, we have to specify boundary and initial conditions. We choose as boundary condition on the boundary $\partial \Omega$ of the whole domain Ω a zero flux condition

$$(k(x/\varepsilon)\nabla \sigma_f^\varepsilon) \cdot \nu = 0, \ x \in \partial\Omega, \ t > 0,$$

where ν denotes the outer normal of the domain. Further, we assume that the flow

$$\sigma_f^\varepsilon = \sigma_m^\varepsilon, \ x \in \partial\Omega_m^\varepsilon, \ t > 0$$

is continuous across the common boundary of the pore and matrix space. The initial conditions

$$\sigma_m^\varepsilon(0, x) = \sigma^{\text{init}}(x), \ x \in \Omega,$$
$$\sigma_f^\varepsilon(0, x) = \sigma^{\text{init}}(x), \ x \in \Omega$$

state that the flows in both pore systems are given at time zero. Again, under appropriate assumptions existence and uniqueness for a unique solution for each fixed ε is well known. In the homogenisation limit which can be performed (Allaire 1992), a coupled system of differential equations is obtained that splits in two parts. For the coarse scale we get for the homogenisation limits σ_m, σ_f

of σ_f^ε and σ_m^ε

$$\frac{\partial \sigma_f}{\partial t} - \operatorname{div}(k^h \nabla \sigma_f) = -\frac{1}{|\Omega_1^m|} \int_{\Omega_m^1} k^h \nabla_y \sigma_m(x,y)\, dy, \quad x \in \Omega,\ t > 0, \qquad (3)$$

$$(k^h \nabla \sigma_f) \cdot \nu = 0, \quad x \in \partial\Omega,\ t > 0,$$
$$\sigma_f(0,x) = \sigma^{\mathrm{init}}(x), \quad x \in \Omega$$

where Ω_m^1 denotes a unit cell that is associated to each macroscopic point, and k^h is the homogenisation limit of the conductivity k. On the fine scale we have

$$\frac{\partial \sigma_m(t,x,y)}{\partial t} - \operatorname{div}_y(k^h \nabla_y \sigma_m(t,x,y)) = 0, \quad y \in \Omega_m^1,\ t > 0, \qquad (4)$$

$$\sigma_m(t,x,y) = \sigma_f(x), \quad x \in \partial\Omega_m^1, \qquad (5)$$
$$\sigma_m(0,x,y) = \sigma^{\mathrm{init}}(x), \quad x \in \Omega_m^1. \qquad (6)$$

We note that we get a global coarse scale equation, which describes the behaviour on the coarser scale in terms of space, as well as a local equation, which describes the mesoscopic behaviour and which is coupled to the coarse scale equation in every point. The local and the global equation belong to the same type; in the above case they are parabolic equations. In order to get the solution of the macroscopic problem, for each x we have to solve a three-dimensional microscopic problem. For that reason, a numerical solution of the coarse equations in this way would be very expensive.

Fortunately, we can make a further reduction of the equations (4), (5),(6) that is associated with a *type change*. In order to do this, we observe that the boundary condition (5) for the equation on the fine scale does not depend on y. Thus we can solve this equation by introducing the solution $r(y)$ of the microscopic problem

$$\frac{\partial r}{\partial t} - \operatorname{div}_y(k^h \nabla_y r) = 0, \quad y \in \Omega_m^1,\ t > 0,$$
$$r(y,t) = 1, \quad y \in \partial\Omega_m^1,\ t > 0,$$
$$r(y,0) = 0, \quad y \in \Omega_m^1$$

as a fundamental solution. We do not need this solution for each y, since only the mean value of the solution on the fine scale must be known for the solution of the equation (3) on the coarse scale. By taking the mean value

$$\tau(t) = \tfrac{1}{|\Omega_m^1|} \int_{\Omega_m^1} r(y,t)\, dy$$

we can compute the right hand side of (3) and obtain eventually an *integro-differential equation*

$$\frac{\partial \sigma_f}{\partial t} - \operatorname{div}(k^h \nabla \sigma_f) = -\int_0^t \frac{\partial \sigma_f}{\partial t}(t-s)\tau(s)\, ds, \quad x \in \Omega,\ t > 0$$

for the coarse scale. It follows that the contribution of the fine scale equation acts as a storage term or *memory term* for the high scale equation. Since the function $\tau(t)$ can be precomputed, we can describe the macroscopic flow by a single integro-differential equation. This means that we have now to solve an equation of a different type.

3 Application

We will now study the problem of fluid flow in porous media involving strong capillary forces, as well as water uptake by roots. As in the preceeding example, we assume that the porous medium is periodic with *double porosity* structure. The domain Ω splits into cubic cells with side length ε, where pore space and matrix space occurs in each cell. By Γ_1^ε we denote the common boundary of the pore space Ω_f^ε and the matrix space Ω_m^ε. The plant is modelled as having a root in each cell with boundary Γ_2^ε.

We assume that the flow of a fluid through the porous medium obeys Darcy's law in pore space as in matrix space. Constitutive relations for the water content S and the conductivity K are given as functions of the hydraulic pressure P, which are shown in Figure 2. The different order of magnitude in pore space and

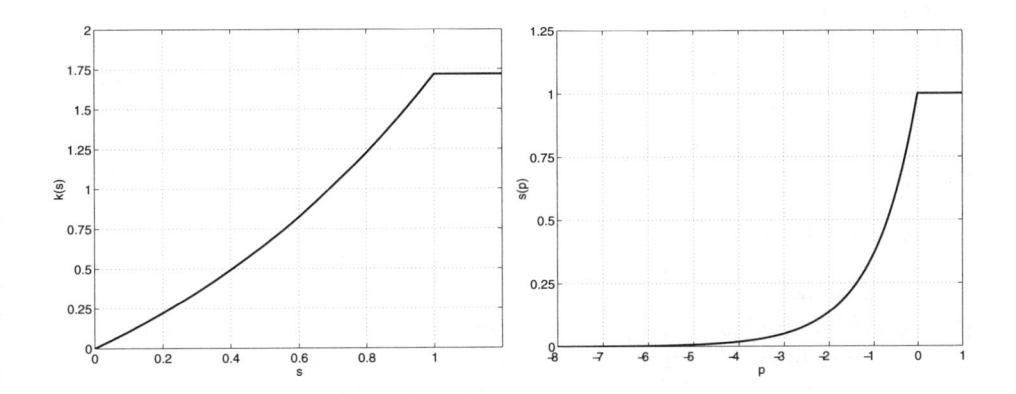

Fig. 2. Hydraulic conductivity and water content

in matrix space of the conductivity is controlled by a scaling parameter α_ε.

From the continuity equation and Darcy's law we get Richards' equation as a mesoscopic description for the hydraulic pressure P_m^ε in the matrix and P_f^ε in the pore space. We assume that the pressure is continuous at the common boundary Γ_1^ε of the pore and matrix space. At the boundary Γ_2^ε of the roots, we impose a continuity condition for the flow. For convenience, the boundary of the whole domain Ω is splitted in two parts Γ_D and Γ_N where the pressure is prescribed on Γ_D and a zero flux condition is given on Γ_N. By $\vec{e_n}$ we denote the unit vector in direction of the gravity. $\nu_{\Omega_m^\varepsilon}$, ν_Ω are the outer normals of Ω and Ω_m^ε, resp. Together with the continuity law, we obtain the equations for the pressure

$$\partial_t S(P_f^\varepsilon) - \text{div}(K(S(P_f^\varepsilon))(\nabla P_f^\varepsilon + \vec{e_n})) = 0 \text{ in } \Omega_f^\varepsilon,\ t > 0,$$

$$\partial_t S(\alpha_\varepsilon^{1/2} P_m^\varepsilon) - \alpha_\varepsilon \text{div}(K(S(\alpha_\varepsilon^{1/2} P_m^\varepsilon))(\nabla P_m^\varepsilon + \vec{e_n})) = 0 \text{ in } \Omega_m^\varepsilon,\ t > 0,$$

$$P_f^\varepsilon = P_m^\varepsilon \text{ on } \Gamma_1^\varepsilon,\ t > 0,$$

$$K(S(P_f^\varepsilon))(\nabla P_f^\varepsilon + \vec{e_n}) \cdot \nu_{\Omega_m^\varepsilon} = \alpha_\varepsilon K(S(\alpha_\varepsilon^{1/2} P_m^\varepsilon))(\nabla P_m^\varepsilon + \vec{e_n}) \cdot \nu_{\Omega_m^\varepsilon} \text{ on } \Gamma_2^\varepsilon,\ t > 0,$$

$$P_f^\varepsilon = P_D^\varepsilon \text{ on } \Gamma_D,\ t > 0,$$

$$K_f^\varepsilon(\nabla P_f^\varepsilon + \vec{e_n}) \cdot \nu_\Omega = 0 \text{ on } \Gamma_N,\ t > 0.$$

We now incorporate water uptake by plant roots by assuming specific outflow conditions on a periodic system of inner boundaries, one in each individual cell. We assume that the maximal outflow at any time is controlled and that a threshold has to be passed.

We take into account the influence of capillarity by the scaling of the saturation. Finally, we have incorporated the outflow of water into roots. We omit a detailed description here and do not give the variational equations, which describe the water uptake.

In order to derive the homogenisation limit, we have to use another important concept, the concept of Young measures. This has been treated in great detail (cf. Ball 1989). It allows us to derive a limit for certain functions, in the sense that we get a statistical description or local distribution for function values (see above). Once having established the existence of solutions for generic initial values, it is possible, by assuming an appropriate relationship between the ε depending parameters, to obtain a homogenisation limit which can be written formally as

$$\partial_t(|Y_f|S(P) + |Y_m|(1 - S_m)) - \text{div}(K^H(S(P))(\nabla P + \vec{e_n})) = \chi_{\{S^* > 0\}} k,$$

where the source/sink term k is determined by an elliptic equation. $|Y_f|$ and $|Y_m|$ are the volumina of the pore space and the matrix space in a cell of length $\varepsilon = 1$. Here S^* is the limit of the external saturation, S_m the limit of the matrix saturation, and K^H is the homogenisation limit of the conductivity K (s. Canarius 1999 for a more detailed discussion). $\chi_{\{S^* > 0\}}$ is a characteristic function, which takes the value 1 if $S^* > 0$ and is equal to 0 otherwise. We arrive at a coupling of a macroscopic/global equation with a mesoscopic, which determines k and which involves also nonlocal effects.

References

Arbogast, T., J. Douglas and U. Hornung (1990): Derivation of the double porosity model of single phase flow via homogenisation theory, SIAM J. Math. Anal., 21, 823–836.

Allaire, G. (1992): Homogenisation and two-scale convergence, SIAM J. Math. Anal., 23, 1482–1518.

Ball, J. (1989): A version of the fundamental theorem for Young measures, in: *PDEs and Continuum models for phase transitions*. Lecture Notes in Physics 344, M. Rascle, D. Serre, M. Slemrod (eds.), Springer-Verlag, Berlin-Heidelberg-New York, 287–315.

Bourgeat A., S. Kozlov, and A. Mikelic (1995): Effective equations of two phase flow in random media. Calculus of Variations and Partial Differential Equations, 3, 385–406.

Bourgeat A., S. Luckhaus, and A. Mikelic (1996): Convergence of the homogenisation process for a double porosity model of immiscible two-phase flow. SIAM J. Math. Anal., 27, 1520–1543.

Bourgeat A., A. Mikelic, and S. Wright (1994): Stochastic two-scale convergence in the mean and applications. J. Reine Angew. Math., 456, 19–51, 1994.

Brümmer, G. W., J. Gerth, and K. G. Tiller (1988): Reaction kinetics of the adsorption and desorption of nickel, zinc and cadmium by goethite. I. Adsorption and diffusion of metals. J. Soil Sci. 39, 37–52.

Canarius, T. (1999): Ein parabolisches Hindernisproblem mit Anwendungen in der mathematischen Hydrologie. Dissertation, Universität Bonn.

Hornung U., W. Jäger, and A. Mikelic (1994): Reactive transport through an array of cells with semi-permeable membranes. Mathematical Modelling and Numerical Analysis, 28, 59–94.

Hornung, U. and R. E. Showalter, Diffusion models for fractured media. Journal of Mathematical Analysis and Applications, 147, 69–80.

Jikov, V. V., S. M. Kozlov and O. A. Oleinik (1994): Homogenisation of Differential Operators and Integral Functionals. Springer Verlag, Berlin, Heidelberg, New York.

Niemeyer, A. (2000): Prozeßanalyse mit Hilfe von mathematischen und statistischen Methoden am Beispiel von Ad- und Desorption im Boden. Dissertation, Universität Bonn.

Peszynska M. (1995): Memory effects and microscale, in: Modelling and optimization of distributed parameter systems: application to engineering, K. Malanowski, Z. Nahorski, M. Peszynska (eds.). Chapman Hall, London-Glasgow-Weinheim, 361–367.

Sattel-Schwind, K. (1988): Untersuchung über Diffusionsvorgänge bei Gelpermeations-Chromatographie von Poly-p-Methylstyrol, Dissertation, Fachbereich Chemie, Universität Heidelberg.

Selim, H. M. and Liwang Ma (1995): Transport of reactive solutes in soils: A modified two-region approach. Soil Sci. Soc. Am. J. 59, 75–82.

Spang (2000): Messung und Modellierung von Sorptionsprozessen in einem Lößboden — Batch- und Säulenperkolationsexperimente mit Stoffen unterschiedlicher Matrixaffinität, Dissertation, Universität Bonn.

Vogt, C. (1982): A homogenisation theorem leading to a Volterra integro-differential equation for permeation chromatography, Preprint 155, SFB 123, Universität Heidelberg.

Upscaling of Hydrological Models by Means of Parameter Aggregation Technique

Bernd Diekkrüger*

Geographical Institute, University of Bonn, Germany

Abstract. The classical approaches to water and solute transport in soils have been developed under the assumption of homogeneity of the properties and boundary conditions. Often, this assumption is not valid for field conditions. Because heterogeneity can not be neglected for water and solute transport, methods for considering scale dependent spatial variability are needed. Different approaches to handling variability and uncertainty are presented here. The concept of effective parameters is discussed using different approaches. As the concept of effective parameters is not applicable in all cases, a method for considering the internal variability of the most important parameter is introduced. It is shown that upscaling can not be performed using a single method. Depending on data quality and data availability and on the problem that has to be solved, an appropriate upscaling method has to be chosen from the different available approaches.

1 Introduction

According to the definition of Blöschl (1996), scaling denotes a change in scale, either in time or in space. He differentiates between a more specific definition in which scaling is interpreted as the change in area or volume, and a broader definition in which scaling refers to all aspects of changing a scale including extrapolation, interpolation, aggregation and disaggregation. In this context upscaling means increasing and downscaling decreasing the scale. This is similar to the definition of scaling given by King (1991) who defines "scaling up" as "the translation or extrapolation of ecological information from small local scales to larger landscapes and regional scales".

In regional analysis two contradictory phenomena have to be considered: the increase of spatial variability with scale and the decrease of knowledge about this variability. On single sites, experiments can be carried out on a limited number of points. However, already for small catchments only soil maps are available for spatial analysis. These soil maps may have an acceptable resolution for small catchments (in Germany 1:5,000) but for larger areas only soil maps at the scale of 1:50,000, 1:200,000, or even 1:1,000,000 are typically available. The use of upscaling techniques is therefore limited by the availability of information. Most of the examples shown below relate to water transport in

---

* b.diekkrueger@uni-bonn.de

soils. Therefore, a brief description of the important equations is given.

Notations

q=$|\vec{q}|$	volumetric water flux density [cm/d]
S	S(x,y,z,t) = source and sink term [cm^3/cm^3/d]
Ψ_m	soil matric potential [hPa]
$|\Psi_m|$	soil suction [hPa]
Ψ_h	hydraulic potential [hPa]
Ψ_p	pressure potential [hPa]
Ψ_z	gravitational potential [hPa]
θ	water content [cm^3/cm^3]
Θ	$(\theta-\theta_r)/(\theta_s-\theta_r)$ = normalised water content [-]
θ_s	saturated water content [cm^3/cm^3]
θ_r	residual water content [cm^3/cm^3]
K_r	normalised hydraulic conductivity [-]
$K(\theta)$	= $K_r K_s$ = hydraulic conductivity [cm/d]
K_s	saturated hydraulic conductivity [cm/d]

One of the basic equations describing water transport in porous media is the mass conservation equation

$$\frac{\partial \theta}{\partial t} = -\nabla \vec{q} + S$$

which states that the rate of temporal change of the water content per unit volume equals the net gain of fluid in space per unit volume plus sinks or sources within this unit. This equation is complemented by a functional relationship between the flow vector q and the water content θ. Darcy's law relates the water flow to the gradient of a potential, the hydraulic potential Ψ_h, which in turn depends on the water content θ. The hydraulic potential is the sum of the gravitational potential, the matric potential and the pressure potential.

$$\Psi_h = \Psi_z + \Psi_m + \Psi_p$$

The gravitational potential is simply the vertical distance from a reference elevation. The matric potential is due to the adsorptive forces of the soil matrix. It depends on the water content of the soil. The pressure potential is related to the weight of the water column above the point under consideration. The latter potential is only important in saturated soils. The water flow is then related to the water potential by

$$\vec{q} = -K(\theta)\nabla\Psi_h.$$

It has to be noted that K is a function of the water content θ. This relationship is referred to as the conductivity curve. Combination of the equation of conservation of mass with Darcy's law yields the Richards equation:

$$\frac{\partial \theta}{\partial t} = \nabla(K(\theta)\nabla\Psi_h) + S.$$

More details can be found in Mendel et al. (2002). In order to solve Richards equations the basic soil hydrological properties, the retention and conductivity curves, have to be specified. It is useful, to introduce normalised hydraulic conductivities and water contents. The conductivity function is written in the form $K = K_s K_r$, where K_s denotes the saturated hydraulic conductivity, and K_r describes the functional relationship. K_r is referred to as the normalised hydraulic conductivity. The normalised water content is defined by

$$\Theta = \frac{\theta - \theta_r}{\theta_s - \theta_r}.$$

Several parameterisations of empirical relationships are in use. One of these approaches is the parameterisation of the retention and conductivity curves according to van Genuchten (1980) which are in wide use:

$$\Theta = \begin{cases} \frac{1}{(1+(\alpha|\Psi|)^n)^m} & \text{for} \quad \Psi < 0 \\ 1 & \text{for} \quad \Psi \geq 0 \end{cases}$$

where $m = 1 - 1/n$ and a and n are fitting parameters. In van Genuchten (1980), the conductivity curves have been parameterised as a function of the water content

$$K_r = \Theta^{1/2}(1 - (1 - \Theta^{1/m})^m)^2$$

Assuming that Richards equation is valid on all scales of interest, the problem is to determine the parameters of the retention and the conductivity curves for each scale. Since the water flux q depends in a nonlinear way on the soil hydrological properties, the expectation of $q(x, t, \Phi)$ cannot be obtained by simply inserting the expectations of the parameters into the equation for q (Smith and Diekkrüger 1996):

$$E[q(x, t, \Phi)] \neq q(x, t, E[\Phi])$$

in which Φ is the vector of the model parameters $\Phi = (\theta_s, \theta_r, K_s, \alpha, n)$. Therefore, other scaling concepts are required.

In this paper different methods for determining effective parameters are described. Depending on the model concept and the discretisation, one or even more soil units have to be considered within a discretisation unit. This requires different aggregation techniques and will be discussed below. The problem is further complicated if only the texture class is known but no measurements of the soil hydrological properties are available. Texture classes relate to classified data on sand, clay, and silt content of the soil. In this case the uncertainty concerning the basic information has to be considered. Due to the non-linear behaviour of the hydrological processes, effective parameters may not exist in all cases. An example is given here of how to deal with this problem.

2 Upscaling using effective parameters

Upscaling is important because the inherent random variability of properties such as soils, relief, and vegetation cannot be neglected if processes such as water and solute transport are to be simulated. Small scale variability means that differences from point to point may occur which results in differences in the state variables such as water content and solute concentration. Often, one is not interested in this small scale variability but in a mean response over a larger area. Furthermore, it may be too time consuming to consider small scale variability if one is only interested in the mean behaviour on a larger scale. Therefore, the parameters have to be aggregated into so called effective parameters. An effective parameter is defined as a parameter that represents the mean behaviour of a certain model result over a given area and time period. It has to be noted that the derivation of effective parameters requires the definition of an objective criterion. Different objective criteria may result in different effective parameters.

This will be illustrated by computing one-dimensional water and solute transport characteristics in porous media. The following concept is based on the notion of an ensemble of uncorrelated soil columns each with a different set of soil parameters from which effective parameters are derived. This concept implies that lateral flow in the soil can be neglected which is applicable if the correlation lengths of the underlying spatial fields are small so that microscale patterns only weakly influence the flow field. The expectation of a model result at any time t and any point x can be computed from the statistical distribution f of the model parameters (Richter et al. 1996):

$$E[q(x,t,\Phi)] = \int_\Phi q(x,t,\Phi)f(\Phi)d\Phi$$

As shown before, the water flux q depends in a nonlinear way on the soil hydrological properties. Therefore, the expectation of q cannot be obtained by simply inserting the expectations of the parameters.

For calculating effective parameters, the following steps have to be performed:

1. Define an objective criterion based on water fluxes, soil water content, solute concentration, or soil suction.
2. Calculate mean profiles or time series of the results of interest considering the ensemble of parameters. This can be performed by simulating all measured soil properties or, if the underlying probability density functions of the parameters are known, by either direct integration or by the Latin Hypercube (LH) method (see below and Richter et al. 1996).
3. Construct percentile bands for the profiles and time series at chosen locations or points in time.
4. Take either any parameter set that results in a profile or time series within a prescribed percentile band as effective parameter set, or search for a parameter set that best describes the mean behavior using inverse modelling techniques (see below).

3 Derivation of effective parameter using the Latin Hypercube approach

The derivation of effective parameters requires the calculation of the mean behavior of the model. An intuitive approach is to generate a random sample of the parameter vector and to derive estimators from the sampling statistics of the model output. This approach is referred to as Monte Carlo method. In many practical situations, the number of replications is limited by the running time of the numerical code. Therefore, sampling techniques have been developed that are more efficient than simple random sampling (McKay et al. 1979). These are stratified random sampling and Latin hypercube sampling. Let S denote the sampling space of the parameter vector Φ. Stratified sampling is performed by partitioning S into disjoint subsets and obtaining random samples from each subset. Latin hypercube sampling is an extension of stratified sampling. Assuming that the components of Φ are uncorrelated, the technical procedure is as follows:

i) The range of each component of the random vector Φ, Φ_i, is partitioned into n intervals. The intervals are spaced such that the probability that Φ_i falls into any interval is $1/n$. The intervals can easily be constructed from the distribution function of Φ_i (cf. Fig. 1). Let k denote the number of components of Φ. The sample space is thus partitioned into n^k cells.

ii) Select n cells such that all intervals are represented for each component. This is achieved by the so called Latin hypercube sampling method. The rationale behind this method is explained by a two dimensional example

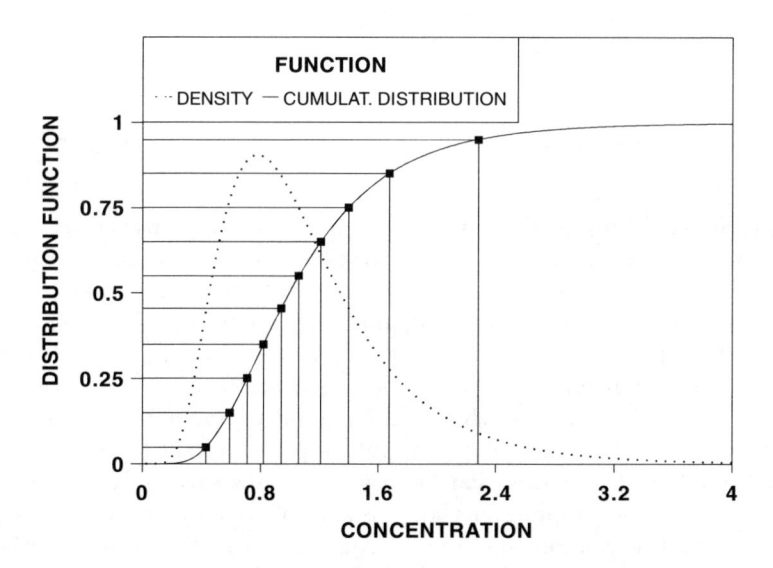

Fig. 1. Generation of realizations of random variables by means of their distribution function.

with n=5. The sample space is thus partitioned into 25 cells. Suppose now that a sample of size n=5 is to be obtained. A sampling scheme with the above properties can be constructed in the following way.

Let the rows and columns of a 5 × 5 square represent the partitions of component Φ_1 and Φ_2 respectively. Each cell is thus labeled by the interval number (i, j). The restriction that each interval must be represented in the sample is equivalent to the condition that each row and each column is sampled once. Let a, b, c, d, e denote the first column of the square. Then the following columns are constructed by permutation. Figure 2 shows a possible arrangement. A feasible random sample is then composed of all cells labeled a. Such an arrangement is called a Latin square. This method ensures that each component of the random vector Φ has all partitions of its sample space represented and that the components are properly matched. Extension to more than two dimensions leads to Latin hypercubes.

Φ_1

	1	2	3	4	5	
	a	b	c	d	e	1
	d	e	**a**	b	c	2
	b	c	d	e	**a**	3
	e	**a**	b	c	d	4
	c	d	e	**a**	b	5

Φ_2

Fig. 2. Latin square arrangement for a two-dimensional sampling problem.

In the above example all components could be easily matched ($n = 25$ replications). In practice, the typical number of parameters is on the order of 5 to 10. Consider the case of 5 parameters with 10 intervals for each parameter. Typical computation times for water transport models in the unsaturated zone including crop growth and evaporation over a vegetation period are in the order of 0.5 to 1 min per simulation on a PC. If for one replication all 10^5 partitions of the parameter space are simulated this would amount to about 70 days of computer time. It is obvious that the computer time is prohibitively large and that only a limited number of partitions can be used. As shown by McKay et al. (1979), the Latin hypercube sampling method has several advantages over random sampling and stratified random sampling. Using the Latin Hypercube method it is possible to consider any distribution function as well as correlated parameters.

The following example is based on a soil survey consisting of 86 sample points on a field of about 3 ha (Diekkrüger et al. 1996, Richter and Diekkrüger 1997).

At each sampling location, soil texture, organic matter content, bulk density as well as the soil water retention curve and the saturated hydraulic conductivity curves were measured at two depths. As an example, the saturated hydraulic conductivity is shown in Fig. 3. Due to the geological, pedological, and hydrological situation it can be assumed that the water flow at this flat site is mainly vertical and the lateral flow is neglected. Therefore, for each sampling location a one-dimensional model can be used. The extrapolation from the local, homogeneous scale to the large heterogeneous area is performed via the expectation of the model results. As in this example the data base is quite large, the known density function can be used for simulating water and solute transport. In order to test the suitability of the Latin Hypercube approach, the results obtained from using measurements were compared with the results from using the LH approach and results from the expectations of the probability density functions.

The vegetation period from March 1 to Oct. 31, 1994 was simulated. Barley was grown as crop on this field. It was assumed that the pesticide chlorotoluron and bromide as a tracer were applied. The pesticide degradation and sorption parameters were derived from the study of Pestemer et al. (1996). They vary with bulk density, organic matter content and water content. The simulated total

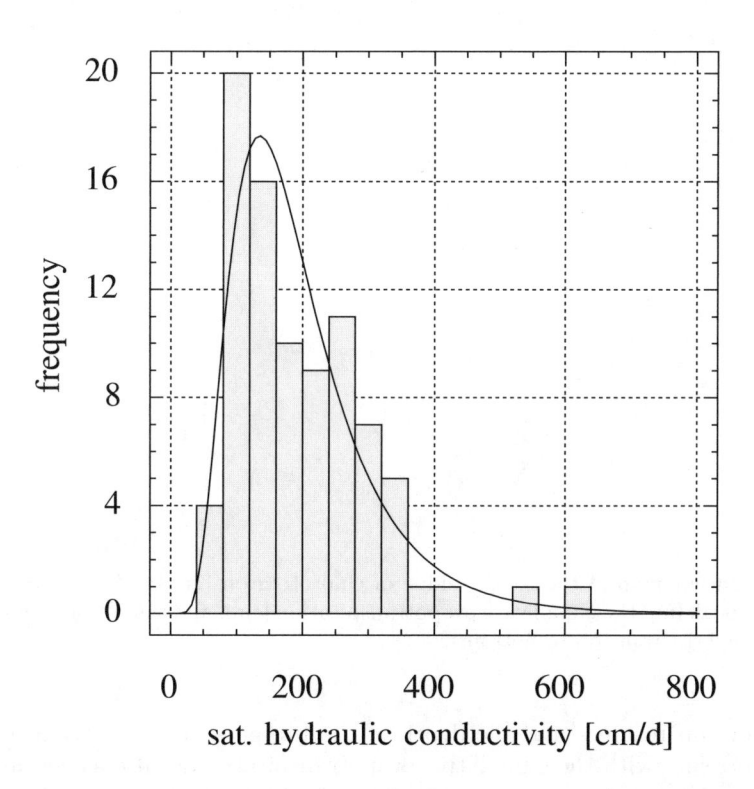

Fig. 3. Log-normal distribution of the saturated hydraulic conductivity K_s.

Fig. 4. Contour map of the total residue of chlorotoluron [mg kg^{-1}], the depth of the bromide peak [m], the cumulative evapotranspiration [cm], and the silt plus clay content [% by weight] (from top to bottom).

chlorotoluron residues and the depth of the bromide peak are given in Fig. 4. They correlate with the cumulative evapotranspiration as also shown in Fig. 4.

As described before, it is important that for deriving effective parameters one has to define criteria such as total pesticide residue or total evapotranspirtion.

The results of the different simulations are given in Tab. 1. For the chosen criteria 'total chlorotoluron (CT) residue', 'depth of bromide peak', and 'cumulative evapotranspiration' the averaged values are comparable with the simulation using mean values. Differences may occur when choosing other criteria. Effective parameters derived from the simulation results are given in Tab. 2. They were defined according to step 4 explained in chap. 2 considering all chosen criteria. That set of parameters was defined as effective set of parameters which best describes the mean behaviour μ_S or μ_{LH}. No inverse modelling technique has to be applied in this case. Although the mean simulation results are comparable, the effective parameters differ significantly.

4 Determination of effective parameters using the inverse modelling technique

Several methods have been developed for the identification of soil hydraulic parameters under laboratory conditions by solving the inverse problem. These methods require measured data such as water fluxes, water content or soil suction. Many authors are using water flux data from soil columns (one-step outflow experiments), where the boundary conditions are held constant (Hornung 1983,

Table 1. Mean values and standard deviation of the criterion measures 'total chlorotoluron (CT) residue', 'depth of bromide peak', and 'cumulative evapotranspiration'. The last column shows the results obtained by a simulation based on the expectations of the model parameters. Notation: LH = Latin Hypercube, S = simulation of sample points, M = mean parameter values, μ_S = mean result obtained from simulating all sample points, σ_S = corresponding standard deviation, μ_{LH} = mean result obtained using the Latin Hypercube method, σ_{LH} = corresponding standard deviation, μ_M = mean result obtained using mean of the parameter values.

criterion measure	μ_S	σ_S	μ_{LH}	σ_{LH}	μ_M
total CT residue [kg ha^{-1}]	0.237	0.163	0.260	0.222	0.236
depth of bromide peak [m]	0.855	0.265	0.843	0.224	0.875
cumulative evapotranspiration [mm]	386.3	19.6	385.5	1.760	382.6

Table 2. Effective parameters derived from the modal classes of the histograms of the criterion measures 'total chlorotoluron (CT) residue', 'depth of bromide peak', and 'cumulative evapotranspiration'. Notation: LH = Latin Hypercube, S = simulation of sample points, u = upper soil layer, l = lower soil layer.

	θ_s	θ_r	K_s	α	n	ρ	C_{org}
LH_u	0.37	0.090	312	0.046	1.50	1.49	0.98
S_u	0.33	0.0	120	0.038	1.20	1.58	0.95
LH_l	0.38	0.055	308	0.098	1.70	1.61	0.36
S_l	0.28	0.0	132	0.057	1.251	1.73	0.33

Parker et. al. 1985 a and b, Valiantzas and Kerkides 1990). Others are using water content and soil suction data measured in time and space (Kool and Parker 1988). All methods have shown that the identification of hydraulic parameters of the Richards equation is feasible although the problem is frequently ill conditioned. Ill conditioned means that the solution of the mathematical problem is not unique. Nevertheless, it is rarely shown that this method is also applicable under field conditions.

Feddes et al. (1993) suggested to use the inverse modelling technique to determine large scale effective soil hydraulic properties. In this study they simulated the 1-d water fluxes of 32 different sites using measurements of soil hydrological properties. The outputs of these simulations were averaged to obtain mean soil water profiles and areal evapotranspiration and infiltration. By using the inverse modelling technique one set of parameters was estimated which resulted in a simulation similar to the mean areal behavior. They were able to simulate water fluxes and water content to a certain accuracy and concluded that, if areal measurements of evapotranspiration and water content are available the inverse technique is able to determine effective parameters.

A similar study was carried out by Richter et al. (1996) who analyzed the applicability of the inverse modelling technique to field measurements of soil suction and water content. Because their field data exhibited small scale spatial variability their study is comparable to the approach of Feddes et al. (1993). However, the situation is more complicated because different soil layers have to be considered.

The Estimation Problem
Given are:
1. a dynamical model in terms of a partial differential equation with parameter vector $\Phi = (\Phi_1, \ldots, \Phi_p)$ and with initial and boundary conditions.
2. time series of measured state variables e. g. water content and/or soil suction, or water fluxes in one or several depth layers.
3. a performance criterion which measures the deviations between model predictions and observations.
4. an error structure.

The mathematical problem then is to find a parameter vector Φ which minimizes the performance criterion chosen.

Performance Criteria
The choice of the performance criterion has a large influence on the convergence of the optimization procedure. The most widely used criterion is the weighted sum of squared deviations between predicted and observed values.

$$L(\Phi|y_{11}, \ldots, y_{kn}) = \sum_{i=1}^{k} \sum_{j=1}^{n} w_{ij}[y_{ij} - \hat{y}_{ij}(\Phi)]^2$$

with weighting factors w_{ij}. The y_{ij} denote measurements taken at time t_j in depth layer i and the $\hat{y}_{ij}$ denote the solutions of the boundary value problem.

In water transport models time lags of several days between observed and simulated time series of the soil suction frequently occur and are acceptable. In this case the classical application of the least squares criterion results in large residual sum of squares and in a poor performance of the optimisation procedure. Convergence is frequently not achieved at all or the procedure converges to a fake minimum yielding parameter values that cause the trajectories to run through the mean values of the data.

It is therefore necessary to use alternative measures of deviations between model predictions and measurements. One possibility is to calculate least square distances in all variables which is also called orthogonal distance regression. The disadvantage is that in cases where large amplitudes at a small temporal scale are measured this method may try to minimise the deviation between an increase in measurements and a decrease in simulation results which is obviously not appropriate.

A third criterion is the difference of the areas under the simulated and measured time series where the time series of measured data is approximated by a polygon. This criterion is often called l_1-approximation and captures a possible bias. In the study of Arning (1994) it was shown that this criterion gives the best results when time lags occur.

The identification problem is thus to find a parameter vector Φ such that

$$L(\hat{\Phi}|y_{11},\ldots,y_{kl}) = \min_{\Phi \in U}[L(\Phi|y_{11},\ldots,y_{kl})]$$

where U denotes the admissible parameter space and, in this example, $\Phi = (K_s, \theta_s, \theta_r, \alpha, n)$. If data sets of water content θ and matrix potential Ψ are used simultaneously for the estimation, the total deviation is calculated as a weighted sum of both performance criteria. In this case, the weighting factors are necessary to compensate for different orders of magnitude between water content and soil suction. Parameter estimation was carried out using a combination of a deterministic one-dimensional simulation model which solves Richards' equation and the statistic software BMDP. The program package BMDP offers an interface for a user-written FORTRAN subroutine for derivate free non-linear regression problems. The algorithm is based on the Gauss-Newton scheme. BMDP delivers estimates of the variables, their standard deviation as well as the correlation matrix.

One objective of this investigation was to study the influence of different measurement periods on the estimates. The quality of the estimated sets of parameters was verified by comparing long-term simulations of about five years with field measurements. For all calibration periods it was possible to find a set of parameters which produce an acceptable agreement with the measurements. This is illustrated for one example in Fig. 5. When the estimated set of parameters is applied to a long-term forecast, often reliable simulation results are obtained (Fig. 6) for the upper and lower part of the soil profile. In the root zone a large range of soil suction and water content was measured while in the lower part of the profile wet conditions due to a perched water table occur during the whole

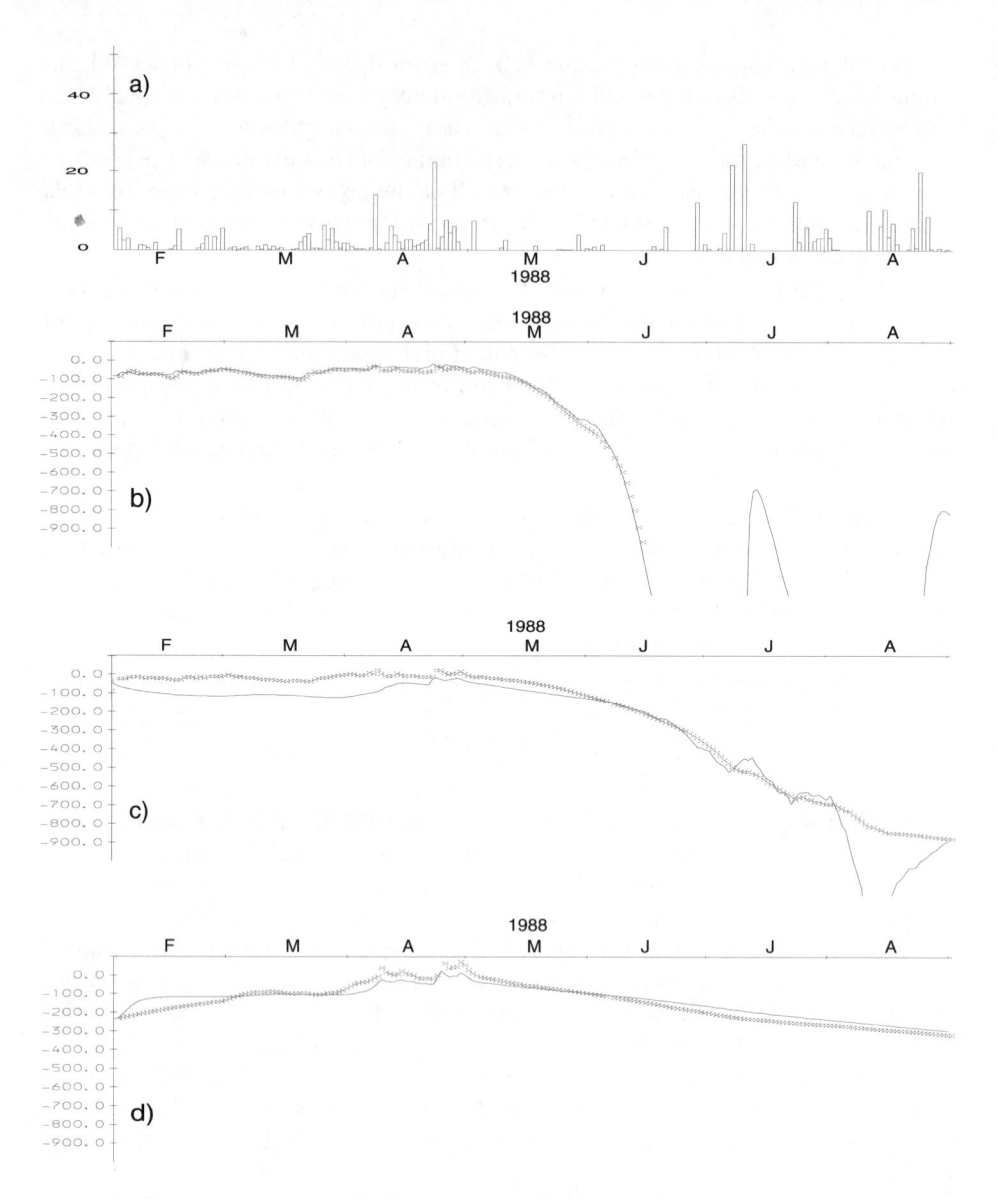

Fig. 5. Precipitation data [mm] (a), and simulated and measured soil suction [hPa] (b-d) for the calibration period Feb. to Aug. 1988 for the depths of 40 cm (b), 100 cm (c) and 210 cm (d).

period. As the temporal variability of the state variables is low below the root zone, the estimates for this layer exhibit high standard deviations.

This demonstrates that it is indispensable to estimate the saturated properties of the soil in a different way. One way is to use so called pedotransfer functions to estimate the soil hydraulic properties from soil texture. This would

Upscaling of Hydrological Models

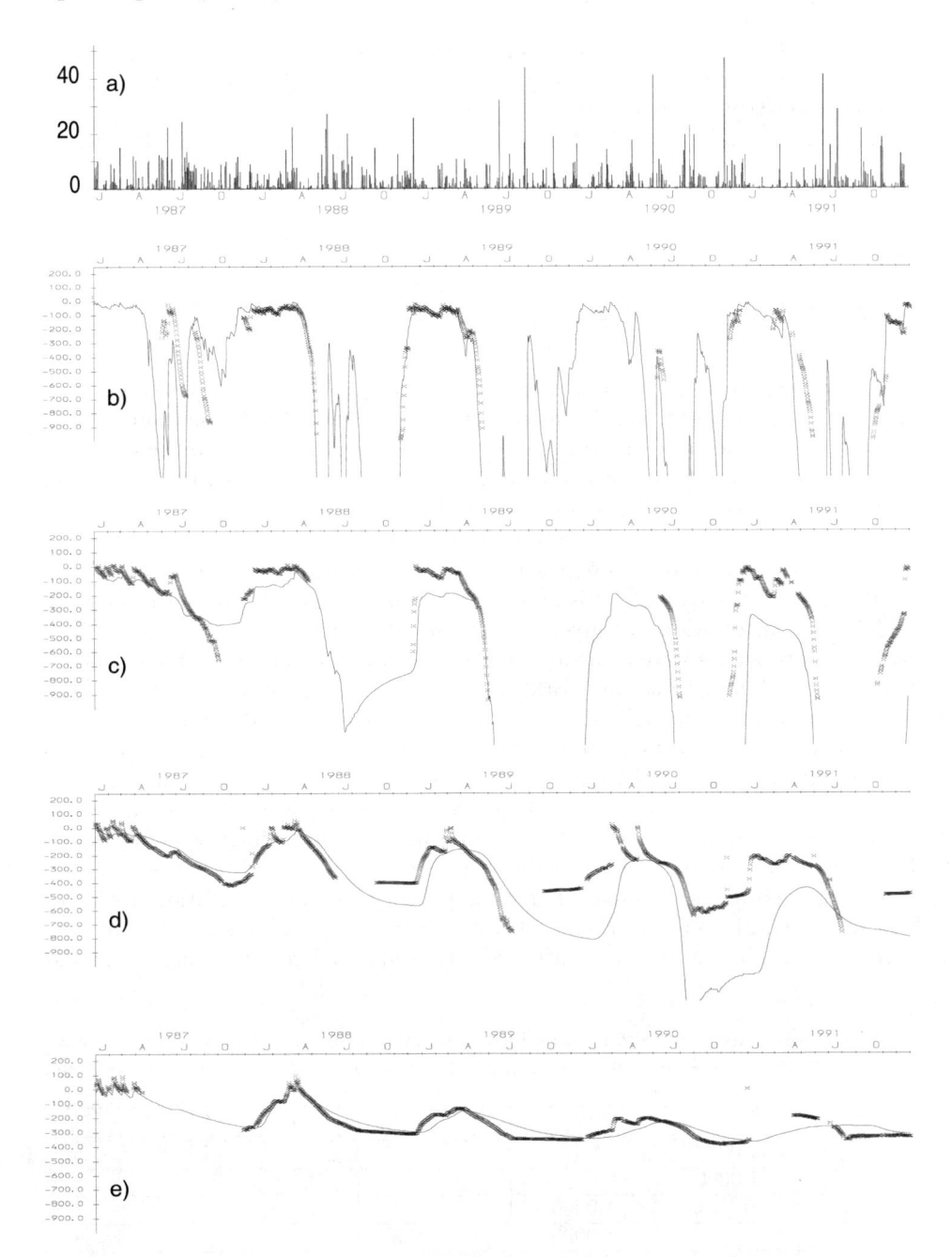

Fig. 6. Precipitation data [mm] (a) and simulated and measured soil suction [hPa] (b-e) for a five year period for the depths of 40 cm (b), 80 cm (c), 150 cm (d) and 240 cm (e). The parameters were estimated by solving the inverse problem.

Table 3. Simulated water balances (mm) for the period Jan. 1987 to Dec. 1991 using parameters measured in the laboratory, derived from a pedotransfer function and by solving the inverse problem for two different periods.

	Lab	PTF	P2 2/88–9/88	P5 12/90–6/91
groundwater recharge	579	538	605	510
evapotranspiration	2343	2397	2498	2519
evaporation	927	1066	1080	1060
transpiration	1416	1331	1418	1459

help to determine the wet part of the conductivity and retention curve in order to obtain a unique solution. It should further be noticed that if measured state variables are to be used, the range of e.g. soil suction is limited to the measurement range using a given measurement technique. If one measures soil suction using tensiometer cups this is limited to about 800–900 hPa which means that for the soils no measured soil suction data are available.

The performance of different sets of parameters can be evaluated by comparing the water balances. This is shown in Tab. 3 where groundwater recharge and evapotranspiration are given for the simulation period of five years. The hydraulic properties were measured in the laboratory, estimated using the pedotransfer function of Rawls and Brakensiek (1985) and estimated by solving the inverse problem for two different calibration periods. The differences between these simulations are small. It has to be noted that evapotranspiration exerts the largest influence on the model outcomes. As a good simulation of the state variables does not necessarily result in good simulated water fluxes, the water fluxes are uncertain. This is illustrated in Tab. 4 for an artificial one-layer situation using "measured" state variables. As can be seen, the correlation between some of the parameters is high. For some parameters, this correlation decreases and for others it increases if the saturated water content θ_s is fixed to a measured value. This means that although the simulated results using estimated

Table 4. Correlation matrix for an artificial one-layer situation. Upper values θ_s estimated using the inverse modelling technique, lower values θ_s fixed the measured value.

	θ_s	θ_r	K_s	α	θ_r
θ_s	1.0000				
	1.0000				
θ_r	-0.1756	1.0000			
	–	1.0000			
K_s	0.9939	-0.1323	1.0000		
	–	0.2152	1.0000		
α	0.9927	-0.1514	0.9992	1.0000	
	–	0.3289	0.8274	1.0000	
n	-0.9496	-0.8803	0.2783	-0.8988	1.0000
	–	-0.1562	0.8963	-0.6840	1.0000

effective parameters are reliable, the uncertainty concerning the parameters is quite high. Continuous measurements of evapotranspiration may help in addressing this problem. In this study it is assumed that the evapotranspiration model is calibrated correctly because the most important parameters were determined in lysimeter studies. However, it is difficult to measure e.g. temporally and spatially variable root lengths. This problem is more complicated in areas that are characterised by different land uses as can often be found in raster based models using large grid sizes (see below).

Nevertheless, considering the spatial and temporal variability of soil properties and boundary conditions, the results of this study are promising because they showed that plausible estimates of hydraulic properties are feasible. This means that the derivation of effective parameters using inverse modelling technique is practical.

5 Upscaling using aggregated parameters

Kabat et al. (1997) differentiate between effective and aggregated parameters. While in their approach effective parameters are related to areas with a single soil texture class, aggregated parameters refer to areas consisting of several soil texture classes. While, in regional simulations, effective parameters are required when the simulation units are derived from the intersection of soil map, land use map, relief information, etc. within a GIS, aggregated parameters are required for grid-based simulations in which the grid size is larger than the mapping units. For deriving effective parameters Kabat et al. (1997) applied the inverse modelling technique (see above).

The problem of deriving aggregated parameters often arises when atmospheric models are used which include approaches for simulating soil water dynamics. As grid size is often large, the inhomogeneities within a grid cell influence the simulated water fluxes significantly. Using areally dominant parameters may result in acceptable results over a number of grids but may also be completely wrong for a single grid. Following the idea of Noilhan and Lacarrere (1995), Kabat et al. (1997) derived a mean soil water retention curve and a mean hydraulic conductivity curve from a pedotransfer function using averages of the sand and the clay fraction. When applying the aggregated soil parameters they found that the simulated water fluxes were predicted within an accuracy of about 10 %.

Assuming that a clay, a loam, and a sand soil can be found within one grid cell with the same probability, one is interested in the mean effective (or aggregated) parameters. Therefore, in the first step one has to derive mean values of θ_s, θ_r, K_s, α and n from soil texture using e.g. the pedotransfer function of Rawls and Brakensiek (1985). As pointed out by Kabat et al. (1997), the consideration of the dominant soil type is not a feasible solution in this case. Even if 60 % of the soil within the area is a clay soil the mean behaviour may differ significantly from that of the clay soil. Therefore, the standard approach often chosen in GIS applications is not feasible.

Different approaches to calculating effective values are possible:

a) Calculate mean parameters of the soil hydrological properties. This can be done by considering the expected type of probability density function (normal – lognormal)
b) Calculate the ensemble retention $\theta_e(\Psi)$ and conductivity $K(\theta_e)$ curves by averaging the individual curves. As shown by Smith and Diekkrüger (1996) the resulting curve is not necessarily described by the same functional relationship as the original curves.
c) Calculate the mean texture from the given textures and apply the pedotransfer function to the mean value.
d) Calculate the integral of the most important property used in the simulation and search for a mean characteristic that fits to the integral property. If one is mainly interested in mean evapotranspiration the integral property for a simulation of a vegetation period would be the plant available water content θ_{paw}.

An example for calculating aggregated soil properties according to approaches a and c is given in Tab. 5. It can be seen that in this case the mean values differ only slightly. For different objective criteria chosen, no significant differences between the simulations may be obtained if both parameter sets were applied in a simulation model. As shown in the study of Smith and Diekkrüger (1996), soil water content profiles may differ significantly if different methods for obtaining effective parameters were compared. As discussed before, differences in soil water content may not necessarily lead to differences in water fluxes. Nevertheless, because solute transport and solute transformation may be more sensitive to changes in water content this is not a general finding in deriving effective parameters.

Often, only the texture class is known for a given site. If one has to derive model parameters from the texture class one has to relate a distinct sand, silt, and clay content to the area of interest. The uncertainty related to this disaggregation may cause significant differences in the simulation results. As is shown in Fig. 7 the ratio of fast (runoff + interflow) and slow (groundwater) flow components is significantly influenced by the methods of disaggregating the soil

Table 5. Soil texture derived soil hydrological properties and mean properties for the aggregated soil. [1]mean parameters derived from averaging soil texture. [2]mean parameters calculated from averaging soil properties. *lognormally distributed

soil type	sand content	clay content	porosity	θ_s	θ_r	K_s	$1/\alpha$	n-1	θ_{paw}
clay	5	60	0.55	43.4	12.0	1	80	0.16	17.6
sand	10	10	0.45	40.7	4.2	4.6	53	0.36	28.7
loam	70	5	0.35	29.8	5.0	46.4	16	0.49	12.0
mean[1]	28.3	25	0.45	35.9	9.1	5.0	39	0.28	18.5
mean[2]	-	-	-	38.0	7.1	6.0*	41*	0.34	19.4

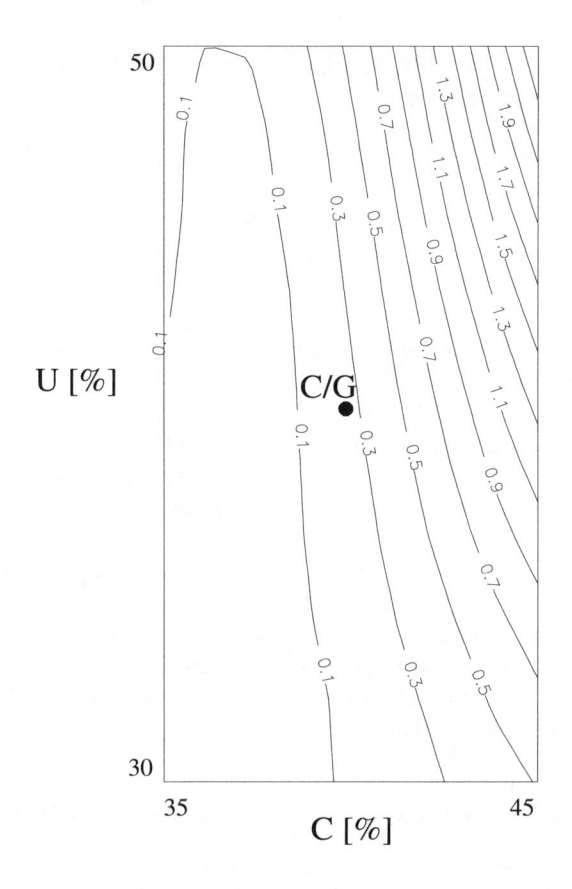

Fig. 7. Ratio of fast (runoff+interflow) to slow (groundwater) flow components within the texture class clayey loam (German classification). Interpolation was performed using the Kriging method. C/G = centre of gravity. C = clay content, U = silt content (redrawn from Bormann 2000).

texture class. Therefore, the definition of effective parameters has to consider uncertainties related to the data base. In this case, the centre of gravity (ratio of about 0.3) was not representative of the analysed texture class (mean ratio of about 0.6).

All approaches for deriving aggregated parameters are applicable if the boundary conditions and all other parameters such as vegetation and relief are spatially uniform. As land use patterns in general are independent from soil type distribution, and evapotranspiration is directly influenced by slope and aspect, this concept has to be further modified. Furthermore, as discussed below, runoff can not be computed correctly using mean or effective parameters.

6 Upscaling considering explicitly the subgrid variability of the important parameters

In some cases the concept of effective parameters fails. This will be illustrated by analysing the infiltration-runoff process. Numerous studies have pointed out that the saturated hydraulic conductivity is the most important parameter controlling infiltration. Let us assume a soil in which the inhomogeneity of the saturated hydraulic conductivity K_s can be described by a log-normal probability density function with the expectation of $10\,\text{cm}\,\text{d}^{-1}$ and a standard deviation of $10\,\text{cm}\,\text{d}^{-1}$. Effective K_s-values are defined here as those values that result in the same runoff volume as the detailed K_s-values. The runoff volume of a storm of a given intensity and a duration of 2 hours is computed using the Smith and Parlange (1978) infiltration equation. Runoff volume and effective K_s values are given in Tab. 6. As can be seen from the results, the effective parameter K_s is a function of the rainfall intensity which means that the definition of a constant effective parameter fails in this case.

As we have seen above the mean soil water behaviour could be described by effective soil hydrological parameters including K_s. Calculating the water fluxes and solute transport in the soil is rather time consuming compared to the calculation of the infiltration rate using an infiltration equation. Therefore, the following concept is proposed and realized in the Soil-Vegetation-Atmosphere-Transfer-model SIMULAT (Diekkrüger 1996):

a) Calculate the mean infiltration rate considering the variability of the most important parameter (here K_s) by means of the Latin Hypercube method.

Table 6. Results of the runoff production simulations. Parameters of the probability density function are $\mu_{ks} = 10\,\text{cm}\,\text{d}^{-1}$ and $\sigma_{ks} = 10\,\text{cm}\,\text{d}^{-1}$. The design storm had a uniform rainfall intensity and a duration of 2 hours.

rainfall intensity [cm d^{-1}]	runoff production [cm d^{-1}]	runoff coefficient [-]	runoff production using expected value of K_s	runoff coefficient using expected value of K_s [-]	effective K_s value [cm d^{-1}]
12	0.24	0.0150	0	0	2.80
18	1.32	0.0757	0	0	4.60
24	3.84	0.1614	0.36	0.0135	5.90
30	7.44	0.2491	3.48	0.1163	6.80
36	11.88	0.3290	8.04	0.2248	7.30
42	16.68	0.3972	13.32	0.3164	7.80
48	21.72	0.4533	18.84	0.3918	8.10
54	27.12	0.5027	24.48	0.4525	8.35
60	32.76	0.5457	30.24	0.5046	8.50
72	44.28	0.6144	41.88	0.5821	8.57
84	55.92	0.6661	53.52	0.6369	8.63
96	67.80	0.7058	65.40	0.6819	8.68
108	79.68	0.7377	77.40	0.7172	8.73
120	91.56	0.7635	89.40	0.7455	8.80

b) Calculate mean soil water fluxes using effective parameters and mean infiltration rates.

This concept is applicable if the averaging-power of the vegetation is smoothing out the variability in soil properties. This is frequently observed in soils where the plant available water is the limiting factor for evapotranspiration.

In grid based simulation models, a combination of methods for aggregated soils and methods for calculating the subgrid variability of the most important parameters has to be applied. Actually, this concept is further developed by considering the relief effects (slope, aspect, elevation) on the potential evapotranspiration. The combination of methods for calculating effective parameters with methods for calculating mean boundary conditions may lead to appropriate simulation results on the larger scale, while due to non-linear modul behaviour any of the two methods alone will not.

7 Conclusion

As shown in the examples upscaling can not be performed by using a single method. Depending on the problem, the data availability and the data quality different approaches are feasible. It is quite clear that soil properties can not be measured on the scale of interest. Therefore, one has to calculate properties in a scale dependent way. The rules for deriving effective parameters depend on the problem that has to be solved. In the case of pure water flow, different objective criteria are important than in the case of solute transport. Although it is often possible to define effective parameters, one has to carefully check under which conditions they are applicable as shown by the inverse modelling example. Because the effective parameter are often not only a property of the porous media but also of the boundary conditions, they may not always be time invariant as shown in the example considering the internal variability of runoff production.

References

Arning, M. (1994): Lösung des Inversproblems von partiellen Differentialgleichungen beim Wassertransport im Boden, Doctoral dissertation, Institute of Geography and Geoecology, Technical University of Braunschweig, Germany, 137 p.

Bormann, H. (2000): Hochskalieren prozeßorientierter Wassertransportmodelle — Methoden und Grenzen. Doctoral dissertation, Geographical Institute, University Bonn. Bonner Geographische Abhandlungen (in press).

Blöschl, G. (1996): Scale and scaling in hydrology. Habilitationsschrift. Wiener Mitteilungen, Band 132, 346 p.

Diekkrüger, B. (1996): SIMULAT — Ein Modellsystem zur Berechnung der Wasser- und Stoffdynamik landwirtschaftlich genutzter Standorte. In: Richter, O., Söndgerath, D., and B. Diekkrüger (eds.): Sonderforschungsbereich 179 "Wasser- und Stoffdynamik in Agrarökosystemen". Landschaftsökologie und Umweltforschung 24, pp. 30–47.

Diekkrüger, B., Flake, M., Kuhn, M., Nordmeyer, H., Rühling, I., and D. Söndgerath (1996): Arbeitsgruppe "Räumliche Variabilität". In: Richter, O., Söndgerath, D., and B. Diekkrüger (eds.): Sonderforschungsbereich 179 "Wasser- und Stoffdynamik in Agrarökosystemen". Landschaftsökologie und Umweltforschung 24, pp. 1232–1306.

Feddes, R. A., Menenti, M., Kabat, P., and W. G. M. Bastiaanssen (1993): Is large scale inverse modelling of unsaturated flow with areal average evaporation and surface soil moisture as estimated from remote sensing feasible? J. Hydrol. 143:125–152.

Hornung, U., 1983: Identification of nonlinear soil physical parameters from an input-output experiment In: P. Deuflhard and E. Harrier (eds.), Workshop on Numerical Treatment of Inverse Problems in Differential and Integral Equations, Birkhäuser, Boston, Mass., pp. 227–237.

Kabat, P., Hutjes, R. W. A., and R. A. Feddes (1997): The scaling characteristics of soil parameters: From plot scale heterogeneity to subgrid parameterization. J. Hydrol. 190:363–396.

King, A. W. (1991): Translating models across scales in the landscape. In: Turner, M. G. and R. H. Gardner: Quantitative methods in landscape ecology. pp. 479–517. Springer Verlag.

Kool, J. B. and Parker, J. C. (1988): Analysis of the inverse problem for transient unsaturated flow, Water Resour. Res., 24 (6), pp. 817–830.

McKay, M. D., Beckman, R. J., and W. J. Conover (1979): A comparison of three methods for selecting values of input variables in the analysis of output from a computer code. Technometrics. 2(21):239–245.

Mendel, M., Hergarten, S., and Neugebauer, H. J. (2002): Water uptake by plant roots — a multi-scale approach. This volume.

Noilhan, J. and. P. Lacarrere (1995): GCM gridscale evaporation from mesoscale modelling. J. Climate 2:206–233.

Parker, J. C., Kool, J. B., and van Genuchten, M. Th. (1985a): Determining soil hydraulic properties from one-step outflow experiments by parameter estimation: I. Theory and numerical studies, Soil Sci. Soc. Am. J., 49, pp. 1348–1354.

Parker, J. C., Kool, J. B. and van Genuchten, M. Th. (1985b): Determining soil hydraulic properties from one-step outflow experiments by parameter estimation: II. Experimental studies, Soil Sci. Soc. Am. J., 49, pp. 1354–1359.

Pestemer, W., Nordmeyer, H., Bunte, D., Heiermann, M., Krasel, G., and U. Walter (1996): Herbiziddynamik im Boden. In: Richter, O., Söndgerath, D., and B. Diekkrüger (eds.): Sonderforschungsbereich 179 "Wasser- und Stoffdynamik in Agrarökosystemen". Landschaftsökologie und Umweltforschung 24, pp. 537–597.

Richter, O., Diekkrüger, B., and P. Nörtersheuser (1996): Environmental fate modelling of pesticides: from the laboratory to the field scale. Verlag Chemie. 281 p.

Rawls, W. J. and D. L. Brakensiek (1985): Prediction of soil water properties for hydrological modeling. In: Proceedings of the symposium watershed management in the eighties. 293–299, Denver.

Richter, O. and B. Diekkrüger (1997): Translating environmental fate models of xenobiotic across scale. Hydrology and Earth System Sciences 4:895–904.

Smith, R. E. and B. Diekkrüger (1996): Effective Soil Water Characteristics and Ensemble Soil Water Profiles in Heterogeneous Soils. Water Resour. Res. 32.7:1993–2002.

Smith, R. E. and J.-Y. Parlange (1978): A parameter-efficient hydrologic infiltration model. Water Resour. Res. 14(3):533–538.

Valiantzas, J. D. and P. G. Kerkides (1990): A simple iterative method for the simultaneous determination of soil hydraulic properties from one-step outflow data, Water Resour. Res., 26, (1), pp. 143–152.

Van Genuchten, M. Th. (1980): A closed-form equation for predicting the hydraulic conductivity of unsaturated soils, Soil Sci. Soc. Am. J., 44, pp. 892–898.

Parameterisation of Turbulent Transport in the Atmosphere

Matthias Raschendorfer[2], Clemens Simmer[1]*, and Patrick Gross[1]

[1] Meteorological Institute, University of Bonn, Germany
[2] German Weather Service, Offenbach, Germany

Abstract. The thermo-hydrodynamic equations describing the state of the atmosphere can only be solved numerically with discrete space and time intervals. Therefore the equations have to be averaged over these intervals. Averaging the equations results in covariance products which represent the sub-grid scale (turbulence) contributions to the mean atmospheric flow. Since these covariance products add additional unknowns leading to a closure problem, they have to be parameterised by use of known quantities. The general concept of parameterisation is described on the basis of turbulent transport theory. In this context the two turbulence parameterisation schemes of the Lokal-Modell (LM), the current meso-scale limited area weather forecast model of the German Weather Service (DWD), are discussed in more detail. Differences in the simulated atmospheric states resulting from these schemes are shown in a case study.

1 Why do meteorologists need parameterisations?

1.1 Basic set of equations

The atmosphere being mainly constituted of the three components dry air, water vapour and liquid water can be described accurately enough by the equation of state for an ideal gas:

$$p = \rho RT \tag{1}$$

with p, total atmospheric pressure, ρ, the air density, T, the air temperature, and R, the specific gas constant of the air mixture. In addition a couple of budget equations are applicable to some properties ϕ of the atmosphere, which all can be written in the following general form:

$$\partial_t(\rho\phi) + \nabla \cdot (\rho\phi\underline{v} - a^\phi \nabla\phi) = Q^\phi. \tag{2}$$

Here $\underline{v} := (v_1, v_2, v_3) = (u, v, w)$ is the wind vector, ∂_t the local time derivative and $\nabla := (\partial_1, \partial_2, \partial_3)$ the Nabla-operator with ∂_i the local derivative in i-direction of a local Cartesian coordinate system, where $i = 3$ belongs to the

---
* csimmer@uni-bonn.de

vertical upward direction for a given point in the atmosphere. Q^ϕ is the source term of a property ϕ and $a^\phi := \rho k^\phi$ is the dynamic molecular diffusion coefficient, assumed to be a constant, rather than the corresponding kinematic coefficient k^ϕ.

In order to get a closed set of equations for the state of the atmosphere, we need such equations for the following seven properties:

- $\phi = v_i$: velocity component in i-direction (momentum density), with the source term

$$Q^{v_i} = -\partial_i p + \rho g_i - 2\rho[\underline{\Omega} \times \underline{v}]_i, \quad (3)$$

with $\underline{g} = (0,0,-g)$ the vector of earth gravity, $\underline{\Omega}$ the vector of the angular velocity of the earth. Further, in (2) for $i = 1, 2, 3$, $\mu := a^{v_i}$ is the kinematic viscosity.

- $\phi = T$: temperature, with source term

$$Q^T = \frac{(Q^h + d_t p)}{c_p} \quad (4)$$

with Q^h the diabatic and isobaric heat source, c_p specific heat capacity of the air mixture and $d_t := \partial_t + \underline{v} \cdot \nabla$ the total time derivative
- $\phi = q_v$: mass fraction of water vapour
- $\phi = q_l$: mass fraction of liquid water
- $\phi = 1$: unity, with source term $Q^1 = 0$ (continuity equation)

We will omit here the explicit presentation of the more complicated source terms for the water phases as well as their laminar diffusion coefficients for brevity. These seven equations are often called the primitive equations in meteorological jargon. Here, the three budget equations for v_i represent the local temporal change of the three-dimensional wind vector, which is controlled by the transport of the wind components as well as the pressure gradient, gravity, and Coriolis forces. The underlying physical principle of the equation of motion is the conservation law for momentum. The conservation of inner energy derived from the first law of thermodynamics is inherent in the equation of the temperature tendency. Besides temperature advection, the source term Q^T is responsible for a temperature change. It contains the pressure work $d_t p$ and the diabatic and isobaric heating Q^h, which includes radiative heating, the convergence of diffusive flux densities, as well as heating due to phase changes of water. The latter sources and sinks also occur in the budget equations for water vapour and liquid water. These budget equations, as well as the continuity equation, are based on the fundamental law for mass conservation. Together with the equation of state the above eight so-called thermo-hydrodynamical equations form a closed set to determine the state of the atmosphere (described by the eight variables $v_1, v_2, v_3, T, q_v, q_l, p$ and ρ), if the source terms are known. These equations, or appropriate approximations form the core of any numerical weather prediction or climate model.

More advanced meteorological models separate the liquid phase into cloud water and rain water and often include another budget equation for the solid water phase. Even more detailed models exist (so-called spectral cloud models), which further separate the different hydrometeor classes into size classes with individual budget equations, which can be even further expanded by additional budget equations for aerosol classes with complicated physic-chemical source terms describing the interaction between the cloud particles.

The prognostic equations — these are the seven equations with temporal change terms mentioned above — are defined in terms of differential operators in space and time. Therefore they are only valid in the limit of infinitesimal temporal interval and spatial increments. In terms of practical applications, however, they are valid only when the volume elements contain a sufficient large number of molecules to apply statistical thermodynamics in order to allow for the definition of a temperature or of pressure. Because most of the basic equations are inhomogeneous partial differential equations of second order, they do not have an analytical solution. Therefore they have to be solved numerically on a grid with discrete space and time scales.

1.2 Averaging the basic equations and the closure problem

Meteorological phenomena cover a wide range of horizontal scales ranging from millimetres to thousands of kilometres. The explicit simulation of atmospheric flow with a numerical model including the smallest horizontal scales would thus require numerical grid spacing in the order of 1 mm, which, at least with todays computer power, is absolutely unrealistic. Thus, it is necessary to average the basic equations over larger space and time intervals. In the framework of numerical simulations, these intervals can be identified by the grid spacing and the time step of a numerical model. In contrast to the original differential equations which describe in principle the whole spectrum of atmospheric motions, the discretised (finite difference) equations of a numerical model resolve only those scales which are larger than at least twice the grid spacing. However, the flow of the smaller scales (sub-grid scale flow) still has to be taken into account, because e. g. a considerable amount of temperature, water vapour and momentum is transported by small scale turbulent motions.

Mathematically a filter operator in space and time has to be applied to the primitive equations before the numerical discretisation can be done. This filter has to be designed in such a way that two constraints are fulfilled. On the one hand, the filtered fields have to be as smooth as needed in order to approximate differentiation in space and time properly within the numerical grid. On the other hand, the grid scale structure of the variable fields should not be changed by the filter, because this structure can be described by the numerical model on the discrete grid. A convenient filter taking into account both of those two requirements is the (moving) averaging operator over the finite time interval Δt and (the air containing part of) the grid volume element G, which for any variable ζ is defined as follows:

$$\overline{\zeta}(\underline{x},t) := \frac{1}{|\Delta t| \cdot |G|} \int_{\Delta t \times G} \zeta(\underline{x}+\underline{r}, t+\tau) d^3 r \cdot d\tau. \tag{5}$$

According to Reynolds (1895) any variable ζ may be decomposed into its average and the deviation:

$$\zeta' := \zeta - \overline{\zeta}. \tag{6}$$

$\overline{\zeta}$ describes the resolvable part of the flow and is also called the grid scale value of ζ. The sub grid scale perturbation ζ' represents the non resolvable part of the flow.

We assume that the mean value of a variable varies much more slowly than the deviation from the average. Than we can assume $\overline{\zeta'} = 0$, which sometimes is called the Reynolds approximation.

Often, people apply an ensemble average instead of that volume average, where the ensemble contains a large number of realisations of the flow. Although that average fulfils the Reynolds approximation exactly, it is only a formal operator which can not be realised physically. In particular, it is not related to the grid spacing and it can not be used to express the influence of the sub grid scale variations of the earth's surface. Thus, in the following, we always think about a volume average. In order to describe formally the effect of the density fluctuations, we use here the weighted averaging operator, which is called Hesselberg mean. For a variable ζ the Hesselberg mean and it's deviation are defined as follows:

$$\hat{\zeta} := \frac{\overline{\rho \zeta}}{\overline{\rho}}, \qquad \zeta'' := \zeta - \hat{\zeta}. \tag{7}$$

According to Boussinesq (see Van Mieghem 1973) it is in close approximation $\frac{\rho'}{\overline{\rho}} \approx -\frac{\theta'_v}{\overline{\theta_v}}$, where $\theta_v := \theta\left[1 + \left(\frac{R_d}{R_v}\right) \cdot q_v - q_l\right]$ represents the virtual potential temperature, with R_d, R_v being the specific gas constants for dry air and water vapour respectively. Thus it is $\overline{\zeta''} = -\frac{\overline{\rho' \zeta''}}{\overline{\rho}} \approx \frac{\overline{\theta'_v \zeta'}}{\overline{\theta_v}}$ rather than $\overline{\zeta''} = 0$. Comparing the magnitude of related terms with the other terms in a equation, shows that in the most cases density fluctuations can be neglected. Only in connection with buoyancy they have to be considered.

Before we can write the so-called filtered (or averaged) budget equations, we have to look, if the filter operator and the differentiation operators do commute. It is easy to be seen that this holds for the time derivative. But in the case of differentiation in space this is not true in general. Both operators can only be commuted if the air containing part of the control volume G does not change during it's shift in space. Though this is no real problem in the free atmosphere, we have to pay attention in the vicinity of the earth surface, where the control volume may contain some fixed bodies (roughness elements). In this case commutation is not valid (see also section 2.2), rather it is $\overline{\partial_i \zeta} = \partial_i \overline{\zeta} + \overline{\partial_i \zeta'}$, where $\overline{\partial_i \zeta'}$ represents the effect of rigid intersections on the averaged gradient of any variable ζ.

Applying the averaging operator to the basic budget equations leads to:

$$\partial_t(\overline{\rho}\hat{\phi}) + \nabla \cdot \underline{\overline{F^\phi}} + R^\phi = \overline{Q^\phi}, \tag{8}$$

where $\underline{F^\phi} = \overline{\rho}\hat{\phi}\hat{\underline{v}} + \overline{\rho \phi'' \underline{v}''} - a^\phi \nabla \overline{\phi}$ is the averaged total flux density of ϕ and R^ϕ is the residual term generated by the averaging of the total flux diverge in the

Parameterisation of turbulent transport in the atmosphere 171

Table 1. Demonstrating the closure problem by means of a simplified compilation of equations and unknowns for various statistical moments of temperature. The full set of equations includes even more unknowns (after Stull, 1988).

prognostic equation for	order of the moment	number of equations	number of unknowns
$\overline{\phi}$	first	1	3
$\overline{\rho v''\phi_i''}$	second	3	6
$\overline{\rho \phi'' v_i'' v_j''}$	third	6	10

case of intersecting obstacles. The three parts of the total flux density are due to advection, sub grid scale (turbulent) motion and molecular diffusion, where the latter can be neglected except very close to rigid surfaces.

1.3 The closure problem

The averaged budget equations now include the additional quantities $\overline{\rho\phi''v''}$, which are called covariances, double correlations, second order statistical moments or (because of their physical meaning) turbulent flux densities.

According to Boussinesq the covariances can be approximated like $\overline{\rho\phi''v''} \approx \overline{\rho\phi'v'}$, which often is used in the literature. These correlation products follow from the averaging of the non-linear advection term and represent the scale interaction. They describe the mean contribution of transport induced by the sub-grid scale motions to the overall flow and thus link the non-resolvable scales to the resolvable ones. Thus, the covariance terms represent the inter-scale exchange processes of the flow. They are often of the same order or even larger than the terms containing the grid scale variables. Especially in the atmospheric boundary layer, which is characterised by highly fluctuating variables, the sub-grid scale terms play a dominant role in the averaged budget equations. The boundary layer is a very important part of the atmosphere, because it communicates the exchange between the Earth's surface and the free atmosphere. This exchange is physically accomplished by eddies of different sizes. The development of turbulent eddies is driven by surface heating and wind shear.

Because the covariances are a priori unknown, the number of unknowns in the system of the averaged equations becomes larger than the number of equations. If we construct budget equations for these covariances from the basic equations, new unknowns in the form of e. g. triple correlations (third moments like $\overline{\rho\phi''\varphi''v''}$ arise). This procedure may be repeated, but each time higher order correlation terms will appear, which lead to an inflating growth of unknowns as shown in Table 1. Therefore it is impossible to close the averaged thermo-hydrodynamic equations, because there will be always more unknowns than equations. This problem is referred to as the closure problem. Besides the advection term, also the averaging of the source terms produce additional unknowns, if they are non-linear functions in the model variables.

A solution to overcome the closure problem is to express the unknown statistical moments (as well as the averaged source terms) with the help of some artificially constructed approximations in terms of known grid-scale variables. This procedure is called parameterisation. Thus, a parameterisation is an approximation to nature which will never be perfect, but should be physically reasonable. The aim of a parameterisation is to be suitable for the considered application. In the context of solving the averaged thermo-hydrodynamic equations a parameterisation is also a closure assumption.

2 Parameterisation of turbulence

Now we look closer at the parameterisation of turbulent processes hidden in the covariances discussed above. In the framework of turbulence the potential temperature $\theta := T(\frac{p_0}{p})^{\frac{R}{c_p}}$, which is the temperature an air parcel would have if it is moved adiabatically (no exchange of mass or energy with its neighbourhood) to a reference pressure p_0 (usually 1,000 hPa), is often used instead of the temperature, because it is a conservative property during adiabatic parcel displacements. Thus we get rid of one mayor source term in the averaged temperature budget equation, namely the adiabatic heating.

After a general description of turbulence parameterisations we present in more detail two turbulence schemes implemented in the "Lokal-Modell" (LM) (Doms and Schaettler, 1999), the operational weather prediction model of the German Weather Service, DWD. The treatment of cloud microphysical processes in atmospheric models, which constitutes the main contributions to the source and sink terms in the scalar budget equations as well as the usual parameterisation of radiation will not be discussed here. A comprehensive description of these parameterisations is given in Pielke (1984).

2.1 General aspects

Many turbulence concepts have been developed in order to close the averaged equations by the construction of closure assumptions (see e. g. Stull 1988). These approximations can be sub-divided into classes, which depend on the highest order of prognostic equations that are retained (see Tab. 1). A closure assumption taking into account the prognostic equations for the mean values (first moments) and approximating the second moments is referred to as a first order closure. A closure of second order additionally considers the budgets of the second order moments but approximates unknown terms in these new equations.

First order closure In a first order closure one has to find an expression for the covariance term $\overline{\rho \phi'' v''}$, which represents turbulent flux densities of momentum (components of the Reynolds stress tensor), temperature, water vapour or cloud water. These quantities are of crucial interest, because they dominate the exchange processes between the Earth's surface and the free atmosphere.

Most of the closure assumptions treat turbulent fluxes like molecular diffusion, which can be described according to Fick's law. This leads to the following flux gradient parameterisation:

$$\overline{\rho \phi'' \underline{v}''} := -\overline{\rho} K^\phi \nabla \overline{\phi} \tag{9}$$

where K^ϕ is called turbulent diffusion coefficient or eddy diffusivity. This parameterisation often is called Austausch- or K-theory as well. In this approach the turbulent fluxes are always directed down the mean gradient of the related property. K^ϕ has dimensions of $\frac{m^2}{s}$, thus, from dimension analysis we expect that it represents the product of an effective size and an effective velocity of coherent patterns (eddies, turbulence elements) within the turbulent flow. So, K^ϕ is a measure of the mixing strength of the turbulent flow. With the first order closure the determination of the turbulent fluxes is shifted to the calculation of K^ϕ, which has to be expressed as a function of the mean atmospheric variables and usually depends on static stability and vertical wind shear.

The validity of the approach is based on the mixing-length theory proposed by Prandtl in 1925 (see e.g. Stull 1988), who claims, however, that K-theory is only valid for neutral and stable conditions. The classical approach to investigate the stability of the atmosphere is the parcel method, in which a small isolated air parcel is vertically displaced from its original position. If the parcel after the displacement is warmer than its environment, the parcel will be accelerated away from its initial position by buoyancy. This is the case if $\partial_z \overline{\theta} < 0$ and the atmosphere is referred to as unstable. But if θ increases with height, the parcel will be accelerated back towards its former position and the atmosphere is stable. A neutral stratification exists, if θ is constant with height. In neutral and stable situations the turbulent transfer is dominated by small eddies of a length scale smaller than the scale of the mean gradient. In highly convective conditions, however, large eddies dominate the turbulent transport. In the real atmosphere these eddies can even account for a transport up the gradient (counter gradient flux), which is not reflected by K-theory.

Second order closure Now budget equations of the second order moments have to be derived. Even a second order closure may result in a flux gradient form. The characterising distinction is, that the parameterisation of K^ϕ does use second order equations. If we have two concentration variables ϕ and φ with the corresponding source terms Q^ϕ and Q^φ, the budget equation of the covariance $\overline{\rho \phi'' \varphi''}$ has the following form:

$$\partial_t \left(\overline{\rho \phi'' \varphi''} \right) + \nabla \cdot \underline{F}^{\phi \varphi} + R^{\phi \varphi} = - \left(\overline{\rho \phi'' \underline{v}''} \cdot \nabla \hat{\varphi} + \overline{\rho \varphi'' \underline{v}''} \cdot \nabla \hat{\phi} \right) \\ - \left(a^\phi + a^\varphi \right) \overline{\nabla \phi'' \cdot \nabla \varphi''} \\ + a^\phi \overline{\varphi'' \nabla^2 \hat{\varphi}} + a^\varphi \overline{\phi'' \nabla^2 \hat{\varphi}} \\ + \overline{\phi'' Q^\varphi} + \overline{\varphi'' Q^\phi} \tag{10}$$

By comparing this equation with the averaged budget of a first order moment (8), we recognise that it has a similar form. Here, $F^{\phi\varphi} := \overline{\rho\phi''\varphi''\hat{v}} + \overline{\rho\phi''\varphi''v''} - \left(a^{\phi}\overline{\varphi''\nabla\phi''} + a^{\varphi}\overline{\phi''\nabla\varphi''}\right)$ is the total flux density of the second moment and $R^{\phi\varphi}$ is the residual term expressing all effects of intersected bodies except those being hidden in the last source term. The triple correlation part of the flux is, as mentioned before, one of the crucial expressions which has to be parameterised. Although mean gradients become considerably large in the laminar layer close to the rigid surfaces, the gradients of the deviations ϕ'' will be much smaller. Thus, the two molecular components of the flux are usually neglected, even close to the rigid surfaces. Strictly speaking, this is not correct in general. For instance, in the case of strong variations of the concentrations at the rigid surface the molecular components of the flux will be responsible for transport of variance into the lower atmosphere. Only in the case of momentum this effect is absent due to the non slip condition at rigid surfaces.

The first source term on the right hand side describes production related to the mean gradients of the model variables and the second term is the non diffusive part of the molecular (laminar) processes called dissipation. As dissipation is important only in the small scale part of the spectrum, where turbulence is assumed to be isotropic, due to a symmetry argument, this term will vanish in the case of ϕ, φ being different if (at least) one is a velocity component. Due to the Boussinesq approximation the third term, being an additional molecular term, is negligible in general.

The fourth term contains the source terms of the original first order budgets and has to be parameterised as well. In order to get as far as possible rid of those source terms, modellers often choose a modified set of variables, by which those source terms can be neglected at least for some of these variables. In that case they are called "conservative" variables. The following two variables are conservative with respect to water condensation or evaporation: $q_w := q_v + q_l$ total water content and $\theta_l := \theta - \frac{L_c \bar{\theta}}{c_p T} q_l$ liquid water potential temperature, where L_c is the specific heat of water fusion. Like θ also θ_l is conserved during adiabatic processes. Thus, if we neglect radiative heating and diffusive fluxes of heat in the second order equations, we can treat θ_l and q_w as conservative in this framework. But momentum can hardly be substituted by a conservative variable, thus we have to live with the momentum sources. Thus in the $\overline{\rho\phi''v_i''}$-equation we get e. g. the following term resulting from the pressure gradient:

$$-\overline{\phi''\partial_i p} = -\overline{\phi''}\partial_i p - \overline{\partial_i \phi p'} + \overline{p'\partial_i \phi} + \hat{\phi}\overline{\partial_i p'} \qquad (11)$$

This term is of great importance and needs to be parameterised. For each term in this expansion there either exist parameterisation formulae, or they are neglected. The latter is the case for the second term, the so-called pressure covariance term. The first term is called buoyancy term and describes in the case of ϕ being a velocity component the buoyant production of sub-grid scale momentum transport. The third term is called return-to-isotropy term, as in the equation for $\overline{\rho v'' \cdot v''}$ the trace of the resulting tensor vanishes, if we assume incompressibility. The fourth term is called drag production and it describes the pressure effect of roughness elements. If there are no roughness elements present, this term vanishes due to $\overline{\partial_i p'} = \partial_i \overline{p'} = 0$. Because of $\overline{\rho \phi''} = 0$, the term in $\overline{\phi'' Q^{v_i}}$, resulting from buoyancy, vanishes. Further, in atmospheric applications the term resulting from the Coriolis force can be neglected in general.

It is beyond the scope of this article to deal with the development of these parameterisations. Thus, it is only mentioned that for all the above terms there exist parameterisations. In the case of triple correlation, which describe turbulent transport (diffusion) of second order moments, usually a gradient approach is used like the following form:

$$\overline{\rho \phi'' \varphi'' v_i''} \propto -\Lambda q \; \partial_i \overline{\rho \phi'' \varphi''} \qquad (12)$$

where Λ is a turbulent length scale and $q := \sqrt{2 \overline{e_{kt}}}$ is the turbulent velocity scale, with $e_{kt} = \frac{1}{2} \underline{v''} \cdot \underline{v''}$ the kinetic energy fluctuation. Since q is needed also in other parameterisations, the budget equation of the turbulent kinetic energy (TKE) $\overline{\rho e_{kt}}$ is of great importance. It has the following general form:

$$\partial_t \overline{\rho e_{kt}} + \nabla \cdot \underline{F}^{e_{kt}} + R^{e_{kt}} = -\sum_{i,j} \overline{\rho v_i'' v_j''} \partial_j \hat{v}_i - \epsilon - \overline{\underline{v''} \cdot \nabla p}, \qquad (13)$$

where $\underline{F}^{e_{kt}} := \overline{\rho e_{kt}} \hat{\underline{v}} + \overline{\rho e_{kt} \underline{v''}}$ is the total flux density of TKE, $R^{e_{kt}}$ is the residual term for intersection effects similar to $R^{\phi \varphi}$ above and $\epsilon := \mu \sum_i \overline{\nabla v_i'' \cdot \nabla v_i''} \geq 0$ is the dissipation of TKE towards inner energy (heat).

The only approximation in this equation is the neglect of all molecular terms except dissipation. The first term on the right hand side describes TKE production by wind shear and the third term, $-\overline{\underline{v''} \cdot \nabla p} = -\overline{\underline{v''}} \cdot \nabla p - \nabla \cdot \overline{\underline{v} p'} + \overline{p' \nabla \cdot \underline{v''}} + \hat{\underline{v}} \cdot \overline{\nabla p'}$, TKE production by pressure work. In this expression the third term, which is the return-to-isotropy term, vanishes if we assume incompressibility. Thus, in the absence of roughness elements, the pressure term consists only of the contribution by buoyant production (first term) and pressure transport (second term), where the latter usually is neglected or assumed to be a part of the sub grid scale flux density $\overline{\rho e_{kt} \underline{v''}}$.

In numerical weather prediction models second order closure usually is done with a lot of additional simplifications in order to minimise computational effort. These are mainly the following approximations:

- Local equilibrium for all second order budgets except the TKE equation thus neglecting the sum of local time derivation, advection and turbulent transport
- Boussinesq approximation: $\overline{\rho \phi'' \varphi''} \approx \overline{\rho} \overline{\phi' \varphi'}$ and $\overline{\phi''} = -\frac{\overline{\rho' \phi''}}{\overline{\rho}} \approx \frac{\overline{\theta_v' \phi'}}{\theta_v}$ (see also 1.1 and 1.2)
- Rotta hypothesis: parameterisation of the return-to-isotropy term according to Rotta (see e. g. Mellor and Yamada 1982)
- Kolmogorov hypothesis: parameterisation of the TKE dissipation according to Kolmogorov (see e. g. Mellor and Yamada 1982)
- Free atmosphere approximation: neglecting the effects of roughness elements in the second order equations

By the use of these approximations the second order equations in the quasi-conservative properties θ_l and q_w together with the three velocity components build a set of 15 linear equations in the 15 second order moments which can be formed by five variables. By application of the boundary layer approximation, that is by neglecting all horizontal gradients of mean quantities, the whole set of second order equations can be reduced to only two linear equations in the two unknowns S^M and S^H, which are the so-called stability functions for momentum (M) and scalar quantities like heat (H). The desired covariances needed in the averaged model equations then have flux gradient form like in the first order closure. The general form reads

$$\overline{\phi' w'} = -K^\phi \partial_z \overline{\phi}, \quad K^\phi = q\Lambda S^\phi, \quad S^\phi := \begin{cases} S^H & \text{for} \quad \phi \in \{q_w, \theta_l\} \\ S^M & \text{for} \quad \phi \in \{u, v\} \end{cases} \quad (14)$$

where ∂_z is used to describe the local derivative in vertical direction. According to the hierarchy of second order closures according to Mellor and Yamada (1974, 1982) such a procedure has the level 2.5. In particular, the so called buoyant heat flux, arising in the buoyancy term of (11) using the Boussinesq approximation, has the following representation: $\overline{\theta'_v \phi'} = K^H \Gamma_v$, where Γ_v is the effective vertical gradient of the virtual temperature. This gradient can be represented as a linear combination of the vertical gradients of $\overline{\theta_l}$ and $\overline{q_w}$.

2.2 Turbulence parameterisations of the Lokal-Modell (LM)

Modellers creating a weather prediction or climate model have to weigh which kind of turbulence parameterisation is the best choice for the scale of the intended model application. They have to take into account both the degree of realism given by a higher order closure and the computational costs connected with them. In the LM two types of turbulence parameterisations are implemented which account for the variety of grid spacing (scales) the model is used with.

The old turbulence parameterisation of the Lokal-Modell The old parameterisation was originally designed for meteorological applications in large-scale forecast models of the meso-β-scale ($\approx 10-100$ km) and larger and is used in the global model (Global-Modell, GME) as well. It was developed also to fit the computer power of the previous generation used at DWD, thus an additional prognostic equation had to be avoided. Consequently the TKE equation was used in the local equilibrium form, which is a diagnostic equation, where buoyant production, shear production and dissipative destruction are assumed to be in balance. That is a level 2.0 closure according to Mellor and Yamada (1974, 1982). This diagnostic TKE-equation can be introduced in the linear equations for the two stability functions. This results in a form, in which the latter can be expressed as (non linear) functions of the gradient Richardson number

$$Ri := \frac{g}{\theta_v} \frac{\partial_z \theta_l}{(\partial_z u)^2 + (\partial_z v)^2}. \quad (15)$$

In this parameterisation sub-grid scale condensation is not taken into account, so, instead of θ_l, here θ is used.

The Ri-dependency of the stability functions gives a limitation for the solution, which is well defined only for $Ri \leq Ri_c$. Ri_c is about 0.19 for the model values used. Thus, for strong stable thermal stratification, as it is the case in the nocturnal boundary layer, the solution of the closure procedure depends on assumed minimum diffusion coefficients, which have to be chosen as additional model parameters.

A special type of parameterisation is needed for the calculation of the surface fluxes at the lower model boundary. The layer between the surface and the lowest model level is assumed to be a constant flux layer, thus, the vertical fluxes can be deduced by integration and have a bulk form like:

$$\overline{\phi'w'}(z) =: -C^\phi \cdot [\overline{\phi}(z_A) - \overline{\phi}(0)], \quad z \in [0, z_A] \qquad (16)$$

Here $z = z_A$ is the height of the first model level and $z = 0$ is the mean height of the rigid surfaces. Then, the turbulent transfer coefficient C^ϕ is a vertical integral of the inverse diffusion coefficient

$$C^\phi = \left(\int_0^{z_A} \frac{dz}{K^\phi(z)} \right)^{-1}, \qquad (17)$$

which is calculated based on the Monin-Obuchov similarity theory (see e. g. Stull 1988). It depends on the so called roughness length of the surface and on thermal stratification of the constant flux layer and can only be determined with the help of empirical stability functions. The whole procedure is implemented according to Louis (1979).

The new turbulence parameterisation of the Lokal-Modell The new turbulence closure scheme is on the level 2.5 according to Mellor and Yamada (1974, 1982) and thus contains a prognostic TKE equation. The most important advantage of this scheme is, that those terms of the TKE budget, which had to be neglected in the diagnostic TKE equation, can now be added. These are in particular advection and turbulent diffusion of TKE.

As there is a great interest in a realistic simulation of the lower boundary layer, the new scheme includes an option to have several model layers in that lowest part of the atmosphere (roughness layer), which may be interspersed with a variety of irregular roughness elements, such as plants of a canopy, buildings of a city or hills of a sub grid scale orography. The fact that volume averaging and spatial differentiation do not commute, if the averaging volume is intersected by obstacles, provides the effect of those bodies mainly in form of surface integrals along the inner boundaries of the obstacles (see discussion above). This concept has been introduced by Raupach and Shaw (1981). But a revised application of that method showed that there has to appear an additional term, describing the reduction of the air volume fraction due to the presence of rigid obstacles in the grid box. These terms arise in the first order budgets and in the second order budgets (such as the TKE equation) as well. Up to now, a parameterisation of them is realised only in the case of the budgets for momentum (form drag force) and TKE (wake production), which both are due to sub grid scale pressure gradients along the surface of a fixed body in a flow.

A specific problem within plant canopies is the short cut of the turbulence spectrum by the action of very small bodies, such as leaves and branches, which transform kinetic energy of large eddies with a big mixing potential into the kinetic energy of very small ones, being on the dissipative scale. Application of band pass filtered 2-nd order equations, which exclude the dissipative part of the spectrum, result in a similar closure scheme. The only difference to the previous scheme of Mellor and Yamada is the appearance of different canopy dependent values of the model constants, representing enhanced dissipation.

Mainly because of the lack of appropriate external parameters describing the intersections statistically, all that described issues connected with rigid intersections are not yet operational in the LM.

Some new development was done by the parameterisation of TKE-production connected with sub grid scale thermal circulation structures, which are forced by sub grid scale patterns of the surface temperature. This effect is hidden in the pressure term as well and has a considerable effect in very stable nocturnal boundary layers over heterogeneous terrain. This production term forces the scheme to simulate more mixing in stable boundary layers and thus removes to some extent a well known shortcoming of traditional turbulence schemes. But single column simulations showed that this positive effect is suppressed to a large extent, if Monin-Obukhov-similarity-theory is used to describe the turbulent surface fluxes of heat and moisture. That is due to the simulation of a strong decoupling between soil and atmosphere, if the surface temperature is much lower than that of the air above. This problem could be solved by applying the TKE-scheme down to the lower boundary of the turbulent Prandtl layer. This was achieved by the evaluation of vertical gradients in that regime with the help of logarithmic Prandtl layer profiles. The main effect of this unified approach is a consistent formulation of the surface layer scheme and the TKE-scheme used above the surface layer.

The additional TKE-production terms all together come from the following expansion of the pressure term:

$$-\overline{\underline{v}'' \cdot \nabla p} \approx P_B + P_C + P_D. \tag{18}$$

On the right hand side, the LM uses the following parameterisations:

- *turbulent buoyancy term*: $P_B = -\frac{g}{\theta_v}\overline{\rho}K^H\Gamma_v$ (which is state of the art),
- *thermal circulation term*: $P_C = \partial_z (H \cdot P_B)$,
- *wake production term*: $P_D = \overline{\rho}\frac{|\hat{v}|^3}{l_{\mathrm{drg}}}$.

Here, $H = l_{\mathrm{pat}}\frac{R_d}{g}\Gamma_v$ is a vertical length scale of coherence, where l_{pat} is the horizontal length scale of sub grid scale thermal surface patterns. Further, l_{drg} is the effective isotropic drag length scale of obstacles intersecting the corresponding grid box.

Errors due to non local effects, which may occur in the convective boundary layer or within a plant canopy (where sometimes counter gradient fluxes are present), are related to simplifications in the second order equations. This is, in particular, the neglect of the third order moments, which describe turbulent transport of second order moments. Those errors are significant, when the length scale of vertical turbulent diffusion at a point is larger than the depth of the vertical interval belonging to this point, where the vertical profiles of the first order moments can be treated as linear. As it is intended to stay within the framework of a level 2.5 scheme, this problem is considered by a modified determination of vertical gradients, using vertical profiles, which are smoothed with the help of a stratification dependent length scale of vertical diffusion. But this method has not yet been tested sufficiently and can only solve for a special group of non local processes.

A further improvement of the new scheme is related to the consideration of sub grid scale clouds and their interaction with turbulence. In order to include condensation effects during sub grid scale vertical movements into the turbulence scheme, the second order budgets are formulated in terms of the "conservative" (see also 2.1) thermodynamic variables θ_l (liquid water potential temperature) and q_w (total water content). The turbulent fluxes of those quantities then have to be transformed in those of the model variables T (temperature), q_v (water vapour content) and q_l (cloud water content), which is done by the use of a statistical sub grid scale cloud model proposed by Sommeria and Deardorff (1976). Their model calculates the volume fraction and the water content of sub grid scale clouds with the aid of the variance of the saturation deficit distribution in a grid box, which can be expressed in terms of turbulent quantities.

Finally, a laminar surface layer is introduced. This makes it possible to discriminate between the values of the model variables at the rigid surfaces (e. g. radiative surface temperature) and values at the lower boundary of the turbulent atmosphere. This innovation results in an increase of the diurnal variation of the soil surface temperature, which is now in better agreement with measurements.

2.3 First Results

The influence of the thermal circulation term on boundary layer development is exemplified in Figs. 1a, 1b, and 2. In order to test this effect without the influence of a presumably not very accurate surface scheme and soil model, single column simulations, forced by observed temperatures and dew points at 2 m height, were performed. The simulations started with vertical profiles obtained by radio soundings at midnight (hour = 0). All measurements are taken from the Lindenberg observatory of DWD. The simulations belong to a high pressure situation in July 1997, when the wind speed was low, so that the single column simulations are easier to compare with data from measurements.

In Fig. 1 measured (mes) and simulated profiles of potential temperature (Tet) are shown after 18 hours (well mixed layer) and 24 hours (stable stratified layer) of simulation time. (a) shows the simulation without using the circulation term ($l_{\text{pat}} = 0$ m) and (b) shows the case with $l_{\text{pat}} = 2{,}000$ m. Obviously, in

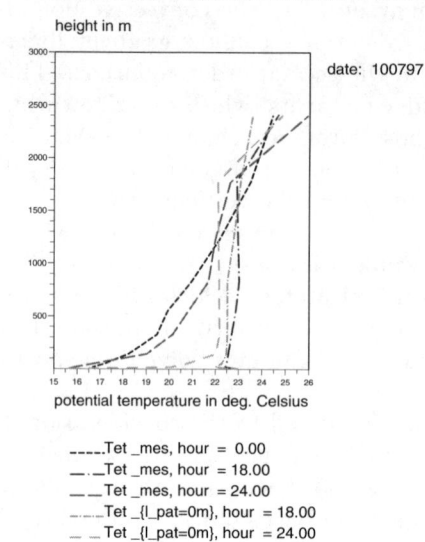

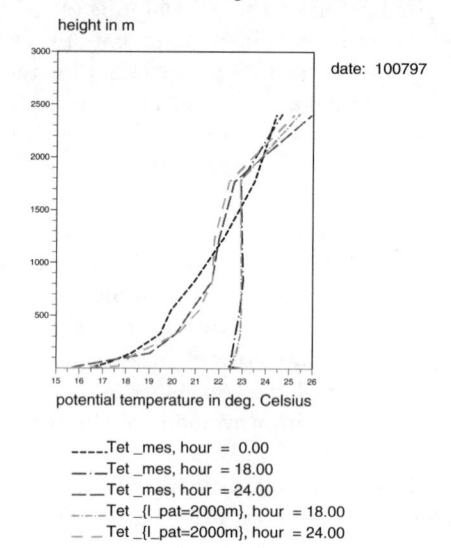

Fig. 1. Measured and simulated profiles of potential temperature in the boundary layer for July 10th, 1997 at the Lindenberg observation site of DWD. (a) show the simulations without the circulation term, while (b) show the simulations with the circulation term in the turbulence closure scheme

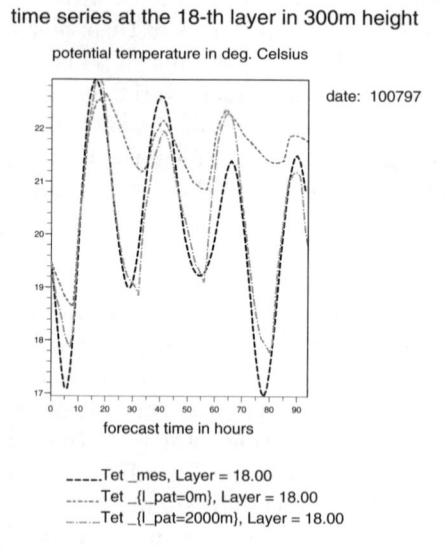

Fig. 2. Measured and simulated time series of potential temperature in the boundary layer beginning July 10th, 1997 at the Lindenberg observation site of DWD in 300 m above ground.

the first case the model gives a much too shallow inversion layer, whereas the circulation term is able to reproduce the measured shape of the inversion fairly well. As seen in Fig. 2, which shows measured and simulated time series at model layer 18 at about 300 m height, only in the case including the circulation term (l_{pat} =2,000 m) the diurnal temperature wave is comparable to the measured amplitude. It is the additional TKE production by this term during nocturnal cooling, that mixes the cold air at the surface more efficiently with higher levels.

Figures 3, 4 and 5 show some results of 12 hour forecast with the LM during the low pressure situation over Germany of 22.07.1999, starting at 00 UTC. Obviously, both, the old model version (ref) and the new version (new) reproduce a jet stream (Fig. 3). Although the new version produces more mixing at stable stratification (Fig. 5), it completely retains the jet stream structure. Fig. 4 shows cross sections for horizontal wind speed and potential temperature at the constant latitude $-5.5°$ with respect to the rotated coordinate system of LM, which is at about 52° geographical latitude (see the y-axis of Fig. 3). The wind speed chart shows two jet maxima corresponding to the western and eastern flanks of the trough. In Fig. 5 the related TKE fields for both LM versions are plotted. Due to the stable stratification (see right picture of Fig. 4) the old scheme gives just the background value for TKE over most of the higher troposphere. Only at some discrete areas, where the Richardson number is less than its critical value, the old scheme is able to calculate a physical solution. But the new scheme is able to simulate some realistic structure of the TKE field even in the higher troposphere. There are two local TKE maxima belonging to the upper and lower edge of each of both jet maxima, where strong vertical gradients of horizontal velocity occur. But at the western jet maximum, especially the upper region of large vertical velocity gradients corresponds to the region of strong stable stratification, damping the shear production of TKE. Thus the corresponding TKE maximum has a rather low magnitude and is shifted somewhat towards regions with weaker thermal stability in the west of the jet. Furthermore, above the mixed boundary layer, which has a height of about 1,500 m, the new scheme produces a more intensive entrainment zone. The sharp upper borders of the boundary layer in both versions are due to the coarse vertical model resolution in comparison with the strong vertical TKE-gradients at that height. Regions, where the boundary layer disappears, belong to sea surfaces of the Atlantic ocean and the northern sea.

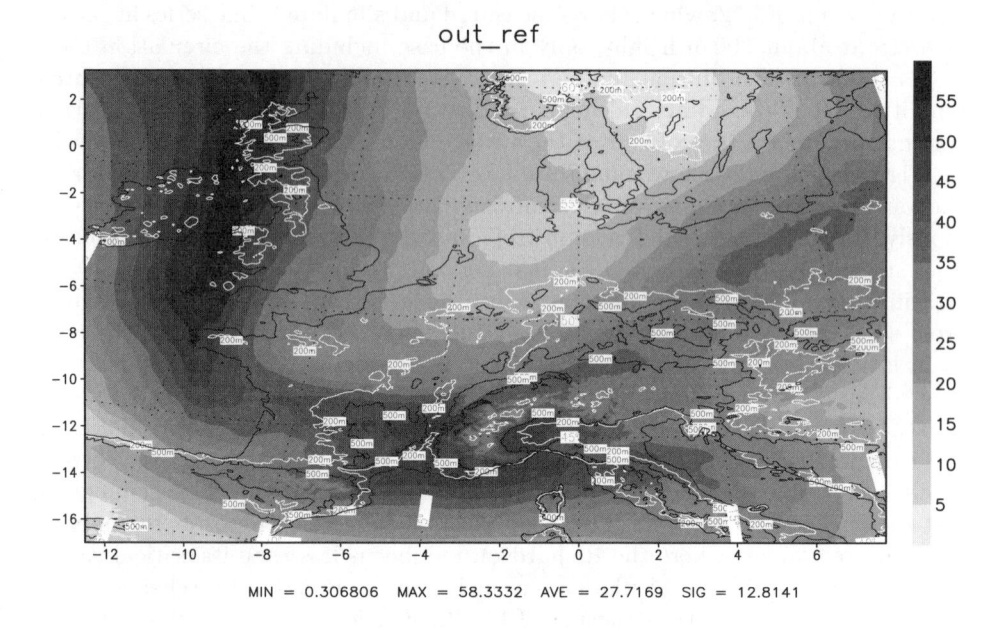

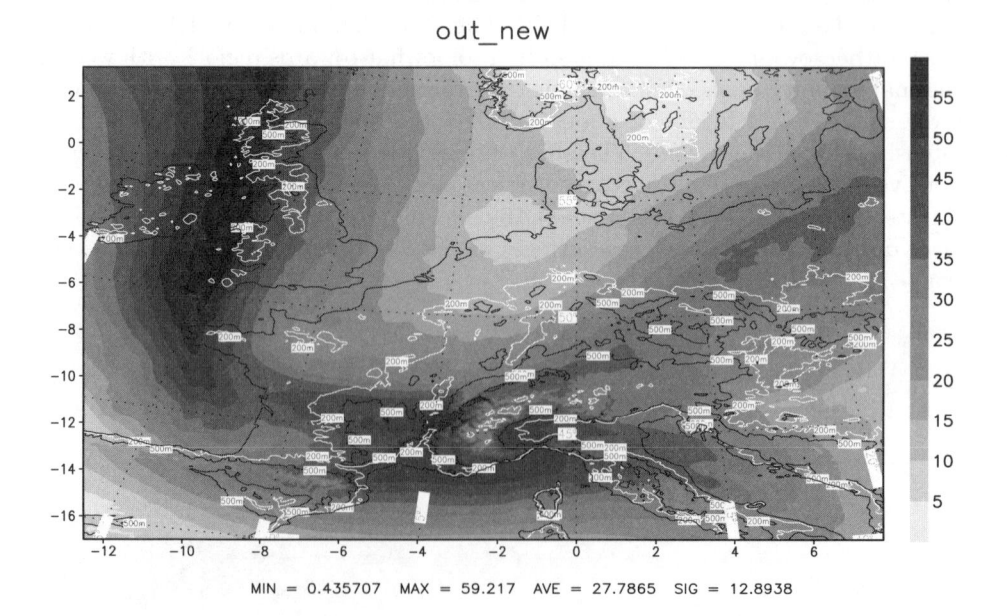

Fig. 3. Wind speed in m/s in 10 km height over central Europe for July 22nd, 1999, 12 UTC, using the current (top) and the new turbulence closure scheme.

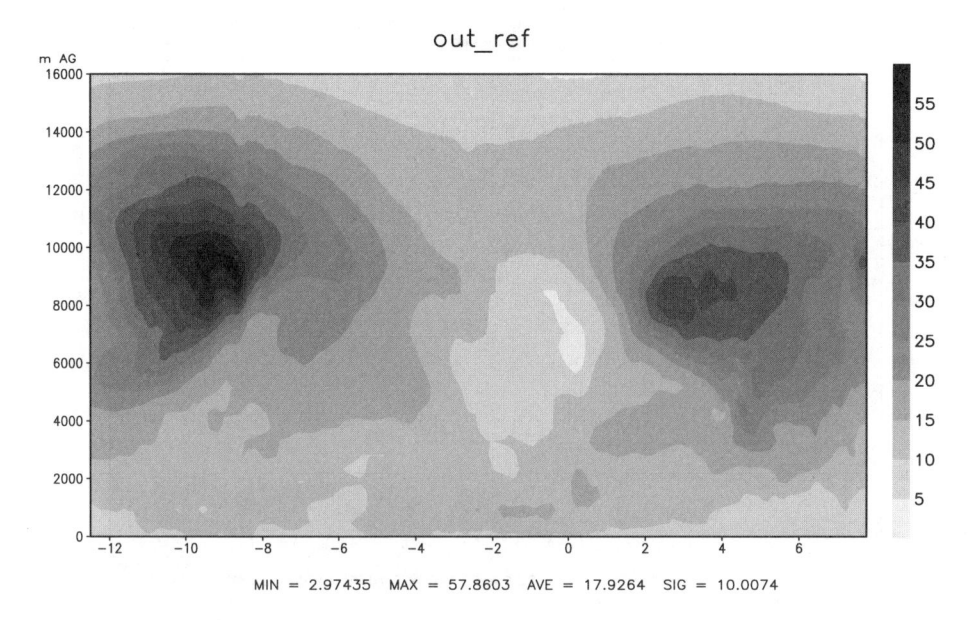

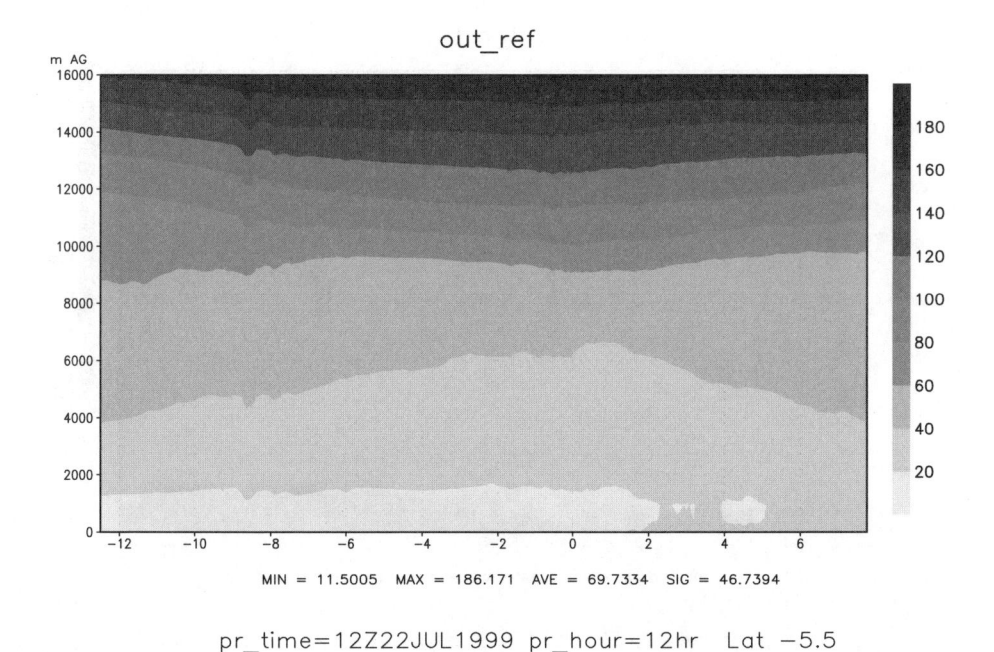

Parameterisation of turbulent transport in the atmosphere 183

Fig. 4. Vertical cross section from West to East at about 52°N on July 22nd, 1999, 12 UTC. Top: horizontal wind speed (m/s) with old turbulence scheme. Bottom: potential temperature (Degree Celsius) with the old scheme. Both are very similar with the new scheme.

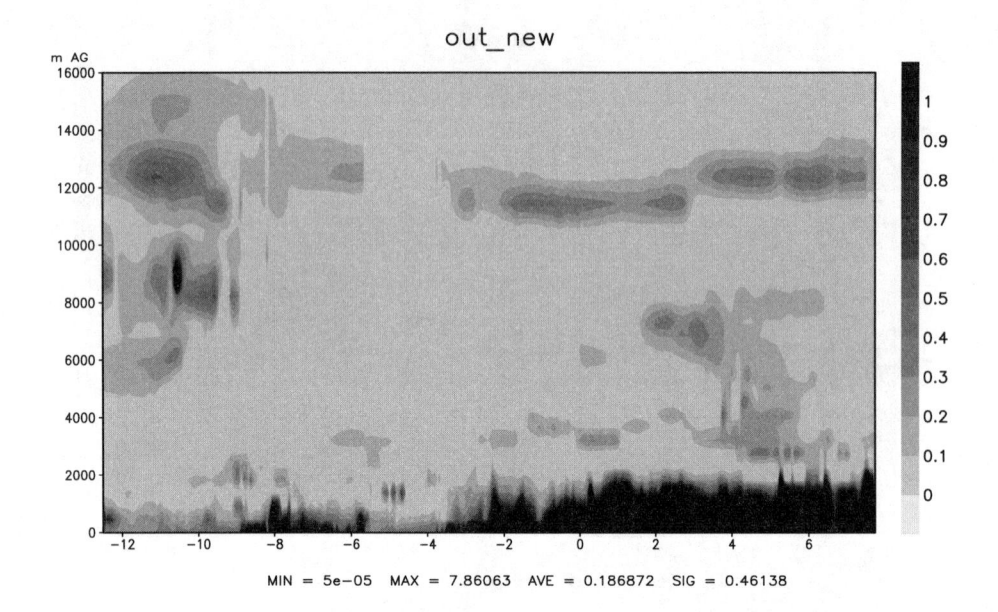

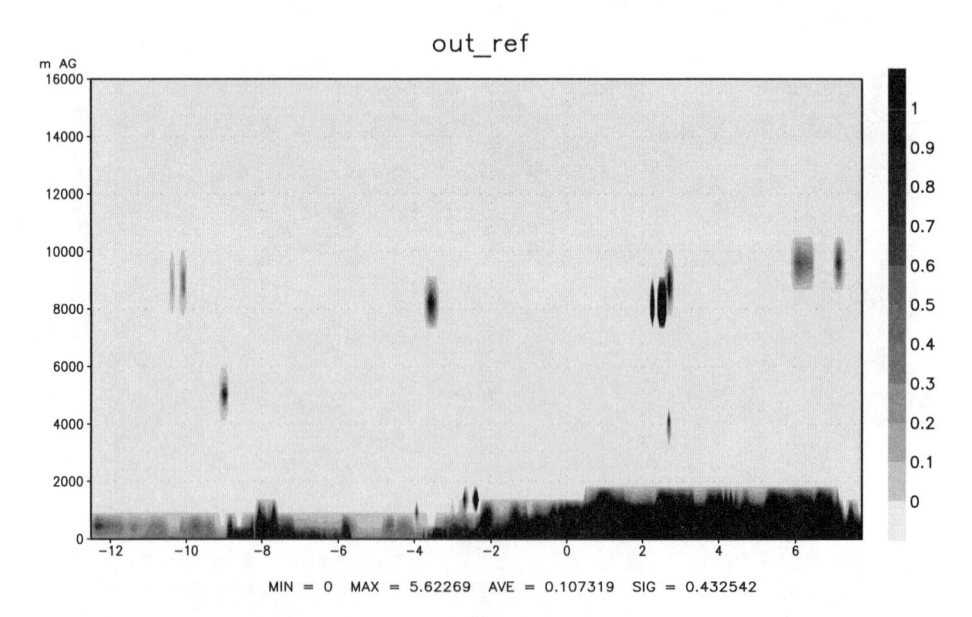

Fig. 5. Vertical cross section from West to East at about 52°N on July 22nd, 1999, 12 UTC. Top: turbulent kinetic energy (m^2/s^2) with new turbulence scheme. Bottom: same with current turbulence scheme.

3 Conclusions

In this contributions, we have tried to describe the basic problem of closing the set of thermo-hydrodynamical equations used in dynamic meteorological models. The necessity originates from the importance of sub-grid scale processes for the description of atmospheric processes. These sub-grid scale processes, produce additional terms and unknowns when the basic set of equations is averaged to appropriate scales in time and space. Adequate equations for describing these additional terms lead to an exploding number of further unknowns, which must be restricted by parameterisations. Basic concepts for these additional equations and parameterisations have been described using the Lokal Modell (LM) of the German Weather Service (DWD) as an example.

References

Doms, G. and Schättler, U. (1999): The nonhydrostatic limited-area model LM (Lokal-Modell) of the DWD. Deutscher Wetterdienst, Offenbach. 180 pp.
Louis, J.-F. (1979): A parametric model of vertical eddy fluxes in the atmosphere. Bound. Layer Metoer., 17, 187–202.
Mellor, G. L. and Yamada, T., (1982): Development of a turbulence closure model for geophysical flow problems. Rev. Geophys. Space Phys., 20, 831–875.
Mellor, G. L. and Yamada, T., (1974): A hierarchy of turbulence closure models for the planetary boundary layer. J. Atmos. Sci., 31, 1791–1806. (Corrigenda (1977) J. Atmos. Sci. 34, 1482)
Miegham van, J., (1973): Atmospheric Energetics. Oxford Monographs on Meteorology, Clarendon Press. Oxford.
Raupach M. R. and Shaw R. H., (1981): Averaging procedures for flow within vegetation canopies. Bound. Layer Meteo., 22, 79–90
Pielke, R. A. (1984): Mesoscale meteorological modeling. Academic Press, Orlando. 612 pp.
Sommeria G. and Deardorff J. W., (1977): Sub grid scale condensation in models of non precipitating clouds J. Atmos. Sci., 34, 344–355.
Stull R. B., (1988): An Introduction to Boundary Layer Meteorology. Kluwer, Dordrecht, 666 p.

Precipitation Dynamics of Convective Clouds

Günther Heinemann[1]* and Christoph Reudenbach[2]

[1] Meteorological Institute, University of Bonn, Germany
[2] Geographical Institute, University of Bonn, Germany

Abstract. The dynamics of precipitation processes represent a typical example for a multi-scale geo-system. The gaps between the scales of the cloud microphysics, resolutions of cloud models and remote sensing methods have to be closed by upscaling or parameterisation techniques. The section gives an overview over the governing processes for the generation of precipitation and the importance of the cloud physics for the modelling and remote sensing of precipitation processes. An example of a cloud model simulation is presented, and a possible approach for parameterisations of convective precipitation processes for infrared remote sensing is given.

1 Introduction

The dynamics of regional convective precipitation processes comprises a wide interval in space and time. It ranges from scales of the cloud microphysics (0.1 μm) to organised convective systems (100 km). The interaction between the microphysical scales (release of latent heat) and the scales of the cloud dynamics (order of 1 km, up- and downdrafts, interaction within cloud clusters) is of high importance for the precipitation processes. Fig. 1 gives a schematic overview over the structure of a thunderstorm cloud (supercell storm), while Fig. 2 shows an image of a thunderstorm from space. The essential elements with respect to the modelling and remote sensing of precipitation processes are a) up- and downdraft areas, where condensation and rain formation as well as evaporation of cloud liquid water (CLW) occur; b) the formation of a cirrus anvil at the tropopause, which is characterised by a deformation of the tropopause (overshooting top) in the area of strongest updrafts; c) rain at elevated levels that evaporates before it reaches the surface (virga). Both, the schematic sketch in Fig. 1 and the real thunderstorm in Fig. 2, show the overshooting top and the anvil, which are key features for the precipitation retrieval discussed in Section 3.

The latter characteristics of the three-dimensional structure of convective precipitation leads to problems using the term "precipitation" in the frame of interdisciplinary research. While the vertical distribution of ice and rain rates is of main interest for questions of the precipitation processes, the surface precipitation is the main need for hydrological applications, since it triggers processes in vegetation and soil. This aspect is also relevant for methods of remote sensing

---

* gheinemann@uni-bonn.de

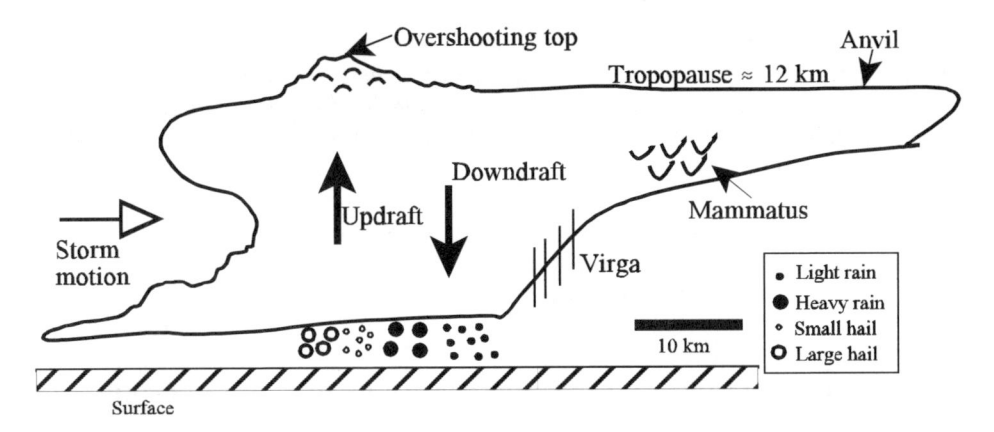

Fig. 1. Schematic structure of a supercell thunderstorm depicting essential elements with respect to the modelling and remote sensing of precipitation processes (inspired by Houze, 1993).

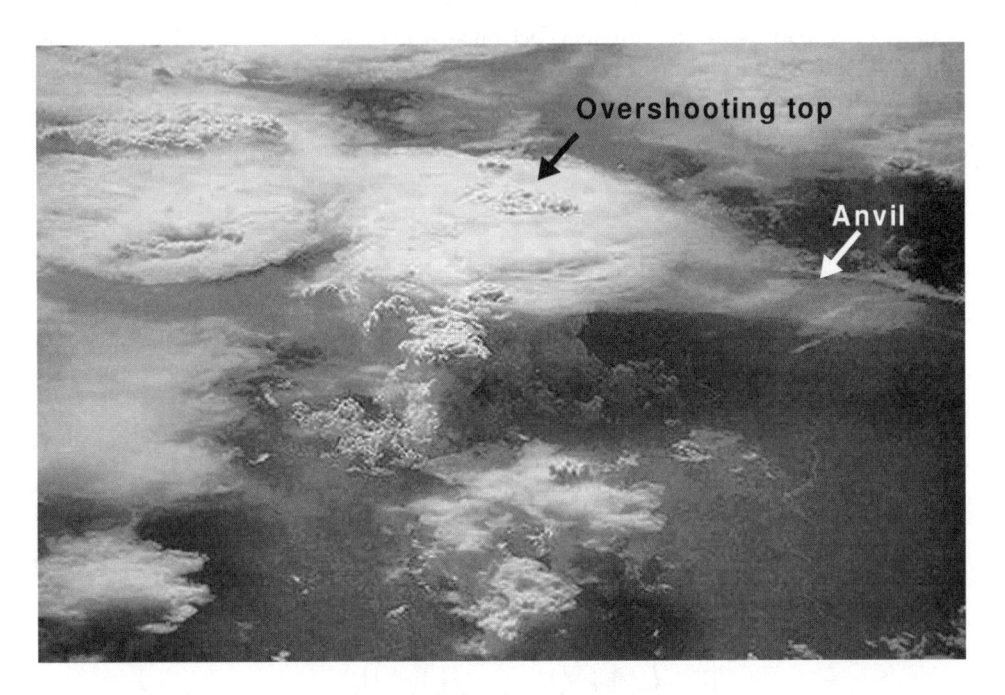

Fig. 2. Image of a supercell thunderstorm made from the space shuttle (courtesy of NASA) with the regions of the anvil and overshooting top marked.

of precipitation, which assume a coupling of structure and height of cloud tops with the surface precipitation (algorithms using infrared satellite data), as well as for rain retrievals using weather radars, which assume a coupling of the drop size distribution (DSD) at the cloud base with the surface precipitation.

Convective precipitation processes occur on a continuum of scales, but usually a separation into discrete subscales is being made when models are used to simulate these processes. The transition of scales of the cloud microphysics up to the thunderstorm scale need to be parameterised either in a physical and/or in a statistical way. The treatment of the various spatial/temporal scale interactions represents therefore an essential problem for the remote sensing, modelling and measurements of convective precipitation. Some aspects of problems and solutions of the scale interactions and the parameterisations of convective precipitation processes are discussed in this chapter.

2 Scale interactions and parameterisations in cloud microphysics

The size distribution of precipitation particles is important for the interaction processes between the particles and for the precipitation process. The scales of the drop size distribution (i. e. the DSD) range from 0.1 μm (size of the condensation nuclei) to those of raindrops of up to several mm. Cloud drops and rain are usually classified using the fall speed (Houze, 1993), which is related to the drop radius. Particles exceeding a radius of 100 μm (1 m/s fall speed) are attributed to rain (drizzle for 100–250 μm). In raining clouds a tendency of a bimodal DSD can be found, which justifies the separation between cloud droplets and rain

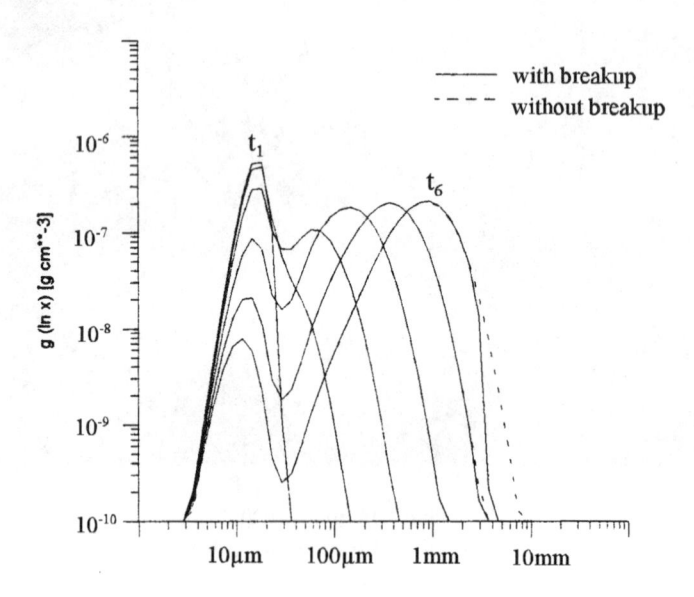

Fig. 3. Simulations of the drop size distribution (expressed as the mass distribution function g) of a convective cloud. Six time steps are shown ($t_1 = 0$ s, $t_6 = 2000$ s), the dashed curve indicates a simulation without the breakup process (from Issig, 1997; modified).

(Fig. 3). The following processes are involved in the formation of cloud drops (generation of an initial DSD) and the formation of raindrops with a sufficient large fall speed (for more details see e.g. Pruppacher and Klett, 1978):

1. Condensation and evaporation: Because of the increase of the saturation water vapour pressure with decreasing drop radius, the initial drop formation for supersaturations occurring in the atmosphere requires condensation nuclei with a scale of 0.1 μm, which are supplied by the atmospheric aerosol. Depending on the size and chemical composition of the aerosol particles an efficient further growth by condensation can take place up to the scale of cloud droplets (10 μm). These processes can be parameterised by experimentally obtained functions, if the aerosol composition and its size distribution is known.
2. Coagulation (growth by collision): Further growth is possible by the process of coagulation, i.e. the collection of smaller drops by (faster falling) larger drops. This process is governed by the DSD and the effective collection cross-section. The latter depends on the product of the geometric cross-sections of the colliding drops and an efficiency factor, which can be larger or smaller than unity. This collision efficiency can be described by experimentally derived functions, but is taken as unity as a first approximation in many cloud models.
3. Breakup of drops: Aerodynamical instability leads to the breakup of drops exceeding a radius of about 2.5 mm. The breakup and the size distribution of the new drops is parameterised statistically by suitable probability functions. The effect of this process is demonstrated in Fig. 3.
4. Ice/water phase interactions: If the ice phase is also present, all processes of the liquid phase have to be considered in analogy also for the pure ice phase (processes 1–3). In addition, the processes between the ice and water phase are important, particularly the growth of ice crystals at the cost of supercooled droplets as well as the coagulation processes between ice and water particles (leading to graupel and hail).

Cloud models, which include all processes in detail and simulate the developments of the size spectra by integrating the relevant differential equations, are called spectral models. Since the computation of the detailed microphysical processes is very time consuming, simpler models are used for many applications, which consider the same processes in form of parameterisations of the exchange between the reservoirs of water vapour, cloud water, rain water, cloud ice, snow and graupel (bulk models). As an example, the bulk parameterisation of the coagulation processes reads as (Kessler, 1969; Houze, 1993):

$$\left.\frac{dq_r}{dt}\right|_{\text{coag}} = k_c \left(q_c - q_{c,0}\right) \quad (1)$$

q_c cloud liquid water content
q_r rain water content
$q_{c,0}$ threshold value (1 g/kg)

k_c^{-1} autoconversion time ($\cong$ 1000 s for $q_c > q_{c,0}$; $= 0$ for $q_c \leq q_{c,0}$)

The main advantage of spectral models is their ability to simulate the DSD and hence all moments of the DSD. This allows the computation of the Rayleigh radar reflectivity (Z) and the rainrate (R) as important quantities for the remote sensing of the precipitation using radar measurements:

$$N = \int_0^\infty f(D)\,dD \qquad \text{drop number} \tag{2}$$

$$q_l = \frac{\pi \rho_w}{6} \int_0^\infty D^3 f(D)\,dD \qquad \text{liquid water content} \tag{3}$$

$$Z = \int_0^\infty D^6 f(D)\,dD \qquad \text{Rayleigh radar reflectivity} \tag{4}$$

$$R = \frac{\pi \rho_w}{6} \int_0^\infty D^3 V(D) f(D)\,dD \qquad \text{rain rate} \tag{5}$$

D is the drop diameter, f(D) the DSD, ρ_w the density of water and V(D) the fall speed.

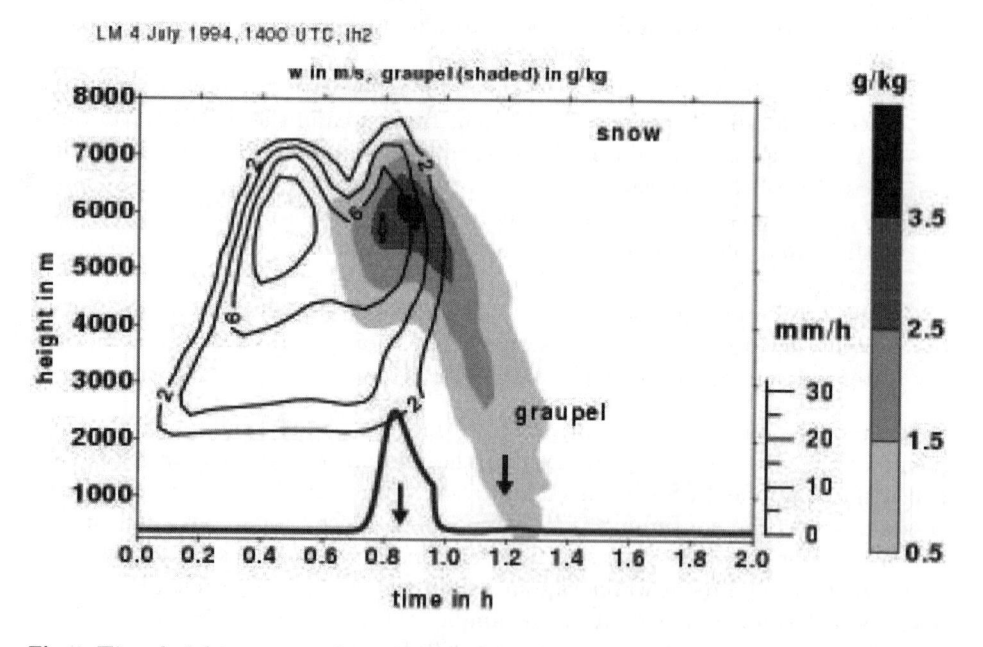

Fig. 4. Time-height cross-sections simulated by the spectral cloud model of Issig (1997) for the situation of 4th July 1994 near Köln for the vertical velocity (isolines every 2 m/s), graupel content (shaded, in g/kg), and snow content (stippled for values exceeding 0.01 g/kg). The time series of the surface rain rate (in mm/h) is shown as a thick line, arrows indicate downdraft areas.

An example of a simulation using a 1.5D cloud model (Issig, 1997) is presented in the following. This model treats the liquid phase in a spectral way, the ice phase is bulk-parameterised. The initial aerosol distribution is taken from Clark (1974). The simulation shown in Fig. 4 uses as initial conditions temperature and humidity profiles of the "Lokal-Modell" of the German weather service (DWD) valid at 14 UTC, 4th July 1994, which is the case of the "Köln hailstorm" (Drüen and Heinemann, 1998). The cloud develops within 30 min, the maximum updraft area lies between 5,000 and 7,000 m. The conversion from cloud droplets to rain water starts after 30 min at these heights. The growth of the drops continues until their fall speed exceeds the updraft, and after 0.7 h rain starts to reach the ground. The maximum surface rain rate is 25 mm/h. The main rain event at the surface lasts about 15 min, and around 1.2 h simulation time a very weak secondary rain maximum is present, which is associated with graupel melting before it can reach the ground. This rain water from melted graupel particles is also the reason that moderate values of the Z field are present after the main surface rain event (see below). A large fraction of the rain water evaporates in the atmosphere.

The simulated Z values as a function of the near-surface rain rate (Z/R relation) reflect the temporal development of the DSD (Fig. 5). Prior to the maximum rain rate, the rain consists of larger drops at the same rain rate when compared to the situation after the maximum. For the derivation of rain rates from radar (Z) measurements this implies, that the Z/R relation depends on the development stage of the convective cloud, which is in turn triggered by the actual atmospheric conditions leading to the thunderstorm development. Two standard Z/R relations are shown additionally in Fig. 5. The first is the relation used operationally by the DWD ($Z=256\ R^{1.42}$), which is close to the well-known Marshall and Palmer (1948) distribution ($Z=296\ R^{1.47}$), and the second is a relation for a DSD of thunderstorms (GDSD; Ulbrich, 1983; Fujiwara, 1965). The GDSD is in better agreement with the simulations compared to the standard DWD relation, and both relations show large differences for times prior to the near-surface rain maximum. The application of these standard Z/R relations for the development phase of the rain event would result in an overestimation of rain rates with a factor of 4–8.

The cloud model simulations can also be used to investigate the height dependence of the Z/R relation. The Z/R relation valid at 1,875 m above the ground is similar to that at 125 m after the rain maximum (except the secondary rain event associated with the melting of graupel), but larger differences can be seen prior to the near-surface rain maximum, where higher Z values at the same rain rate indicate that larger rain drops occur more frequently compared to the surface level. Breakup processes seem to be responsible for the modification of the DSD between these model levels.

3 Scale transitions and parameterising techniques used in satellite remote sensing

Numerical modelling approaches combined with remote sensing data promise a more physically based linkage between the recorded surface observations and

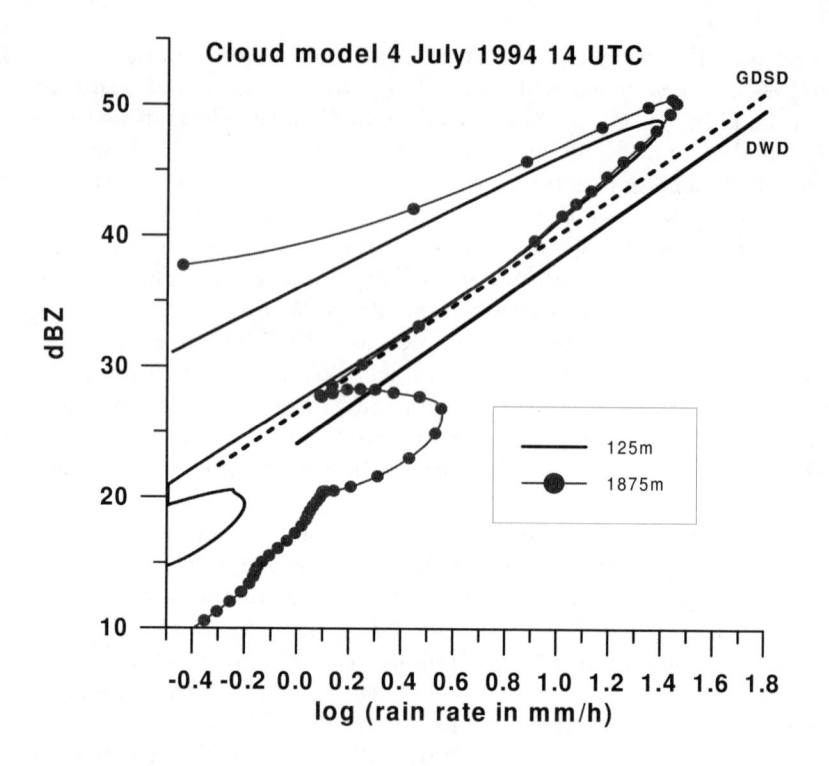

Fig. 5. Temporal development of the Z/R relation for the simulation of Fig. 4 for the model levels at 125 and 1,875 m. Z/R relations of DWD (solid) and GDSD (dashed) are shown (see text). Z is shown as dBZ= 10 log Z. The upper branches of the curves correspond to the first phase of the rain event.

the appropriate processes. On the other hand, they lead to new problems in connection with the handling of the occurring scale transitions. In spite of this problems, the application of temporal high-resolution remote sensing data offers the only reliable observation basis for the detection of convective process patterns. The achievable accuracy of quantification is crucially dependent on the handling of the appropriate scale transitions.

In this Section, an algorithm for the quantitative estimation of precipitation using infrared geostationary satellite data is outlined. The algorithm is based on the technique introduced by Adler and Negri (1988), but was modified substantially. In the new algorithm, which is described in detail in Reudenbach et al. (2001), observed Meteosat infrared (IR) cloud surface temperatures are coupled with the modelled temperatures and precipitation rates of a 1D cloud model with bulk parameterisation of cloud microphysics (Zock et al. 1995). Fig. 6 explains schematically the basic relations and proposed solutions which are discussed in detail in the following.

As Fig. 6 shows, a conceptual model has to be developed which is able to connect the cloud microphysics and dynamics in a suitable way with satellite observations. Due to the physical constraints of IR-sensors the only possibil-

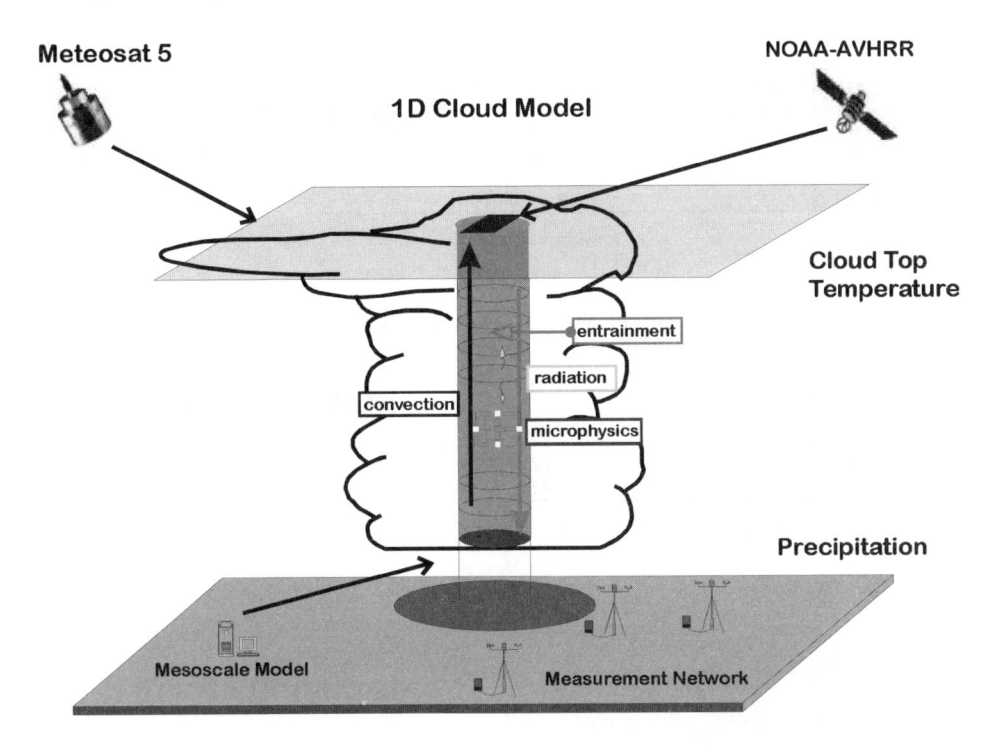

Precipitation Dynamics of Convective Clouds 193

Fig. 6. Basic scheme of numerical model, remote sensing and ground measurement data integration corresponding to the spatial and temporal scale transitions.

ity to link observation and simulation appears to be the top cloud surface. A numerical cloud model (see sec. 1) will be employed to calculate the necessary meteorological fields (cloud top temperature, precipitation) in a physically consistent way. However coupling the simulation results with satellite observations is only possible if one takes into account the occurring spatial and temporal scale transitions. The diverging scales of the cloud model and the Meteosat pixel resolution (Fig. 6) have to be linked up with a suitable upscaling procedure.

For the adaptation of the Meteosat spatial resolution to the cloud model, simultaneous NOAA AVHRR data (1.1 km resolution) have been used, to utilise an empirical correction of Meteosat IR brightness temperatures of the convective cores. The retrieval algorithm provides an estimation of the stage of cloud development which is used to fit the temporal scales of cloud model run and Meteosat observation together.

As a result the highly parameterised microscale processes of the cloud-dynamics and microphysics of the cloud model provides a physically based estimate of the convective response, (i.e. precipitation rate and maximum cloud height) in a given atmosphere, which can be implemented in a satellite based retrieval technique.

The geostationary satellites of the Meteosat series are scanning the Earth's disk completely at half-hour intervals (resolution of approx. $6 \times 10 \, \text{km}^2$ for Central

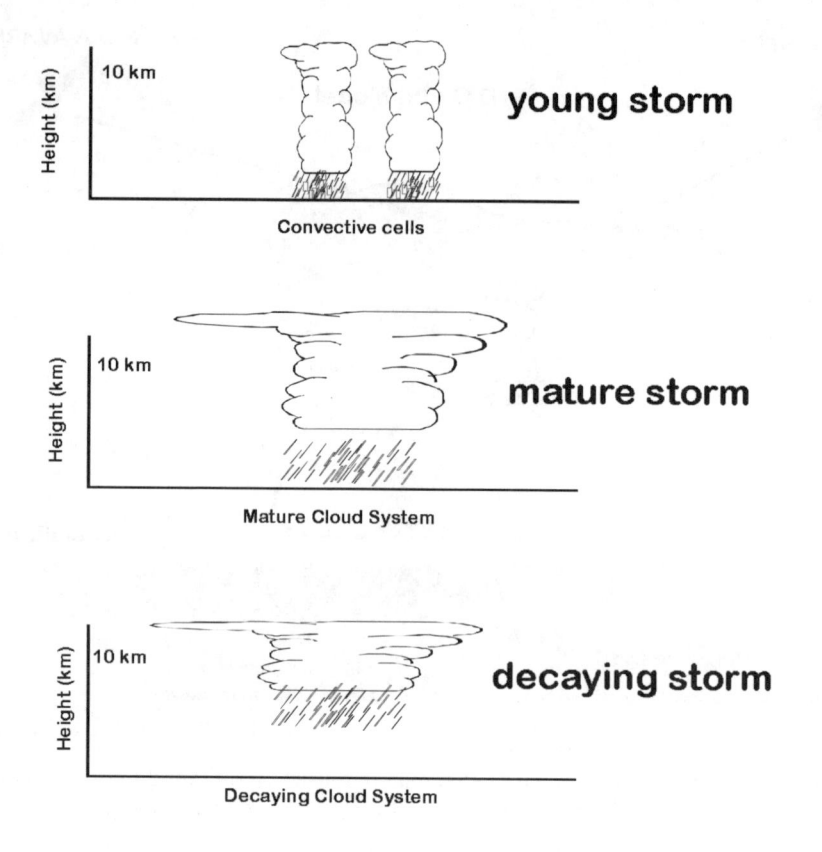

Fig. 7. Scheme of convective life cycle stages (after Johnson and Young, 1983).

Europe). By means of a suitable algorithm the convective minima are isolated from the temperature patterns of each picture sequence and assigned to a defined stage of the convective life cycle. The cloud system is classified into early, mature and the decay stages. Appropriate precipitation regimes are assigned to the respective stage (Fig. 7): the early stage (young storms) is associated with convective precipitation; the mature stage (mature storms) is associated with convective and stratiform precipitation and the stage of decaying storms provides no precipitation. In this way it is determined whether the detected spatial/temporal signal (Meteosat pixel) corresponds to the model conceptions of convective precipitation genesis.

The rain algorithm for active convective cores is based on the 1D parcel-type cloud model. The linkage of remote sensing and model data is established by using the Meteosat infrared temperatures and the simulated cloud top temperatures. The model is adapted to the regional atmospheric conditions by using actual atmospheric profiles of humidity and temperature simulated by the LM of the DWD (see previous section). In order to derive the regression between cloud top temperature and rain rate, the cloud model is operated with a variable diam-

Precipitation Dynamics of Convective Clouds

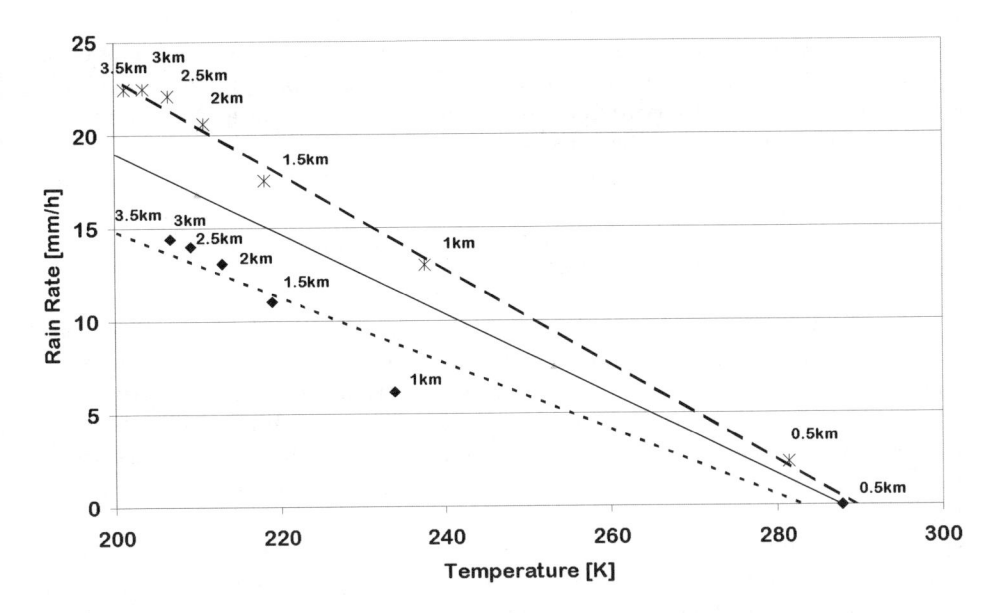

Fig. 8. Cloud model-derived rain rate/cloud top temperature relationships for two actual profiles of 4th July 1994 (dashed) and the averaged linear regression (solid line) for about 30 profiles. The parameter varied is the diameter of the convective updraft area.

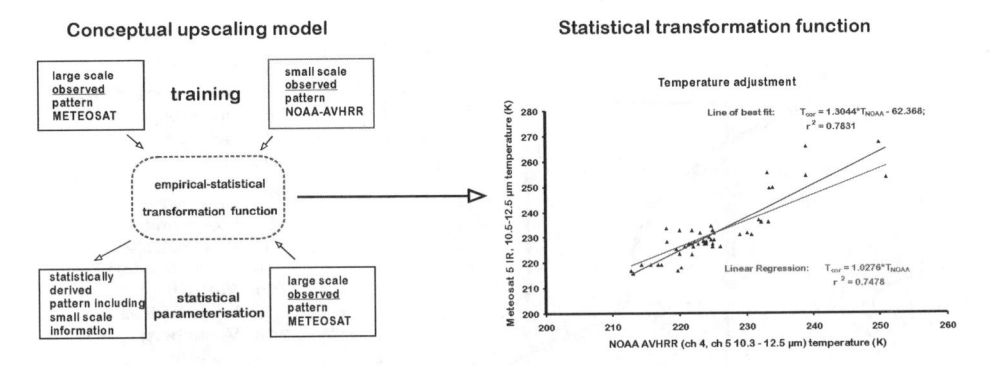

Fig. 9. Statistical upscaling of the spatial cloud model resolution to Meteosat viewing geometry.

eter of the updraft area. Fig. 8 shows two examples of relations between mean rain rate (Rmean) and the modelled cloud top temperature Tct. In order to obtain an averaged statistical relation, Rmean/Tct regressions have to be computed for LM profiles in the area of observed convection. The rain rate can then be approximated by linear regression analysis:

In a preliminary step it is necessary to find a suitable solution for the adjustment of the spatial scale of the 1D cloud model according to the viewing geometry of Meteosat. The used cloud model can be applied up to a diameter of approx. 4 km. The spatial resolution of the modelled precipitation event deviates thus clearly from the observation geometry of the Meteosat data. A linear statistical upscaling procedure is used to decrease the resulting parameterising errors which are caused by cloud surface temperatures that were detected too warm. Simultaneous Meteosat IR temperatures and spatially better resolved NOAA AVHRR temperature data ($1 \times 1 \, \text{km}^2$) of identical minima are used (Fig. 9) to carry out a linear regression analysis.

The derived transformation function serves as statistical scale adjustment of the Meteosat temperatures in reference to the temperatures calculated by the cloud model. To validate the precipitation estimation point measurements of ground stations are used. However, the extreme variability of the observed convective precipitation events prevents the use of individual point observations for validating purposes. In order to derive representative statements about spatial precipitation based on ground measurements, rain gauge networks with high spatial and temporal resolution within the area of the City Bonn were used (Fig. 10). Rain rates of 5 minutes enables a spatial mean value of the ground measurements according to the spatial resolution of the remote sensing data. The kriging technique was used to interpolate the spatial means. It calculates a weighted sum of known observations to estimate an unknown value (Pebesma and Wesseling, 1998).

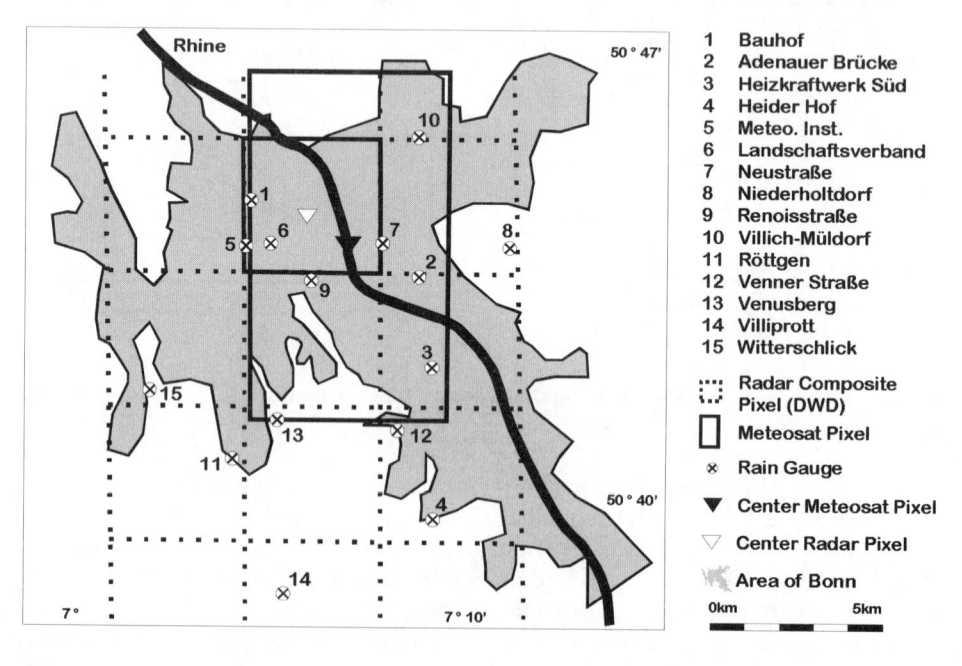

Fig. 10. Ground measurement network (City Bonn) with corresponding remote sensing viewing geometry areas.

Precipitation Dynamics of Convective Clouds 197

In Fig. 11 all rain gauge data during the "Köln hailstorm" event (4th July 1994, 10–14 UTC) of the 15 stations are spatially interpolated and averaged over the size of the Meteosat pixel. Due to the temporal resolution of Meteosat mean rain rates are computed as 30 minutes averages. The interpolated rain rate applies significantly better to the Meteosat derived rain rates. Due to the spatially and temporally aggregated and interpolated ground data according to the Meteosat viewing geometry it is possible to make quantitatively more valid predictions about the area integral sensor signal as this would be possible from the individual point measurements.

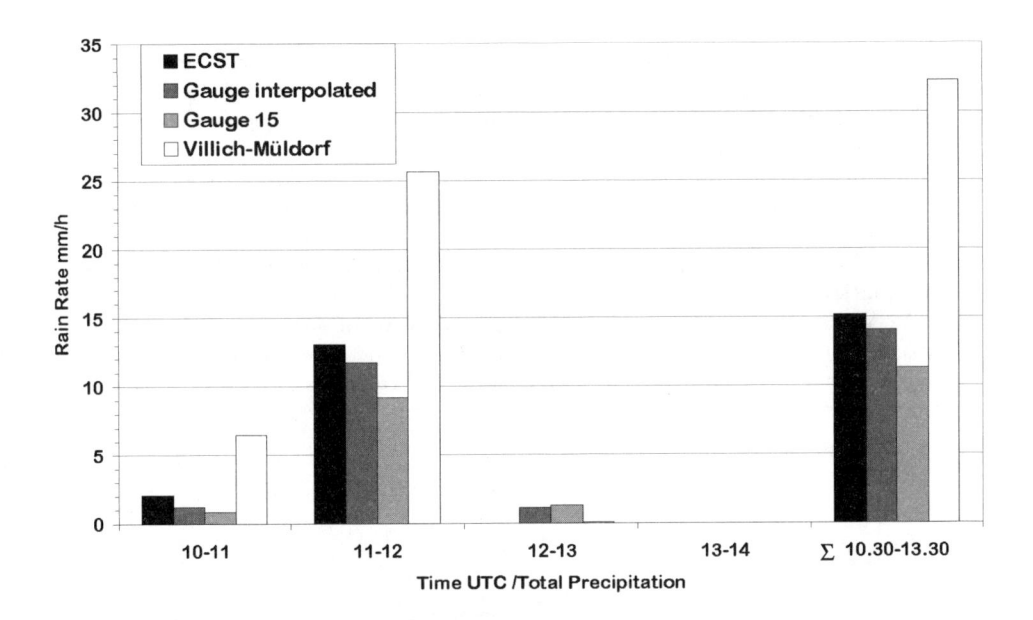

Fig. 11. Comparison of rainfall rates (30 minutes averages and the 4 h average) during the heavy precipitation event 10–14 UTC, 4th July 1994, at Bonn. ECST is the satellite derived estimate for the Meteosat pixel. Gauge 15 is the arithmetic mean of the 15 rain gauge stations shown in Fig. 10; Gauge 9 is the arithmetic mean of the 9 gauges within the Meteosat pixel; Gauge interpol. marks the spatially interpolated and averaged data over the area of the Meteosat pixel.

References

Adler, R. F., Negri, A. J., 1988: A satellite infrared technique to estimate tropical convective and stratiform rainfall. Journ. Appl. Meteor. 27, 30–51.
Clark, T. L., 1974: A study in cloud phase parameterization using the Gamma distribution. J. Atmos. Sci. 31, 142–155.
Drüen, B., Heinemann, G., 1998: Rain Rate Estimation from a Synergetic use of SSM/I, AVHRR and Meso-Scale Numerical Model Data. Meteorol. Atmos. Phys., 66, 65–85.

Fujiwara, M., 1965: Raindrop-size distribution from individual storms. J. Atmos. Sci., 22, 585–591.
Issig, C., 1997: Ein spektrales Wolkenmodell mit integriertem Strahlungsübertragungsmodell zur Unterstützung von Niederschlagsalgorithmen aus Fernerkundungsdaten. PhD Dissertation, Bonn, 104 p.
Houze, R. A., 1993: Cloud dynamics. Academic Press, San Diego, USA, 570 pp.
Johnson, D. H., Young, G. S., 1983: Heat and moisture budgets of tropical mesoscale anvil clouds. J. Atmos. Sci. 40, 2138–2147.
Kessler, E., 1969: On the distribution and continuity of water substance in atmospheric circulations. Meteor. Monogr. 32, Amer. Meteor. Soc., Boston, USA, 84 pp.
Marshall, J. S., and W. M. Palmer, 1948: The distribution of raindrops with size. Journal of Meteorology, 5, 165–166.
Pebesma, E. J. and C. G. Wesseling, 1998. 'Gstat: a program for geostatistical modelling, prediction and simulation.' Computers and Geosciences. 24 (1): 17–31.
Pruppacher, H. R., Klett, J. D., 1978: Microphysics of clouds and precipitation. Reidel Publishers, Dordrecht, 714 pp.
Reudenbach, C., Heinemann, G., Heuel, E., Bendix, J., Winiger, M., 2001: Investigation of summertime convective rainfall in Western Europe based on a synergy of remote sensing data and numerical models. Meteor. Atmosph. Phys. 76, 23–41.
Ulbrich, C. W., 1983: Natural variations in the analytical form of the raindrop size distribution. J. Clim. Appl. Meteorol., 22, 1764–1775.
Zock, A., Menz, G. and Winger, M. 1995: Regionalisation of rainfall models in Eastern Africa using Meteosat Real-Time-Window data. Proceedings of the International Geoscience and Remote Sensing Symposium (IGARSS'95), 10–14 July 1995 Florence, Italy (New York: I. E. E. E.), 250–252.

Sediment Transport — from Grains to Partial Differential Equations

Stefan Hergarten[*], Markus Hinterkausen, and Michael Küpper

Section Geodynamics, Geological Institute, University of Bonn, Germany

Abstract. We illustrate that sediment transport phenomena cover a wide range of spatial and temporal scales and point out the need for continuum models of sediment transport. We show, however, that continuum approaches to sediment transport only result in feasible models in some special cases such as rolling, saltation, and suspended load and discuss the capabilities and the limitations of these approaches with examples from modelling soil erosion processes and ripple formation.

1 Phenomena, Processes, and Scales

Large parts of the earth's surface are shaped by sediment transport processes. For this reason, modelling landform evolution is mainly modelling sediment transport processes.

The structures formed by sediment transport processes cover a wide range of spatial scales. It starts at the microscopic roughness of the surface and continues with structures on the centimetre to meter scale, such as gullies and ripples. The next larger elements are river banks, landslides, and dunes; they may extend to some hundred meters. Finally, the largest scale of sediment transport phenomena consists of basins formed by fluvial erosion.
Sediment fluxes may be driven by

- wind,
- water (in oceans, lakes, and rivers, or Hortonian runoff during heavy rainfall events) or
- gravity (landslides, debris flow).

Unconsolidated sediments are granular matter; their grain sizes vary from less than 1 μm for clay particles to about 1 cm for gravel. Even large boulders of more than 1 m size may participate in sediment transport. But their probability of being moved is considerably lower than for smaller particles. Thus, their direct contribution to landform evolution may be negligible, but their presence can affect transport of smaller particles strongly by armouring and reinforcing surface materials.

In many cases, the sizes of the particles contributing to sediment transport are indeed much smaller than the characteristic scales of the pattern generated

---

[*] hergarten@geo.uni-bonn.de

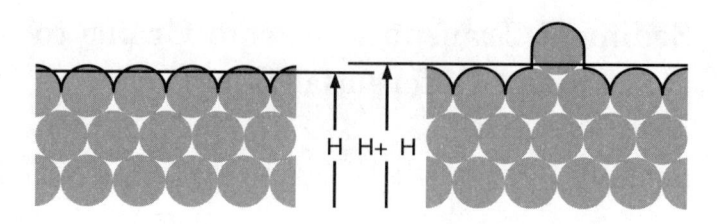

Fig. 1. Definition of an average surface height (left) and illustration of the effect of adding one grain (right).

by the transport process. If we, for instance, consider a typical sand of 0.3 mm grain diameter, a volume of $(10\,\text{cm})^3$ consists of 4×10^7 particles. Thus, modelling transport processes by simulating individual particle motions is only feasible for investigating phenomena on the grain scale, but not for modelling effects on larger scales. Therefore, mostly particle densities (concentrations) are introduced in order to describe the sediment as a continuum.

The first step towards a continuum description is introducing an average surface height $H(\vec{x}, t)$. The vector $\vec{x} = (x_1, x_2)$ describes the coordinates on the model scale that is much larger than the grain scale. Thus, $H(\vec{x}, t)$ can be seen as an average surface height of a sediment pile that consists of many particles; it can be defined as illustrated in the left hand part of Fig. 1. For simplicity, we restrict our considerations to identical grains for the moment, but most of the ideas can easily be extended to sediments of multiple particle sizes.

As shown in the right hand part of Fig. 1, the average surface height increases by an amount δH if a grain is added. This effect can be quantified by introducing a representative particle volume V in such a way that

$$\delta H = \frac{V}{A} \tag{1}$$

if one grain is added to an area A. As sediments contain a certain amount of void space (pores), V is greater than the volume of a single sediment grain; V can be computed by dividing the particle volume by the fraction of pore space (porosity).

As a result of Newton's principle, the state of a moving particle cannot be characterised by its position $\vec{x} = (x_1, x_2, x_3)$ in space alone; the velocity $\vec{v} = (v_1, v_2, v_3)$ is necessary, too. Thus, a particle concentration $c(\vec{x}, \vec{v}, t)$ which depends on six independent variables (plus time) has to be defined within the transporting medium (air or water). In the general case, this procedure will result in a six-dimensional partial differential equation.

In computational quantum mechanics, such high-dimensional problems are common subjects, but they can only be solved if most properties of the solution are a priori known. Thus, such a general continuum approach to sediment transport does not have any advantages compared to simulating individual grain motions. Feasible formulations are only obtained if further assumptions about the transport mechanisms are made.

The complexity of the resulting equations depends on the interactions of the transported particles; these interactions may be

- interactions with the driving fluid (air or water),
- interactions with the sediment surface or
- interactions with other particles being transported.

The latter kind of interactions mostly leads to quite complicated, nonlinear equations. Neglecting them is reasonable as long as the particle concentration is sufficiently small. At higher concentrations, the relevance of particle-particle collisions increases. This effect is ubiquitous in landslides and debris flows where particle-particle interactions are at least as important as interactions with the driving fluid, if the latter is present at all.

In the following we focus on three established transport mechanisms which neglect particle-particle interactions and their transfer to continuum models. The more complicated case with high particle densities leads to the theory of granular flow; discussing its numerous approaches goes beyond the scope of this introduction, so that we recommend the review articles of Iverson (1997) or Jan and Shen (1997).

2 Sediment Transport Mechanisms and their Continuum Representation

2.1 Rolling

Heavy particles are unlikely to be lifted up by water or air, so that they tend to move along the sediment surface. In each river, a certain amount of sediment is transported by rolling. The toppling of sand grains from steep sandpiles or at the lee side of ripples and dunes is a similar mechanism.

Since particle motion is restricted to the sediment surface, rolling can be considered as a two-dimensional process in a planar view. Thus, the particle concentration describes the number of rolling particles per area instead of volume; $\vec{x}$ and $\vec{v}$ are two-dimensional vectors.

In this case, the mass balance equation that describes the evolution of the surface height can be written in the form

$$\frac{\partial}{\partial t} H(\vec{x}, t) = -\mathrm{div}\vec{j}(\vec{x}, t) \tag{2}$$

with the stream density

$$\vec{j}(\vec{x}, t) = V \int c(\vec{x}, \vec{v}, t)\,\vec{v}\,dv_1 dv_2 \,. \tag{3}$$

However, this is the trivial part. Quantifying the driving forces leads to a four-dimensional, time-dependent partial differential equation for $c(\vec{x}, \vec{v}, t)$. Thus, further assumptions about the mechanism are necessary.

A simple, in some cases sufficient idea is to allow only motion with a fixed velocity (that may, for instance, depend on the surface gradient) and keep only the number of rolling grains variable. Such a two state approximation was introduced by Bouchaud et al. (1995) for simulating avalanche dynamics on sandpiles.

2.2 Suspended Load

Suspended load is the opposite limiting case to rolling. If the particles are light and the flow is turbulent, an apparently random pattern of particle motion occurs. Each particle is exposed to a nearly random acceleration many times; so it performs a large number of small scale movements. This may look as if the particles were floating; but this is — strictly speaking — not correct. No matter how small the particles are, their density is higher than that of water, so that they sink in the mean. As the settling velocity decreases with decreasing particle size, the random motion may govern the behaviour, so that the particles may stay in suspension for a long time.

Since the particles are exposed to high accelerations, effects of inertia are negligible for suspended load. Thus, there is no need to keep track of the particle velocity, so that considering a concentration $c(\vec{x}, t)$ where $\vec{x} = (x_1, x_2, x_3)$ is appropriate. The particle concentration in the fluid is then determined by the balance equation

$$\frac{\partial}{\partial t} c(\vec{x}, t) = -\mathrm{div}\, \vec{j}(\vec{x}, t) \tag{4}$$

where $\vec{j}(\vec{x}, t)$ is the particle stream density. Assuming that the velocity of the particles is the sum of the fluid velocity $\vec{v}(\vec{x}, t)$, the settling velocity $\vec{v}_s$, and a random component, the stream density is

$$\vec{j}(\vec{x}, t) = c(\vec{x}, t) \left(\vec{v}(\vec{x}, t) + \vec{v}_s \right) - D(\vec{x}, t)\, \vec{\nabla} c(\vec{x}, t). \tag{5}$$

The first term at the right hand side describes an advective transport, while the second term acts as a diffusive component.

Finding an expression for the diffusivity $D(\vec{x}, t)$ is the nontrivial part; it quantifies the effects of turbulence and eventually of particle collisions. Approaches can be derived from the theory of turbulent flow; but as shown later in the example of soil erosion, it may be appropriate to use $D(\vec{x}, t)$ for a calibration with empirically-based formulas for transport capacity, too.

The evolution of the sediment bed is then determined by the sediment flux at the surface:

$$\frac{\partial}{\partial t} H(x_1, x_2, t) = -V\, \vec{n} \cdot \vec{j}(x_1, x_2, H(x_1, x_2, t), t) \tag{6}$$

where $\vec{n}$ is the upper normal vector at the surface.

We started at the wide range of the spatial scales of sediment transport phenomena, but the time scales cover a wide range, too. Compared to the time scales of surface evolution, the time needed for moving individual particles is very short. From the long term point of view, the flow field and the sediment concentration adapt immediately to any changes in surface geometry or boundary conditions. Thus, there is no memory in these rapid processes, so that they can be described by steady-state equations. This step is called separation of time scales; for suspended load, this means that the left hand side of Eq. 4 can be neglected. However, while the separation of time scales is a great progress in many multi-scale phenomena, it does not really make things easier for suspended load.

2.3 Saltation

Saltation is a transport mechanism between rolling and suspended load. The power of the flow is sufficient for lifting up particles, but not high enough for keeping them in the fluid for long times. The acceleration acting on the particles during their motion is not so strong that effects of inertia are negligible. Thus, the particle traces are not as random as for suspended load; variations may result from the process of lifting rather than from the later stages of motion.

This phenomenon can be simulated efficiently if the time scales can be separated and if the number of saltating particles is low, so that they do not interact with each other. In this case, the traces of a particle starting at any point of the surface can be computed, provided that the initial velocity and the forces acting during the saltation step are given. Thus, it is in principle possible to compute the statistical distribution of the saltation lengths, so that the problem can be reduced to a two-dimensional plane view model.

It is convenient to use the two-dimensional master equation for computing the changes of the surface height:

$$\frac{\partial}{\partial t} H(\vec{x}, t) = V \left(\int \nu(\vec{y}, t) \, w(\vec{y}, \vec{x}, t) \, d^2 y - \nu(\vec{x}, t) \right). \qquad (7)$$

Here, $\nu(\vec{x}, t)$ is the number of saltating particles per area and time starting at the location $\vec{x}$ at the surface; $w(\vec{y}, \vec{x}, t) d^2 x$ is the probability that a particle that starts at the location $\vec{y}$ settles in the interval $[\vec{x}, \vec{x} + \delta \vec{x}]$. The first term at the right hand side of Eq. 7 describes those particles which start at any location and settle at $\vec{x}$, while the second term describes the particles starting at $\vec{x}$.

Eq. 7 can in principle be written in flux density form with a particle stream density $\vec{j}(\vec{x}, t)$, but it is quite complicated and has not any advantages.

Transforming Eq. 7 into a numerical scheme is straightforward; it is simply a redistribution of particle packets according to the statistical distribution of the saltation lengths. However, computing this distribution is costly, so that further simplifications may be necessary. For instance, if the saltation lengths are smaller than the model resolution, Eq. 7 can be written in the same form as the equation for rolling grains with a very simple particle stream density (Hergarten and Neugebauer 1996). Another important limiting case is that of a narrow distribution of the saltation lengths; this means that the statistical variation can be neglected. Such an approach is used in the model of Nishimori and Ouchi (1993) and in the ripple model discussed in Section 4.

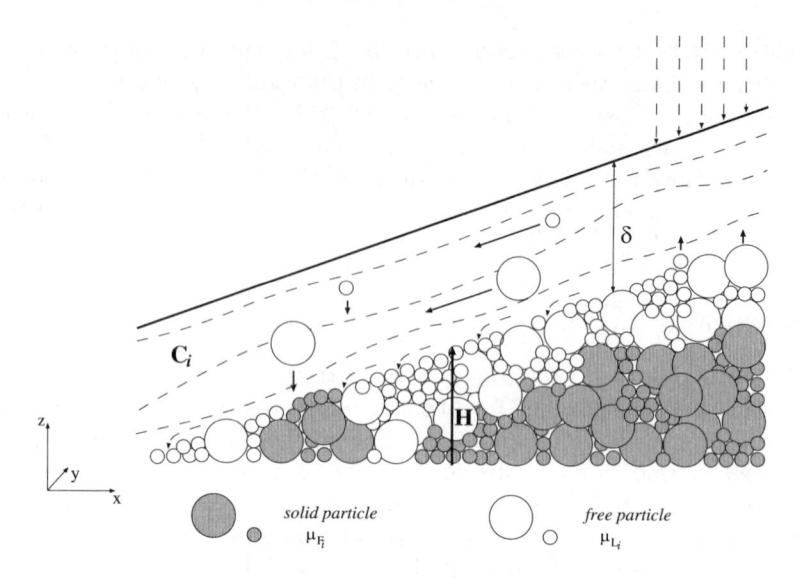

Fig. 2. Illustration of particle motions and soil structure on the grain scale.

3 Soil Erosion

3.1 Particle Motion and Soil Structure

A simplified concept for describing soil erosion processes is illustrated in Fig. 2. It illustrates particle motions and soil structure on the grain scale. The soil contains particles with the same kind of structure but different sizes.

In contrast to sand or gravel sediments as they occur in many fluvial systems, cohesion plays an important part in soil erosion; it prevents the particles from easily being detached by the flow. For simplicity we describe the cohesion by a two-state approximation: Each particle is either connected tightly to its neighbours or can be detached by the flow.

Turbulent surface runoff of water is the major driving force to particle movement. Moreover, stresses on surface caused by the water can disconnect solid particles, so that they can be lifted up afterwards.

For small particle sizes, suspended load is the dominating sediment transport mechanism. So we introduce particle concentrations $C_i(\vec{x}, t)$ in the fluid, where the index i represents a class of particles that is characterised by a grain size. In order to describe the different particles in the soil we consider concentrations $\mu_{L_i}(\vec{x}, t)$ for the free particles and $\mu_{F_i}(\vec{x}, t)$ for the solid particles in the soil, too. While $C_i(\vec{x}, t)$ refers to a number of particles per volume, $\mu_{L_i}(\vec{x}, t)$ and $\mu_{F_i}(\vec{x}, t)$ are relative volume concentrations. With the condition

$$\sum_i (\mu_{L_i}(\vec{x}, t) + \mu_{F_i}(\vec{x}, t)) = 1 \qquad (8)$$

they describe changes in the mixture between the classes and the two states.

3.2 Surface Runoff

Our description of surface runoff is based on a generalisation of Manning's formula (Hergarten et al. 2000); beside the surface height $H(\vec{x}, t)$ it involves the thickness of the water layer $\delta(\vec{x}, t)$ as a second variable:

$$\vec{v} = -\frac{1}{n} \delta^{\frac{2}{3}} |\vec{\nabla}(H + \delta)|^{\frac{1}{2}} \frac{\vec{\nabla}(H + \delta)}{|\vec{\nabla}(H + \delta)|}. \tag{9}$$

The Manning formula describes the average fluid velocity of turbulent flow on a structured surface of a slope. The parameter n considers the microscopic roughness of the surface.

Taking into account that the temporal evolution of the water level δ depends on the rainfall rate R and the infiltration rate I, the balance equation reads:

$$\frac{\partial \delta}{\partial t} + \text{div}(\delta \vec{v}) = R - I. \tag{10}$$

3.3 Erosion and Deposition

Due to suspended load as the dominating transport mode the temporal changes of the particle concentrations $C_i(\vec{x}, t)$ can be described by Eq. 4, and the particle stream density $\vec{j}_i(\vec{x}, t)$ is given by Eq. 5 with the fluid velocity from Eq. 9.

In order to develop an expression for the diffusivity $D(\vec{x}, t)$ we use this for a calibration with the empirically-based formula for the transport capacity from Engelund and Hansen (1967):

$$C_i^{\text{stat}} = \frac{0.05}{\sqrt{g}\, d_i^4} \frac{\rho_w}{\rho - \rho_w} |\vec{v}| \sqrt{\delta} |\vec{\nabla}(H + \delta)|^{\frac{3}{2}} \tag{11}$$

where

d_i = particle diameter,
ρ = mass density of the particles,
ρ_w = density of water,
g = gravitational acceleration.

This formula describes a stationary particle concentration C_i^{stat} in case of steady, spatially homogeneous runoff. Comparing this empirical relation with the solution of the balance equation (Eq. 4) under the same conditions leads to an equation which defines the diffusivity $D(\vec{x}, t)$.

Model calibration requires a second empirically-based formula that determines the rate E of disconnecting solid particles; we choose that of Foster (1982):

$$E = a\,\tau^b. \tag{12}$$

τ is the shear stress at the soil surface; a and b are parameters specified by the soil characteristics. The stress τ is given by the component of gravity force per area of the water parallel to the slope:

$$\tau = \rho_w g \delta \, |\vec{\nabla}(H+\delta)|. \tag{13}$$

The evolution of the slope surface $H(\vec{x},t)$ and of the mixture ratios is determined by the sediment flux in direction of the upper normal vector at the surface (see Eq. 6). Finally, the total particle stream density $\vec{j}(\vec{x},t)$ is the sum of the individual stream densities $\vec{j}_i(\vec{x},t)$.

3.4 Some Results

Figs. 3 and 4 provide a rough overview over the capabilities of the model. The model domain consists of an artificial slope of 16 m×64 m size with a roll-out at the foot. In order to introduce some inhomogeneity, a small noise is superposed. The surface runoff is caused by several rainfall events with a rate of 1.67×10^{-5} m/s (60 mm/hour). There is no inflow at the upper edge, and we assume periodic boundary conditions at the sides of the domain. Fig. 3 shows that a pattern of erosion rills evolves with increasing number of events, so that flow is concentrated in this rills. The description with partial differential equations enables us to trace the evolution on this scale in detail.

Another example is illustrated in Fig. 4; it shows the spatial distribution of the mixture ratio between two particle classes of different sizes. The diameter of the small particles is 0.05 mm, while that of the large particles is 1 mm. In the beginning, the large particles are confined to a layer of 5 mm thickness in the upper region. Through the erosion process, they are transported downslope and the layer becomes smaller. While small particles tend to be transported out of the system, large particles are partly deposited at the foot of the slope. As a result of the inhomogeneous flow paths due to rill formation, deposition is quite inhomogeneous.

However, we should be aware that suspended load is the only transport mechanism in this simplified model. While this assumption is reasonable for the small particles, particles of 1 mm are transported by saltation rather than by suspended load in reality, so that a more comprehensive model which includes at least two different transport mechanisms is required for such simulations.

4 Ripple Formation

Sediment ripples in flumes arise if saltation is the dominating transport mode. In contrast to suspended load the grains are transported in a thin layer over the bed. "The detachment of grains is due to the friction shear stress interacting between the flow and the bed surface; their downstream motion is due to the local flow velocities" (Yalin 1992).

The flow structure over ripples is illustrated in Fig. 5. The turbulent flow detaches at a ripple crest and creates a "roller" at the lee side before it reattaches at

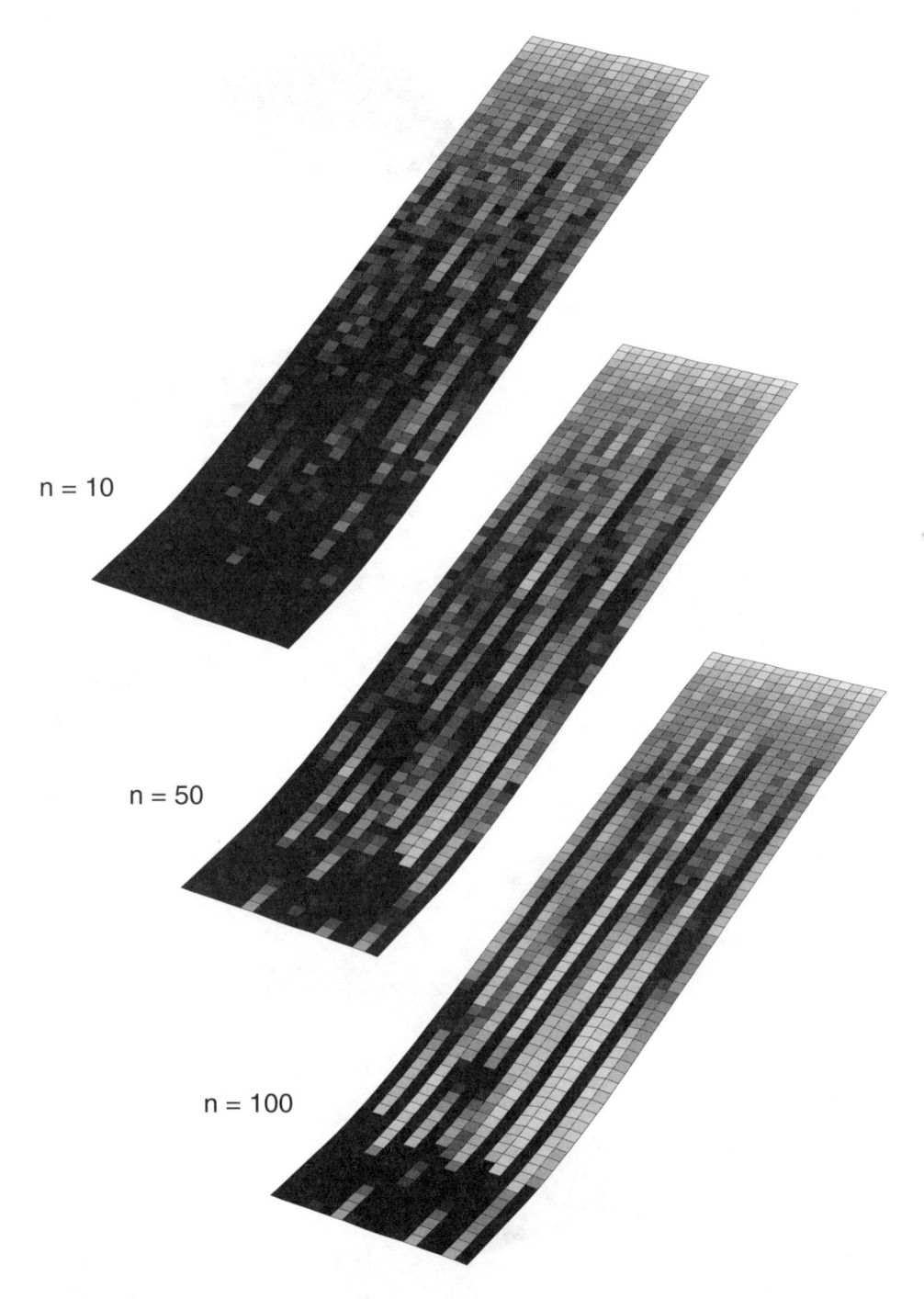

Fig. 3. Spatial distribution of surface runoff during the n-th rain event. The gray scale indicates the thickness of the water layer.

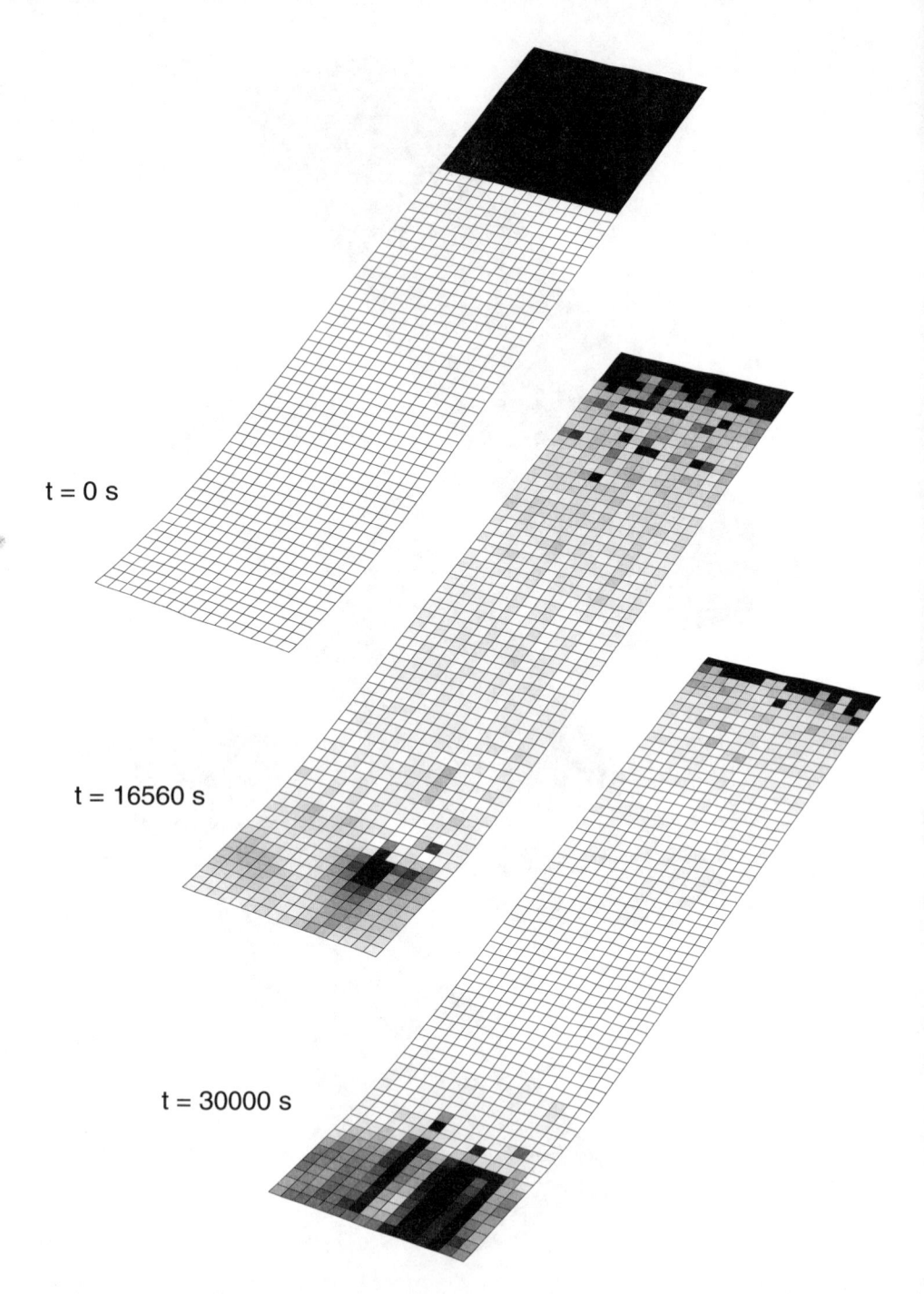

Fig. 4. Mixture ratio at different times. The gray scale corresponds to the average grain size; dark areas represent large particles.

the stoss side of the adjacent ripple. One can observe that the sediment transport ceases in the separation zone and sediment is predominantly deposited. Three regions of different velocity profiles can be distinguished: the internal boundary layer, the wake region, and the outer region (McLean and Nelson 1994).

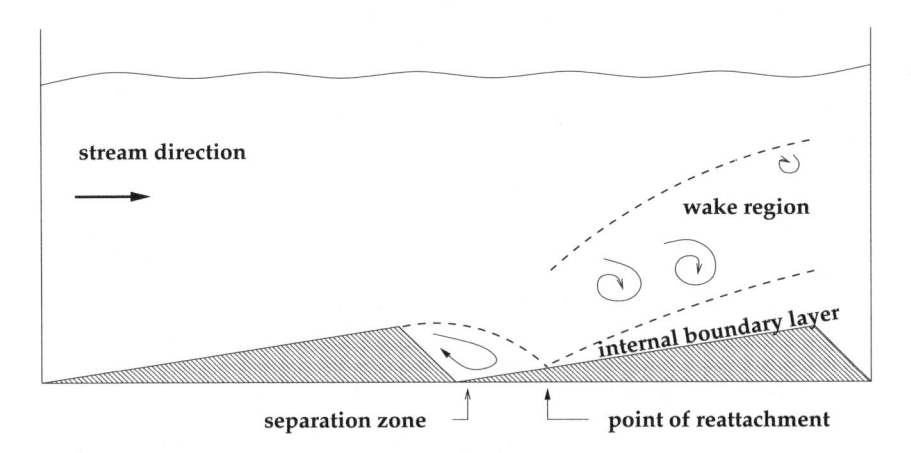

Fig. 5. Schematic plot of the flow structure over a bedform.

Since the transport mode is saltation, Eq. 7 is appropriate for describing the impact of the flow on the sediment bed. Hence the sediment transport varies with the number of entrained particles $\nu(\vec{x}, t)$ and with the distribution of saltation lengths $w(\vec{x}, \vec{y}, t)$. As a first approach we choose the saltation length to be constant and consider the influence of the fluid flow in the term $\nu(\vec{x}, t)$ which is exclusively depending on the current bed topography.

4.1 Lee/Stoss Effect

The basic idea is to describe the number of moved particles along the sediment bed by a kind of shadowing: shadowing is relatively strong in the separation zone where the sediment transport is reduced in consequence of the upstream ripple crest, whereas the shadowing effect ceases behind the point of reattachment.

For the sake of simplicity the average transport direction is chosen to be parallel to the x_1-axis. We assume that due to the shadowing mechanism the number of moving particles over ripples is reduced compared with the number of moved particles over a flat bed. We parameterise this behaviour with a "shadow function" $S(x)$ which varies between zero and one, so that the number of moved particles can be determined as follows:

$$\nu(\vec{x}, t) = \nu_0 \cdot S(\vec{x}, t) \tag{14}$$

where ν_0 is the number of moved particles over a flat bed. An at least qualitatively reasonable approach for the shadow function can be obtained by variational calculus:

$$S(x) = \left[1 - \frac{1}{\Delta} \min_{\substack{u \geq H \\ u(x_{max}) = H_{max}}} \left(\max_{\zeta \in [x^{max}, x]} \{-u'(\zeta)\}\right)\right]_+ \quad (15)$$

with continuous functions u as shown in Fig. 6. The symbol $[\ ]_+$ denotes the positive part of the term in brackets. Δ is the steepness of $u(x)$ where the particles are not moved any longer, so that S is equal to zero then. As a result of this approach, the sand transport decreases at the lee side of a ripple and increases at the stoss side of the adjacent ripple.

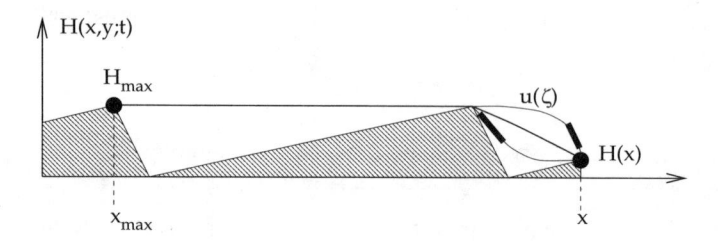

Fig. 6. Illustration of the variational calculus to obtain the shadow function $S(x)$.

As mentioned above, we assume that the saltation length of the particles is constant. Thus, the corresponding probability function $w(\vec{y}, \vec{x}, t)$ has the shape of Dirac's delta function $\delta(\vec{x} - \vec{y} - l\,\vec{e}_x)$, and Eq. 7 turns into:

$$\frac{\partial}{\partial t} H(\vec{x}, t) = V \cdot \left(\nu(\vec{x} - l\,\vec{e}_1) - \nu(\vec{x}, t)\right) = V\nu_0 \cdot \left(\underbrace{S(\vec{x} - l\,\vec{e}_1)}_{\text{Deposition}} - \underbrace{S(\vec{x})}_{\text{Erosion}}\right) \quad (16)$$

The alteration of the sand bed height is given by the difference between the number of incoming and outgoing particles.

4.2 The Diffusional Processes

Eq. 16 describes a process that creates patterns due to the lee/stoss characteristics. However, ripples observed in nature reach a quasi-steady state with a limited height of the crest after a transient process (Baas 1999). Hence, besides constructive processes, there are also destructive influences that smooth the bed's surface.

We identify the following behaviour: the steeper the slope the larger is the number of particles which move downhill due to rolling; moreover avalanching occurs, if the slope exceeds the angle of repose (for sand approximately 32°). Both processes can be described by a diffusional approach for the stream density:

$$\vec{j}_d(\vec{x}, t) = \begin{cases} -D\left(|\vec{\nabla}H| - \tan\varphi\right) \frac{\vec{\nabla}H}{|\vec{\nabla}H|} & \text{if } |\vec{\nabla}H| > \tan\varphi \\ 0 & \text{else} \end{cases}. \quad (17)$$

with φ as the smallest angle where rolling occurs; D denotes the diffusivity. The impact of the diffusional stream on the total sediment transport is shown in Fig. 7.

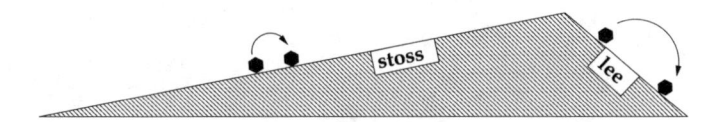

Fig. 7. Influence of diffusion on a constant sediment transport downstream.

Eventually the governing equation of the model reads:

$$\frac{\partial H}{\partial t} = -\text{div}\,\vec{j}_d + V\nu_0 \cdot \left(S(\vec{x} - l\,\vec{e}_1) - S(\vec{x}) \right) \tag{18}$$

4.3 Results

To illustrate the capabilities of the model we perform an exemplary simulation experiment (see Fig. 8): in the initial state we assume a sediment bed which is flat except for small, random fluctuations in height. After some time, ripples begin to grow and migrate downstream, whereby little ones are captured by bigger ones. The patterns are three-dimensional, and the ripple crests evolve towards a lunate shape.

We expect that the system should reach a quasi-stationary state where the growth of the ripple crests ceases. However, we observe in our simulation that the development of the ripples continues until a single ripple occupies the whole model region. Finally, an equilibrium between constructive and destructive processes is achieved, so that the growth process ceases in fact. However, length and height of the resulting equilibrium shape is controlled by the size of the model domain rather than by physical properties in this preliminary model.

Nishimori and Ouchi (1993) have derived a similar model for wind-driven ripples. They also use a diffusional particle stream, but choose a different approach to determine the saltation lengths. Their saltation length l is a function of the surface height:

$$l = L_0 + b\,H(\vec{x}). \tag{19}$$

In contrast to our model they are able to simulate a quasi-stationary state of the ripples. If we compare their results with ours, we find following differences:

- Our bedforms are asymmetric and migrate downstream in accordance to observation; however the growth is only limited by the model region.
- Nishimori and Ouchi's model produces ripples which are limited in height, but their shape is symmetric and they migrate in the wrong direction (upstream).

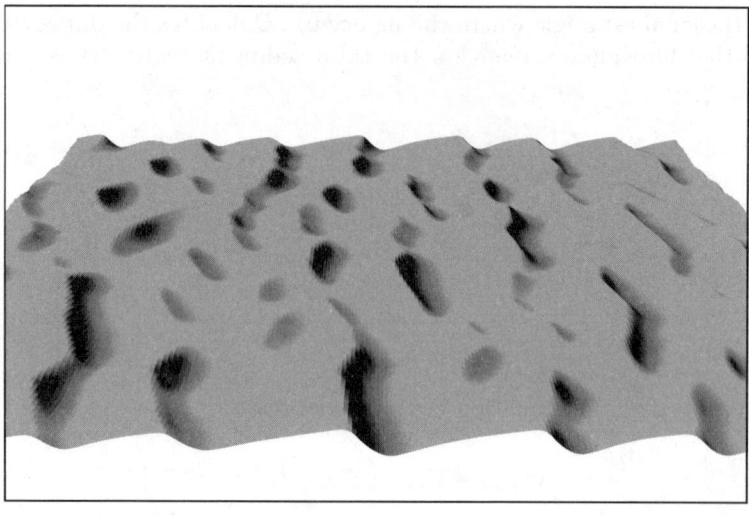

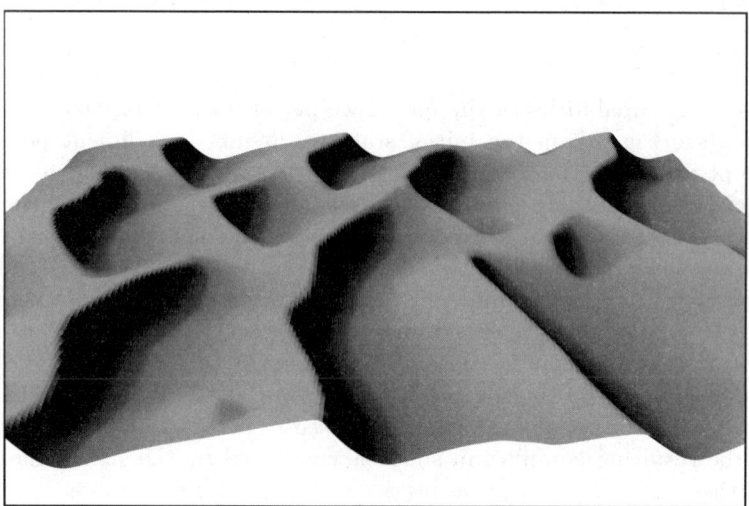

Fig. 8. Ripple development; $l = 1 \times$ grid spacing, $D = 10$, $V\nu_0 = 1$, $k = 1$, $\varphi = 0.1$, periodic boundary conditions.

4.4 Next Steps

The evolution from little ripples structures to bigger ones appears to be realistic in our model, although we do not reach a natural equilibrium state yet. It will be necessary to find out how a variable saltation length affects the results, and if this could lead to a final state that is independent of the boundary conditions. Furthermore, it is well-known, that the ripple development is strongly dependent on the sediment size. Yalin (1992) found the empirical relation that the ripple wave length in equilibrium state is about 1000 times the particle size. This indicates that the scale of particles is important for the final state and has to be considered in a continuum approach.

5 Conclusions

Sediment transport phenomena include a wide range of spatial and temporal scales. Simulations of pattern formation on large scales (larger than a few centimetres) can only be performed on a continuum basis. In the general case, the continuum formulation leads to high-dimensional partial differential equations that cannot be solved efficiently. Feasible models can only be developed for some special cases, such as rolling, saltation, and suspended load. Thus, there is not a general theory that can replace the search for an adequate approach for the phenomenon under consideration.

References

Bass, H. J. (1999): Equilibrium morphology of current ripples in fine sand. Sedimentology, 46, 123–138.

Bouchaud, J.-P., M. E. Cates, J. Ravi Prakash, and S. F. Edwards (1995): Hysteresis and metastability in a continuum sandpile model. Phys. Rev. Lett., 74, 1982–1985.

Engelund, F., and E. Hansen (1967): A Monograph on Sediment Transport in Alluvial Streams. Tekniks Verlag, Copenhagen.

Foster, G. R. (1982): Modeling the erosion process. In: Hydrologic Modeling of Small Watersheds, edited by C. T. Haan. ASAE Monograph 5.

Hergarten, S., and H. J. Neugebauer (1996): A physical statistical approach to erosion. Geol. Rundsch., 85(1), 65–70.

Hergarten, S., G. Paul, and H. J. Neugebauer (2000): Modelling surface runoff. In: Soil Erosion — Application of Physically Based Models, edited by J. Schmidt. Springer, Environmental Sciences, Heidelberg, 295–306.

Iverson, R. M. (1997): The physics of debris flow. Rev. Geophys., 35(3), 245–296.

Jan, C.-D., and H. W. Shen (1997): Review dynamic modelling of debris flows. In: Recent developments on debris flows, edited by A. Armanini and M. Michiue. Springer, LNES 64, 93–116.

Mclean, S. R., J. M. Nelson, and S. R. Wolf (1994): Turbulence structure over two-dimensional bed forms. J. Geophys. Res. 99(C6): 12729–12747.

Nishimori, H., and N. Ouchi (1993): Formation of ripple patterns and dunes in wind-blown sand. Phys. Rev. Lett., 71, 197–200.

Yalin, M. S. (1992): River Mechanics. Pergamon Press Ltd., Oxford, 220 pp.

Water Uptake by Plant Roots — a Multi-Scale Approach

Markus Mendel, Stefan Hergarten*, and Horst J. Neugebauer

Section Geodynamics, Geological Institute, University of Bonn, Germany

Abstract. The soil-root-stem water pathway is a major component of the subsurface hydrological system. We model water uptake and water transport through the soil and through the root system as coupled processes. The model is based on a differential equation approach. The coupled non-linear equation system is solved numerically. On the macroscale, fluid flow through the soil matrix within and outside of the rooted area is modelled using a two-dimensional, axial-symmetric Richards equation. Flow through the root system is described by a Darcy equation. Water uptake and transport through the root system are considered to be passive processes; osmotic effects are neglected. In order to obtain the source and sink distribution of the root system, local soil water potentials (near a single root) are computed in a one-dimensional approximation in a further step. The model is capable of simulating *hydraulic lift*, which is the transport of water from moist into drier soil layers through plant root systems. The calculated amount of passively shifted water is consistent with experimental data. We discuss under which conditions hydraulic lift could be an optimised strategy for the plant and how water uptake patterns are modified by the multi-scale (local) approach.

1 Multi-Scale Concept

Vegetated areas constitute an important part of the earth's hydrologic system. Approximately 70 % of the water falling on the soil surface is returned to the atmosphere through evapotranspiration (Molz 1981). Water transpired by plants is taken up by thin rootlets of the root system so that a water potential gradient at the root soil interface emerges. On a larger scale water flows by diffusion or through macropores (Germann 1990) into the rooted soil domain in order to balance the water deficit induced by the transpiration. Evidently an interaction exists between the water flow through the soil, water uptake by rootlets, water flow through the root system and the transpiration.

Quantitative studies of water uptake by plant roots date back to studies by Gardner (1960). Later many comparable models followed; for an overview see Molz (1981) and Clothier and Green (1997). In all these approaches, water uptake is modelled by a Richards equation choosing a special term for the right hand side (sink term), in this context also called the extraction function.

---
* hergarten@geo.uni-bonn.de

The first aim of the presented article is to investigate the interaction of water flow through the soil (described by the Richards equation) and through the root system. Describing the latter process we need a further differential equation. Experimental data show that water flow through single rootlets can be described in a first approximation by Poiseuille's law (Zimmermann 1983). Using this approach, Doussan et al. (1998) model water flow through a system of individual roots. Passing over to a continuum level Poiseuille's law leads to a saturated Darcy equation that describes water transport through the whole root system. Water flow through the root system shall be driven by the transpiration, i. e. we do not consider osmotic water transport; the transpiration rate is therefore used as the upper boundary condition for the Darcy equation (see Fig. 3).

Restricting to these two macro-scale equations local soil water potential gradients around individual roots cannot be resolved. Since the fundamental studies of Gardner (1960) there has been an ongoing debate under plant physiologists if gradients in the rhizosphere (the soil at the root surface) are sufficiently large to be of physiological importance to the plant. In the literature, you can find support for both opinions. For an overview see Oertli (1996).

As the problem is still under discussion, the second aim of this article is to regard meso-scale water flow towards the individual roots as well. This is managed technically by separating the two scales in the following way: For each cell of a global grid, where the macro-scale equations are solved, an averaged catchment area of an individual root is calculated. The radius of the cylindrical catchment area is a function of the root density; it is approximately 1 cm. Within this cylinder the diffusion of water towards the root is computed in a one-dimensional approximation. A discussion of the applicability of meso-scale soil water diffusion follows in section 3.1. We will investigate under which conditions water uptake patterns are modified significantly by the multi-scale approach.

2 Physiology of Water Uptake and Transport through Plant Tissue

In the following we shall summarise some basics of plant physiology that are important for the understanding of the model assumptions.

Roots take up water from the soil to meet the requirements of the transpiring crown. A generally accepted theory describing the mechanism of sap ascent is the *cohesion theory*. It states that water is pulled through the plant by tension in xylem vessels created by the evaporation of water from the leaves; that means water transport is driven by the difference in water potential of soil and dry air without consumption of any metabolic energy. The cohesion theory was first proposed 100 years ago (Böhm 1893).

As a corollary, a continuous column of water exists between soil and leaves, since water subjected to a negative pressure is in a metastable state. Homogeneous cavitation (embolism) in water theoretically occurs at pressures between -80 and -200 MPa (Green et al. 1990). The stability of water in xylem vessels above this limit depends on the absence of heterogeneous nucleating sites where water contacts xylem walls or inclusions, and on the exclusions of air by capillary forces in cell-wall pores (Pockman et al. 1995). Despite the fact that the cohesion theory was occasionally questioned, it represents the generally accepted theory for explaining sap ascent (Tyree 1997).

Water uptake by plant roots is determined by the *radial conductivity* K_r (root hydraulic conductivity) from the root surface to the root xylem and the *axial conductance* K_a (axial conductivity). Axial conductance and radial conductivity can be measured (e. g. Doussan et al. 1998, Tyree et al. 1998); axial conductance can often be calculated approximately using Poiseuille's law. K_a, calculated with Poiseuille's law, is usually two to five times larger than the measured value due to irregularities of xylem walls (Zimmermann 1983). K_a of roots has been measured for relatively few species; it ranges from 5 to $42 \times 10^{-17} \mathrm{m}^4 \mathrm{s}^{-1} \mathrm{Pa}^{-1}$ for the main roots of maize, sorghum and desert succulents. The values for lateral roots are one to two orders of magnitude lower (Moreshet et al. 1996). K_r also varies with plant species, ranging from 0.1 to $70 \times 10^{-14} \mathrm{m} \mathrm{s}^{-1} \mathrm{Pa}^{-1}$ when measured by applying a hydraulic pressure gradient to the roots (Huang and Nobel 1994). Furthermore, K_r depends on soil saturation, soil temperature, root order (main roots, lateral roots, root hairs), root maturation and the position along the root axis (Doussan et al. 1998, Steudle and Meshcheryakow 1996, North and Nobel 1996).

Alm et al. (1992) showed by using a finite-element model of radial conductivity and axial conductance for individual roots that K_a may substantially limit water uptake by causing large gradients in xylem water potential P_x. Doussan et al. (1998) extended this model by coupling a model of root architecture of maize with local information on axial conductance and radial conductivity ("Hydraulic Tree Model"). Neither of both approaches contains an equation describing water flow through the whole soil matrix on the scale of the root system (macro-scale) and towards the single root (meso-scale).

Since the studies of Caldwell and Richards (1989), it is known that some trees and shrubs perform a process termed *hydraulic lift*. Hydraulic lift is the transport of water from moist into drier, in general upper, soil layers through plant root systems. Plant roots thus do not only take up water but may also lose water if the soil is dry. Emerman (1996) discusses two theories of hydraulic lift. In the first theory, hydraulic lift promotes the uptake of nutrients and thus provides woody plants with a competitive advantage. The second theory assumes that hydraulic lift is a stress response, which results from the inability of tree roots to perform the osmoregulation that would be necessary to prevent shallow roots from leaking water. Caldwell et al. (1998) claims that hydraulic lift facilitates water movement through the soil-plant-atmosphere system. Hydraulic lift can be measured by analysing the stable hydrogen isotope composition of water in understory plants around trees (soil water and groundwater vary in isotope composition), or by collecting intact soil cores from the vicinity of a tree at the same distance at which thermocouple psychrometers have been placed (Emerman and Dawson 1996). Using the second method, Emerman and Dawson estimate the quantity

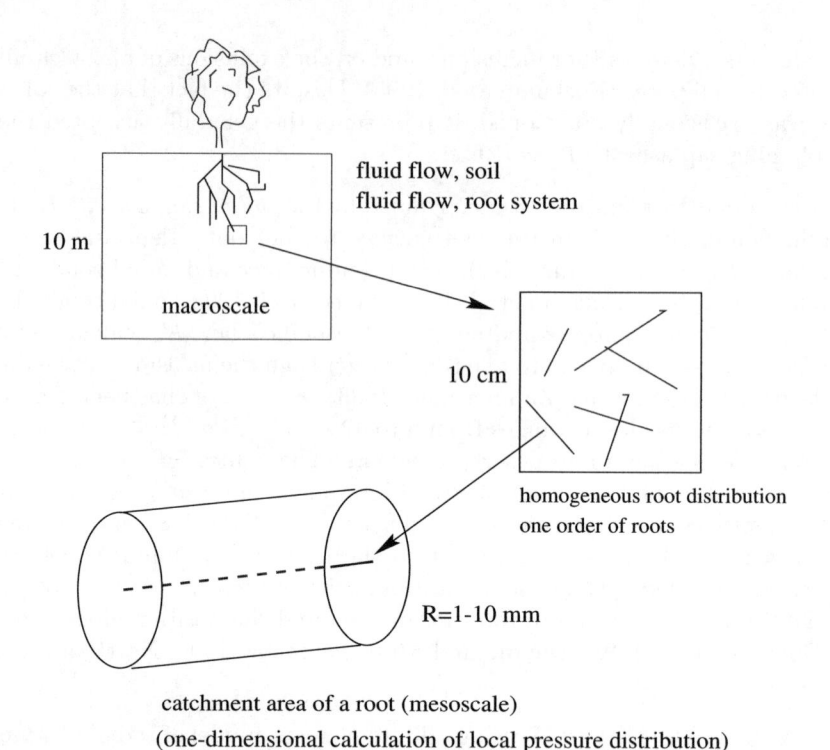

Fig. 1. Macro-scale and meso-scale: On an axial-symmetric, two-dimensional domain (macro-scale) soil water potential P_s and xylem water potential P_x are computed using Eqs. 1 and 2. The "local" soil water potential distribution P_{loc} is calculated within a cylindric domain around a single root that is chosen exemplarily for each "global" grid cell.

of *HLW* (hydraulically lifted water) of a mature sugar maple tree to 102 ± 54 litres per day.

3 Model Assumptions and Equations

In this chapter we explain the differential equations. The coupled nonlinear differential equation system consists of two Richards equations (Eqs. 1, 3) and a saturated Darcy equation (Eq. 2). The domains Ω_1 and Ω_2, where Eqs. 1, 2 are solved, are illustrated in Fig. 1. The catchment area of the single root (Ω_3) is shown in Figs. 3 and 2.

Soil macro-scale (Ω_1)

$$\phi \frac{\partial S(P_s)}{\partial t} - \text{div}\left[K_f(S)\{\nabla P_s - \rho_f \tilde{g}\}\right] = -2\pi R_r\, a_r\, K_r\, (P_{\text{rhi}} - P_x)\, S_{\text{rhi}} \qquad (1)$$

Water Uptake by Plant Roots – a Multi-Scale Approach

Root system (Ω_2)

$$\phi_x \beta_P \frac{\partial P_x}{\partial t} - \mathrm{div}\,[\,\breve{a}_r\,K_a\{\nabla P_x - \rho_f \tilde{g}\}] = 2\pi R_r\, a_r\, K_r\,(P_\mathrm{rhi} - P_x)\, S_\mathrm{rhi} \qquad (2)$$

Soil meso-scale (Ω_3)

$$\phi \frac{\partial S_\mathrm{loc}(P_\mathrm{loc})}{\partial t} - \mathrm{div}\,[K_f(S_\mathrm{loc})\{\nabla P_\mathrm{loc}\}] = \mathrm{div}\,[K_f(S)\{\nabla P_s - \rho_f \tilde{g}\}] \qquad (3)$$

The variables and parameters are:

S = saturation, (macro-scale)
P_s = soil water potential, (macro-scale) [Pa]
P_x = xylem water potential, (macro-scale) [Pa]
a_r = root density [m^{-2}]
K_f = (unsaturated) hydraulic conductivity [m^2 Pa^{-1} s^{-1}]
K_r = radial conductivity [m Pa^{-1} s^{-1}]
K_a = axial conductance of one root [m^4 Pa^{-1} s^{-1}]
R_x = xylem radius ($\approx 10^{-4}$ m)
R_r = root radius (5×10^{-4} m)
ϕ = porosity
ϕ_x = xylem volume per cell/cell volume ($\approx 10^{-4}$)
β_P = compressibility of water (5×10^{-10} Pa^{-1})
$\vec{g}$ = vector of acceleration ($9.81 \cdot \vec{e}_z$ m s^{-2})
ρ_f = fluid density (10^3 kg m^{-3})
∇ = gradient operator

The symbol $\breve{a}_r$ denotes the *root density* tensor; it regards that the root density may depend on the direction for each grid cell. P_loc is the local soil water potential on the meso-scale (Ω_3, see Figs. 1 and 2). P_rhi is the local soil water potential value in the rhizosphere. The same applies to S_loc and S_rhi. P_rhi and S_rhi are important for computing the water uptake amount (see the sink terms of Eqs. 1 and 2).

3.1 Definition of Local Approach and Global Approach

In the following, we speak about the *local approach* if we mean numerical simulations based on the coupled differential equation system (Eqs. 1,2,3). Here, water uptake is assumed to be proportional to the difference of P_x and P_rhi. Water diffusion on the meso-scale is calculated on a constant grid spacing of 0.5 mm (Fig. 2). In contrast to the pore-scale, the meso-scale shall include a "sufficient" number of soil pores. What does sufficient mean in this case? The application of a diffusion equation is only physically reasonable if the grid cell width used is greater than a variable called *Lagrangean length scale L* (Fischer et al. 1979). L is the length required to arrange all pores participating in the flow process according to their size and relative abundance (for an exact definition of L see Fischer et al. 1979). L depends strongly on the soil type and on the saturation. For a saturation $S = 0.8$, the Lagrangean length scale ranges from 9 cm (sand) over 0.3 cm (loam) and 0.2 cm (sandy clay loam) to 0.02 cm (clay loam and clay) (Germann 1990). $S = 0.8$ corresponds to a rather wet soil. If the soil is drier, which is more typical near a transpiring tree, L decreases by up to one order of magnitude (note that larger pores dry first). Thus the computation of water diffusion towards plant roots in the meso-scale should be possible at least for a fine-pored and homogeneous soil.

Common approaches (see Molz 1981) do not use a differential equation on the meso-scale, so that pressure gradients near the root surface are neglected. If we want to use this *global approach*, we have to replace the *local* variables P_{rhi} and S_{rhi} on the right hand side (sink term) of Eqs. 1, 2 with the *global* ones (P_s and S) and remove Eq. 3. We will see later how water uptake patterns are modified using the multi-scale approach.

3.2 Flow through the Soil on the Macro-Scale

Fluid flow through the soil may temporarily occur in two domains. One domain comprises the structural pores, referred to as the macropore system. Here water is primarily driven by gravity and is affected only little by capillary forces. On the other hand, we observe flow through the soil matrix (matrix flow) that is usually described using Richards' equation (Richards 1931).

Matrix Flow The Richards equation (Eqs. 1,3) can be derived from the equation of continuity (mass conservation)

$$\frac{\partial \rho_b}{\partial t} + \text{div}\,\vec{j} = q \qquad (4)$$

where $\rho_b = \phi S \rho_f$ is the bulk density of the fluid, $\vec{j} = \rho_b \vec{v}$ is the volumetric flux, $\vec{v}$ the pore velocity, i.e. the velocity of the fluid in the pores, and q is the sink term, and Darcy's law that may be written as

$$\vec{v}_{Da} = -K_f(S) \cdot (\nabla P_s - \rho_f \vec{g}). \qquad (5)$$

The Darcy velocity $v_{Da} = \phi S \vec{v}$ is the velocity of the fluid moving in the porous medium. The Richards equation consists of 3 terms. It states that the saturation of a soil element increases (described by the first term) if fluid enters through the wall of the element (second term) or a source exists inside the element (sink term). That is why water uptake by plant roots can be modelled by choosing a right hand side of Richards' equation. For the saturation and permeability functions we use the relations of van Genuchten (1980):

$$S(P) = \frac{1}{\left(1 + \left(-10^{-4}\frac{1}{h_b}P\right)^{\lambda+1}\right)^{\frac{\lambda}{\lambda+1}}} \tag{6}$$

$$K_f(S) = K_{f,\max}\sqrt{S}\left(1 - \left(1 - S^{\frac{\lambda+1}{\lambda}}\right)^{\frac{\lambda}{\lambda+1}}\right)^2 \tag{7}$$

with

$$\lambda = \text{pore size index} \in [0.03, 1.1]$$
$$h_b = \text{bubbling pressure} \in [0.013\,\text{Pa}, 1.87\,\text{Pa}]$$

The values of pore size index and bubbling pressure depend strongly on soil texture. We neglect effects of hysteresis as well as water vapour transport. The latter would only play an important role if we considered a large temperature gradient in the soil (in general in the upper soil layer) or a very small saturation. Since the studies of Gardner (1960), many models have been built around *extraction functions* (sink terms) in the Richards equations. In a review, Molz (1981) discusses the applications, relative advantages and limitations of several of such functions. We have chosen an extraction function most similar to that of Herkelrath et al. (1977), but we point out two important modifications: In our approach P_{rhi} means the soil water potential in the rhizosphere (soil near the root surface). Correspondingly, S_{rhi} is the saturation in the rhizosphere. Both variables are solutions of Eq. 3. In Herkelrath's approach, the soil water potential and saturation variables of the sink term of Eq. 1 are global, that means, averaged values. The "local computation" is reasonable since roots only "feel" the pressure at the membrane surface. Furthermore, P_x is not assumed constant, but is also the solution of a differential equation (Eq. 2).

Thus our model describes the whole physical process of water uptake without the need for fitting parameters (apart from the van Genuchten parameters) or the need for xylem pressure values that are very difficult to obtain experimentally.

Macropore Flow Macropores are e. g. animal burrows, such as worm and insect tunnels, as channels formed by living or dead roots or desiccation cracks or joints. The depth limit of these macropores ranges between 0.5 and 3 m in humid climates and may reach more than 20 m in the tropics (Germann 1990). In many field experiments it has been shown that macropore flow supplies a substantial, often predominant contribution to the total flow (e. g. Germann and Levy 1986). Flow velocities in macropores are of up to several cm/s (e. g. Beven and Germann 1981). Modelling macropore flow, one usually starts with an empirical relationship between macropore flow density j_{ma} and macropore water content S_{ma}:

$$j_{ma} = \alpha_{ma} S_{ma}^{\beta_{ma}}$$

The parameters α_{ma} and β_{ma} describe the geometry of the macropore system. Including mass conservation, this leads to a differential equation of first order in space and time (approach of the kinematic wave) that can be solved numerically. Fluid flow from the macropore to the matrix can be described using Richards' equation. Merz (1996) compares measured infiltration rates and saturation fronts with 3 model approaches:

1. only matrix diffusion
2. neglect of the exact hydraulic flow process into the macropores.
   Macropores will be filled uniformly with w ater after precipitation.
3. macropore infiltration and matrix diffusion

Model 3 is in best agreement to the measured values. The pure matrix model is capable of fitting experimental saturation fronts reasonably well too, if K_f-values are increased by a power of ten. Since it is not the main aim of our model to give an optimised calculation of infiltration rates, we consider macropore flow only according to the above mentioned approach 1 with an increased vertical K_f-value by a power of ten.

3.3 Flow through the Root System

Modelling water flow through the whole root system, i.e. including all its rootlets, is not possible for computational reasons. One possibility is to proceed on an artificial root system with a strongly reduced quantity of root segments. The volume of water Q that flows per second through a root segment in x direction can be computed as

$$Q = -K_a \frac{\partial P_x}{\partial x} \qquad (8)$$

K_a is the axial conductance of the root segment. In our model, a root density consisting of *one order of roots* and expressed either in root surface/matrix volume or in root length/matrix volume, takes the place of several single roots. Including mass conservation, this yields a saturated Darcy equation (Eq. 2). Note that xylem vessels are always saturated (if not cavitated), thus K_a is not a function of a saturation variable like K_f is in Eq. 5. Obviously the root system is described strongly simplified compared to a natural root system that consists of several root orders (main roots, lateral roots and root hairs) with different physiological properties (see chapter 2).

Xylem diameter, which is related to axial conductance (Hagen-Poiseuille), horizontal and vertical root density and radial conductivity of roots can be varied within the global grid, but are assumed to be uniform in one (global) grid cell. The usage of a saturated Darcy equation requires that the root system is more or less "regular" in structure. Unlike in a porous media, where pores are connected and flow is commonly described using the Darcy equation, connections between different root segments do not necessarily exist within one cell of the grid. The sink term of Eq. 2 is identical with that of Eq. 1, but differs in sign, since sink and source are reversed. The Darcy equation we use is linear. Cavitations within the root system are assumed to be negligible. The shape of the root system (the direction of the root segments) is expressed by a variable distribution of axial conductance and horizontal and vertical root density. Since xylem vessels are very rigid (Roth et al. 1992) and the compressibility of water is small, the first term of Eq. 2 makes little effect. We consider water uptake to be passive. Modelling osmotic water uptake is quite complicated. Osmotic and hydraulic flow into the root use different paths. For a detailed description see Steudle and Peterson (1998). On the other hand, osmotic (active) water uptake often is stated to be slight during the summer season (e. g. Mohr and Schopfer 1992).

3.4 Flow through the Soil on the Meso-Scale

Pressure and saturation distributions near a single root are calculated in a one-dimensional approximation. In chapter 3.1 we have discussed under what assumptions the computation of fluid diffusion on the meso-scale is physically reasonable. On the meso-scale the grid cell width is set constant $= 0.5$ mm (see Fig. 2). The first grid cell corresponds to the root. The radius of the cylinder ranges between 0.2 cm to 5 cm, depending on root density (see Figs. 1 and 3). Fluid flow into the root is determined by the radial conductivity K_r.

In the case of modelling hydraulic lift K_r is set to be independent of the direction of the pressure gradient. Without hydraulic lift, K_r is approximately 0 in the case of $P_{\text{rhi}} < P_x$, i.e. fluid flow from roots to the soil is assumed to be insignificant.

3.5 Coupling Macro-Scale and Meso-Scale

Fig. 2 shows how to couple the diffusion equations of macro- and meso-scale. The radius of the "catchment area" of a root (Figs. 1 and 2) is computed in a first approximation as a simple function of the root density tensor $\breve{a}_r(r,z)$. Xylem pressure P_x (solution of Eq. 2) is used as a Dirichlet boundary condition for Eq. 3. P_{rhi}, the solution of Eq. 3, is the corresponding boundary condition for Eqs. 1 and 2. (Note that P_{rhi} is not identical with P_{loc}, but $P_{\text{rhi}} := P_{\text{loc}}[\text{cell } 2]$). How does the macro-scale soil flow affects the meso-scale? In our approach, the global flow (divergence term of Eq. 1) "enters" uniformly into this cylinder (Ω_3, Fig. 2). An alternative approach would be to let the global flow enter the meso-scale through the cylinder wall using a Neumann boundary condition. Eq. 3, integrated over the cylinder, leads to Eq. 1, i.e. the mass balance is guaranteed.

4 Numerical Methods

For the numerical realization a finite-difference scheme with consistently staggered grids is used. Both global equations are two-dimensional, the local equation is one-dimensional. We use a radial-symmetric discretisation. For boundary and initial conditions used for the equations see Figs. 1 and 2. Multigrid-methods are

$$\nabla[K_f(P_s)\{\nabla P_s - \rho_f \vec{g}\}]$$

$K_f(1) = S_{\text{rhi}} R_r K_r$

$K_f(2)$ ... after van Genuchten

P_x

S_{rhi}
P_{rhi}

equations (1,2)

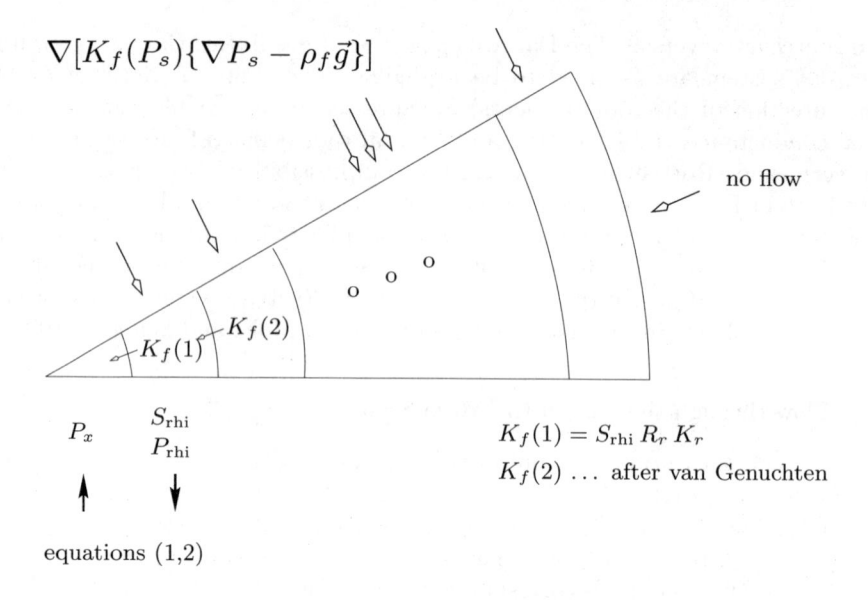

Fig. 2. "Catchment area" of the root: The water potential variables defined for the grid cells in the meso-scale are termed P_{loc}. The central grid cell corresponds to the root. Here P_x is used as a Dirichlet value ($P_{\text{loc}}[\text{cell }1] = P_x$). Grid cell width is set constant = 0.5 mm. The radius of the cylinder ranges between 0.2 cm to 5 cm, depending on root density (see Figs. 1 and 3). $P_{\text{loc}}[\text{cell }2]$ is called P_{rhi} and is needed for the sink terms of Eqs. 1 and 2.

applied to solve the two-dimensional equations, the one-dimensional equation is solved by Gauss elimination. For detailed information about multigrid-methods see the article of Gerstner and Kunoth in this volume. The $K(S)$-function is approximated by a polynomial function within S=0.9 and 1 for numerical reasons. The stability of the numerical code has been tested by investigations on different grid sizes (spatial resolution) and different time step widths.

5 Simulation Example

Fig. 3 shows the axial-symmetric soil domain (Ω_1) with the initial and boundary conditions of Eqs. 1 and 2 used in the simulation example. Radius and depth of Ω_1 are 10 m. Initial condition of soil and xylem water potential is hydrostatic. Ground water level is located in 5.5 m depth. Radius and depth of the rooted area (Ω_2) amount to 5 m. Within this domain, Eq. 2 is solved. Root density decreases linearly in radial and vertical direction from the stem. At the upper, central boundary of the rooted area (radius < 0.31 m) transpiration rate T [m^3 s^{-1}] is set as a Neumann condition ($T(t) \sim \sin t$ and $T(12 \text{ p.m.}) = 0$). Alternatively, P_x or a combination of P_x and T could have been used as a boundary condition. The loss of water in Ω_1, due to transpiration, is compensated by precipitation that is

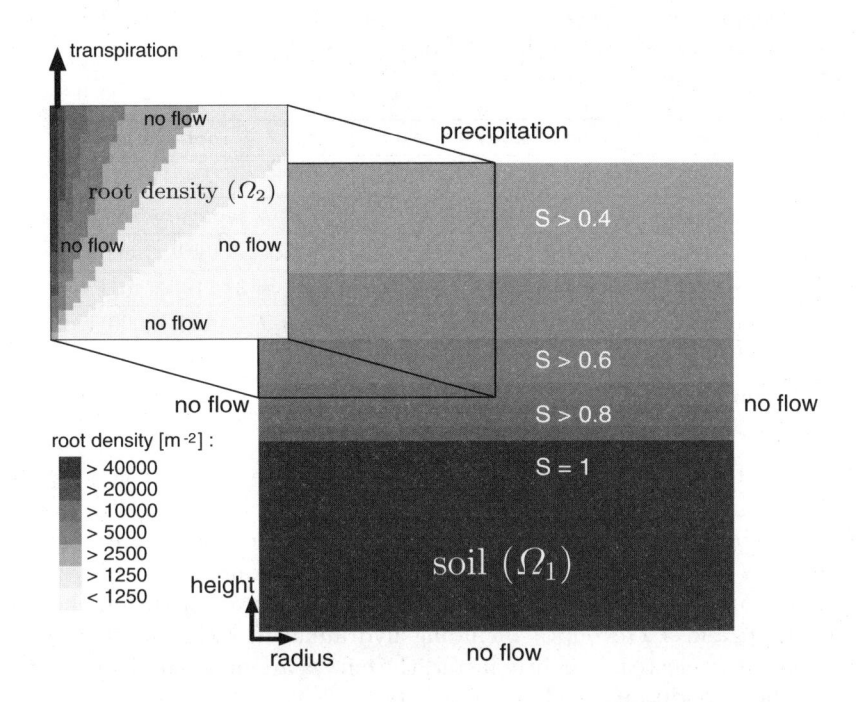

Water Uptake by Plant Roots – a Multi-Scale Approach

Fig. 3. Root density a_r [m^{-2}], boundary and initial conditions (Eqs. 1 and 2) used in the simulation example. The initial distribution of water potential in soil and root system is hydrostatic. Radius and height of domain Ω_1 (soil) are 10 m. Radius and height of domain Ω_2 (root system) is 5 m. The ground water level is located in 5.5 m depth.

assumed to be constant in time, but is reduced by 50 % for $r < 5$ m because of the interception of the shoot. Using this boundary conditions, we expect the system to approach a stationary state (apart from diurnal oscillations). Precipitation is assumed to be constant in time, because it is impossible to resolve infiltration fronts, due to realistic rain events, without an adaptive grid refinement in vertical direction.

In order to compare simulated *HLW* and soil water potential oscillations near the stem with experimental data of a sugar maple tree, the values for the transpiration rate (1. value: 400 l/d) and soil parameters (loam to sandy loam according to the USDA classification system) are chosen comparably to measured data of Emerman and Dawson (1996). Few information exists about the distribution of radial conductivity K_r of plant roots with regard to the macro-scale. K_r values are chosen near the upper limit of the range measured for different species (see chapter 2). In so doing the influence of root hairs and mycorrhiza is taken into account, i.e. the K_r-value used should be interpreted as an effective value. Within a distance of less than 2 meters to the stem K_r decreases linearly due to suberation, lignification and bundling of roots. Certainly, this can only be a very rough approximation of a real spatial K_r-distribution. Actually, K_r

Table 1. Model parameters used in the simulation example. Note: in the literature K_f variables are often expressed in $[\text{m s}^{-1}]$. In this case K_f values are by a factor 10^4 greater and h_b is expressed in [m].

soil (loam to sandy loam)	root system
$K_{f,\max,\text{vertical}} = 5 \times 10^{-9}\,\text{m}^2\text{Pa}^{-1}\text{s}^{-1}$	$\rho_{r,\phi} = 10^4\,\text{m}^{-2}$
$K_{f,\max,\text{horizontal}} = 2.5 \times 10^{-10}\,\text{m}^2\text{Pa}^{-1}\text{s}^{-1}$	root radius $= 5 \times 10^{-4}\,\text{m}$
$K_{f,\max,\text{meso-scale}} = 5 \times 10^{-10}\,\text{m}^2\text{Pa}^{-1}\text{s}^{-1}$	$K_{r,\max} = 10^{-12}\,\text{m Pa}^{-1}\text{s}^{-1}$
$\lambda(\text{pore size index}) = 0.3$	$K_a = 2 \times 10^{-16}\,\text{m}^4\text{Pa}^{-1}\text{s}^{-1}$
$h_b(\text{bubbling pressure}) = 0.3\,\text{Pa}$	$T = 400, 3{,}000\,\text{l/d}$
$\phi = 0.5$	$T(t) \sim \sin(t)$
precipitation $= 400, 3{,}000\,\text{l d}^{-1}\Omega_1^{-1}$	$T(12\,\text{p.m.}) = 0$

can be varied within the model as a function of space, time, root density, soil water potential, and transpiration rate. $K_a = 2 \times 10^{-16}\,\text{m}^4\text{Pa}^{-1}\text{s}^{-1}$ is a mean value for main roots, measured for various plant species (see chapter 2). Table 1 summarises the parameters used in the simulation example.

Fig. 4 shows the temporal development of soil saturation within the rooted area in 0.8 m and 4.2 m depth including hydraulic lift (without hydraulic lift saturation values decrease slightly faster, the curves are quite similar). For initial and boundary conditions see Figs. 2 and 3.

Assuming a (unrealistically high) transpiration rate of 3,000 l/d, a stationary solution of the equation system does obviously not exist. This can also be shown by an analytical examination for the one-dimensional case (we dispense with the calculation here): using the relations of van Genuchten, for all soil types $(\lambda, \alpha > 0)$ an upper boundary exists for the transpiration rate that yields to a stationary solution for the diffusion problem. Fluid flow through soil behaves obviously completely different than laminar flow does through a pipe, in which any flow rate can be created assuming an appropriate pressure gradient.

5.1 Water Uptake with and without Hydraulic Lift

In the following we will compare how water is taken up by roots either with or without the concept of hydraulic lift. Fig. 5a,b show diurnal water uptake oscillations in different soil depths near the stem (radius = 0.47 m) in the long term equilibrium, both computed using the local approach.

Including hydraulic lift, roots near the surface take up water during the day and release water in the night. Water uptake maxima are shifted towards the night as a function of depth. The computed "phase velocity" (of the maxima) is about 0.5 m/h. The amplitudes of the water uptake curves become smaller as a function of the depth. Obviously the system behaviour is similar to the propagation of a travelling wave. In deeper soil horizons, roots take up water during the whole day with very little diurnal fluctuations. The root zone with the maximum effective diurnal water uptake (water uptake minus water release) is in approximately 4 m depth.

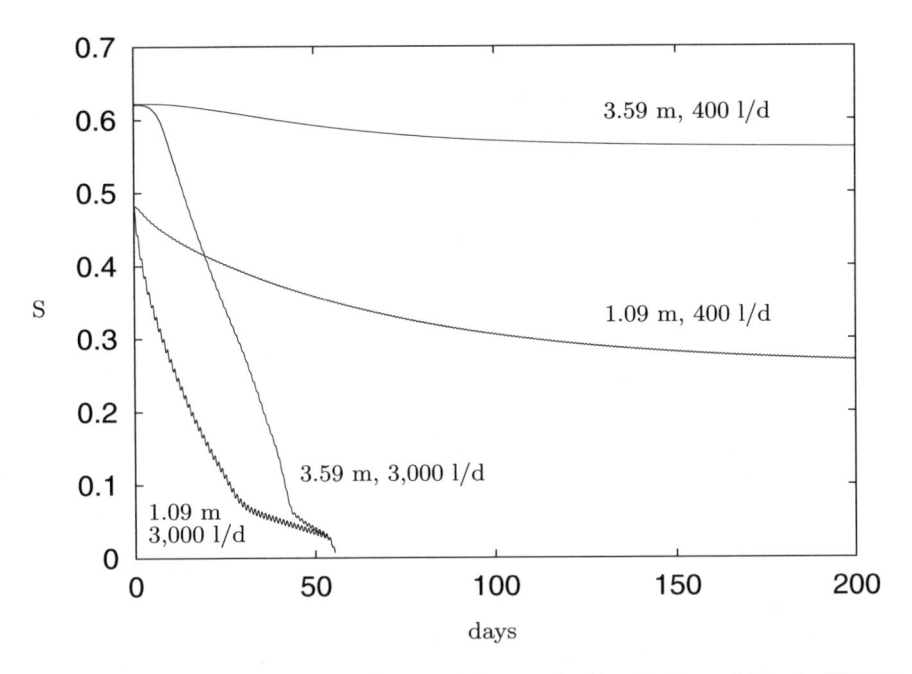

Water Uptake by Plant Roots – a Multi-Scale Approach

Fig. 4. Development of saturation S in 2 different depths (1.09 m, 3.59 m). With hydraulic lift. For initial and boundary conditions see Fig. 3, 2. Transpiration in [l/d]. Saturation values are averaged (in a weighted manner) in horizontal direction within the rooted domain. The system tends towards a long term equilibrium if transpiration is not to strong (i. e. T=3,000 l/d).

Without hydraulic lift, water is taken up almost exclusively in deeper soil horizons. Since roots are not allowed to release water, soil layers near the surface, dried during the day, cannot be refilled by deeper soil water or ground water during the night (apart from matrix diffusion).

Table 2 shows the values of diurnal variations of soil water potential and xylem water potential in the long term equilibrium including and excluding hydraulic lift.

Note the following details:

1. Including hydraulic lift, transpiration can be performed using smaller xylem tensions (greater xylem water potentials). This could mean a *competitive advantage* for the plant and thus a selective value of hydraulic lift because smaller xylem tensions reduce the probability of cavitations (embolism) of xylem vessels within one individual (Hargrave et al. 1994).
2. Soil saturation within the upper rooted area remains higher including hydraulic lift, because water is pumped through the root system from the wetter, outside soil regions into the drier rooted soil. This is in agreement with soil water potential measurements at different distances away from a sugar maple tree that undergoes hydraulic lift (Dawson 1993).

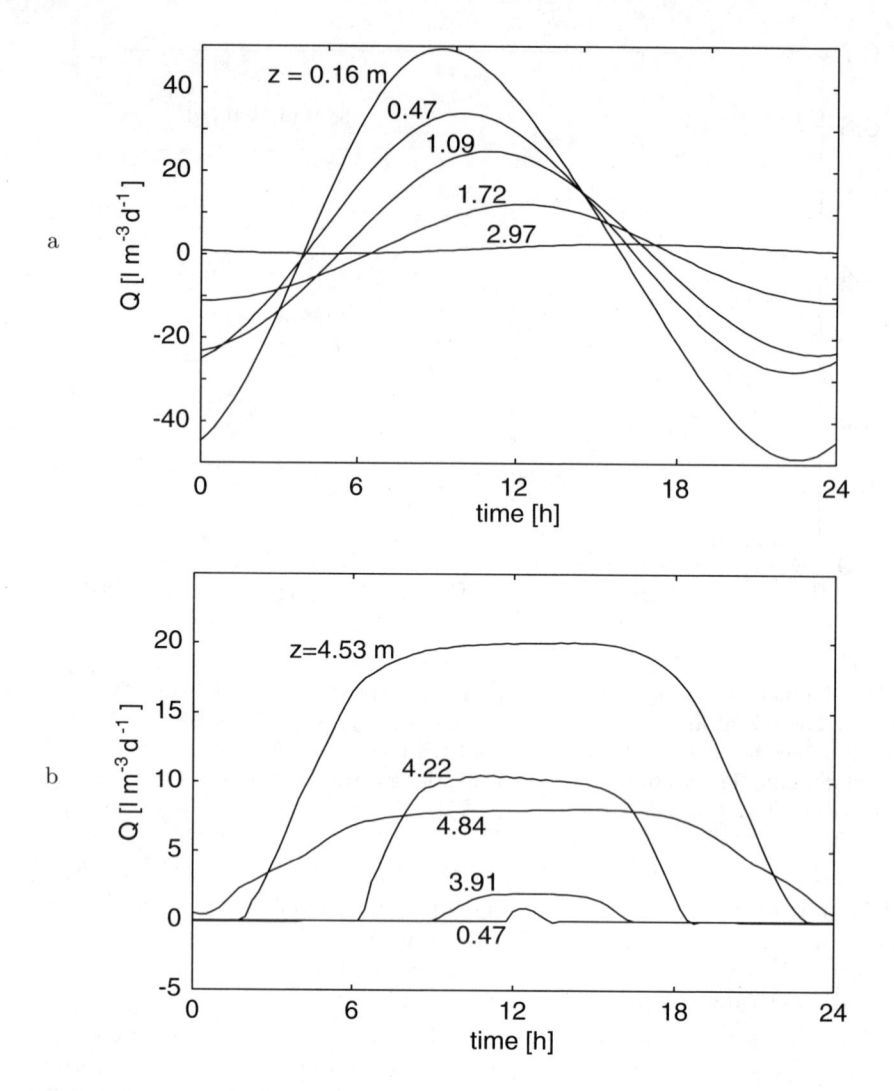

Fig. 5. Water uptake (Q > 0) and water release (Q < 0) of roots. The system is in a long term equilibrium. Local approach. $T = 400\,\text{l/d}$, radius: 0.47 m, different depths. a: with hydraulic lift. b: without hydraulic lift. To improve clarity, some graphs are suppressed.

3. The amount of simulated *HLW* (local approach) is consistent with the value measured by Emerman and Dawson (1996) ($102 \pm 54\,\text{l/d}$, sugar maple tree). Neglecting local pressure gradients (global approach) simulated *HLW* is even above the measured range. Thus we support the thesis that hydraulic lift is a passive process and does not need an active, osmotic component.
4. The amplitude of diurnal soil water potential variations near the surface is higher by two to three orders of magnitude in the case of hydraulic lift.

Water Uptake by Plant Roots – a Multi-Scale Approach

Table 2. Daily oscillations of soil water potential P_s [Pa] and xylem water potential P_x [Pa]. Radius: 0.47 m. Depth: 0.47 m. S(aver.): daily-averaged saturation. LA: local approach. GA: global approach. The system is in a long term equilibrium. T: 400 l/d. HLW is the amount of shifted water by hydraulic lift.

quantity	hydraulic lift	no hydraulic lift
S(aver.)	0.23	0.19
ΔP_s (LA)	1.94×10^5 Pa	530 Pa
P_x(min)	-21.8×10^5 Pa	-23.6×10^5 Pa
HLW(LA)	137	≈ 0
HLW(GA)	157	≈ 0

This supports the experimental method of measuring the amount of HLW with psychrometers or TDR. Emermann and Dawson (1996) have measured daily soil water potential oscillations near the stem of a sugar maple tree to around 1 MPa. The simulated value of ΔP_s at nearly the same distance and depth (r=0.47 m, z=0.47 m) is smaller: $\Delta P_s = 0.19$ MPa (local approach) and $\Delta P_s = 0.22$ MPa (global approach). One can explain this difference as follows: In the simulation the root density is assumed to decrease (only) linearly from the stem in vertical and radial direction. As a consequence the root water potential drops mainly at the upper, central edge of the rooted domain. Simulated daily root water potential oscillations decrease by nearly one order of magnitude within one metre distance to the stem. Due to the maximum principle (i.e. Strauss 1995), soil water potential oscillations are even smaller than root water potential oscillations. Evidently, the root density distribution chosen in the simulation is not very realistic.

Fig. 6 shows the dependence of root and soil water potential oscillations on the effective radial conductivity K_r. Assuming a high value for the radial conductivity, P_{rhi} and P_s as well as the calculated values of HLW differ significantly. This is a direct consequence of Ohm's law. If the radial conductivity is smaller or the soil water potential is higher, e.g. due to a smaller transpiration or a different soil type, local gradients can be neglected in a good approximation (Fig. 6b). The calculated value of HLW decreases assuming a lower value for the radial conductivity. For $K_r = 10^{-14}$ m Pa^{-1} s^{-1} the calculated HLW is 80 l/d (81 l/d using the global approach). This is still within the range measured by Emerman & Dawson (1996) for a sugar maple tree.

6 Conclusions

The main aim of this study was to investigate passive water uptake and passive water transport through the soil and the root system as a coupled process. The calculated amount of passively shifted water in the case of hydraulic lift is

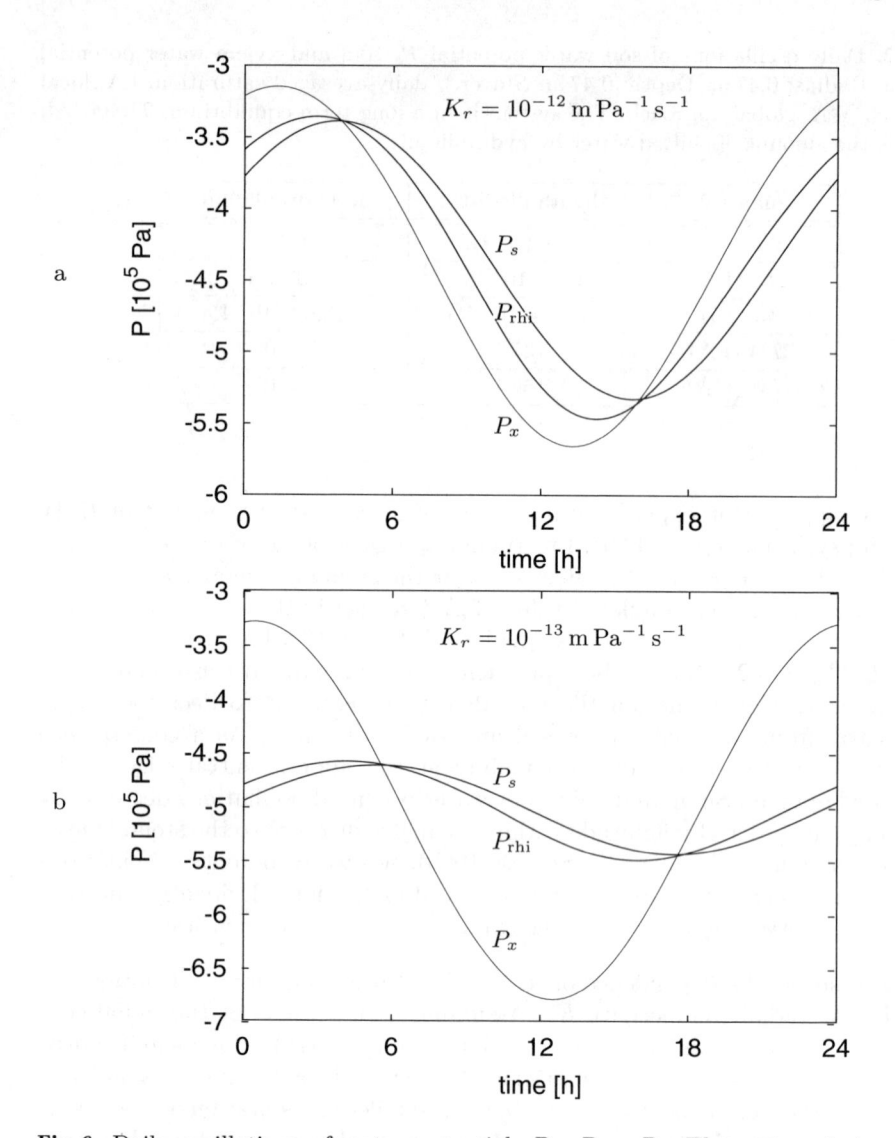

Fig. 6. Daily oscillations of water potential: P_s, P_{rhi}, P_x. The system is in a long term equilibrium. T: 400 l/day. Radius: 0.47 m. Depth: 0.47 m. Local approach: $K_r = 10^{-12}\,\text{m}\,\text{Pa}^{-1}\,\text{s}^{-1}$ (top), $K_r = 10^{-13}\,\text{m}\,\text{Pa}^{-1}\,\text{s}^{-1}$ (bottom). For further parameters see Table 1.

consistent with experimental data; thus our simulation suggests that hydraulic lift is a passive process without any active component. The process of hydraulic lift seems to be an optimising strategy for the plant because smaller xylem tensions reduce the probability of cavitations of xylem vessels.

The equation describing water flow through the root system is quite simple. There is only one order of roots included. With regard to a more detailed interpretation of simulated water uptake patterns, the model must be extended to several root orders. Furthermore in a more detailed discussion of hydraulic lift the influence of understory plants has to be included. Shallow-rooted neighbours of trees and shrubs that are undergoing hydraulic lift are claimed to profit from the shifted water (e. g. Dawson 1993). If the groundwater level is too deep, hydraulic lift could also be a disadvantage for a tree that is influenced by understory plants with regard to the water balance. More information is needed about plants performing hydraulic lift. The calculation of local water potential gradients around the root may be superfluous under certain conditions, e. g. if the saturation is high. But since hydraulic lift is a phenomenon underlying great soil water potential gradients (during the summer season), the local approach should be regarded as a fundamental extension of existing approaches modelling water uptake by plant roots.

References

Alm, D. M., J. C. Cavelier and P. S. Nobel (1992): A finite-element model of radial and axial conductivities for individual roots: development and Validation for two desert succulents. Annals of Botany 69, 87–92.

Beven, K. J. and P. Germann (1981): Water flow in soil macropores. Soil Sci. Soc. Am. J. 32, 15–29.

Böhm, J. (1893): Capillarität und Saftsteigen. Ber Dtsch Bot Ges 11, 203–212.

Caldwell, M. M. and J. H. Richards (1989): Hydraulic lift: water efflux from upper roots improves effectiveness of water uptake by deep roots. Oecologia 79, 1–5.

Caldwell, M. M., T. E. Dawson and J. H. Richards (1998): Hydraulic lift: consequences of water efflux from the roots of plants. Oecologia 113, 151-161.

Clothier, B. E. and S. R. Green (1997): Roots: The big movers of water and chemical in soil. Soil Science 162, 534–543.

Dawson, T. E. (1993): Hydraulic lift and water use by plants: implications for water balance, performance and plant-plant interactions. Oecologia 95, 565–574.

Dawson, T. E. (1996): Determining water use by trees and forests from isotopic, energy balance and transpiration analyses: the roles of tree size and hydraulic lift. Tree Physiology 16, 263–272

Doussan, C., G. Vercambre and L. Pages (1998): Modelling of the hydraulic architecture of root systems: An integrated approach to water absorption — distribution of axial and radial conductances in maize. Annals of Botany 81, 225–232.

Emerman, S. H. (1996): Towards a theory of hydraulic lift in trees and shrubs. In: Morel-Seytoux H. J. (ed.). Sixteenth American Geophysical Union hydrology days. Hydrology Days Publication, Atherton, Calif, 147–157.

Emerman, S. H. and T. E. Dawson (1996): Hydraulic lift and its influence on the water content of the rhizosphere: an example from sugar maple, Acer saccharum. Oecologia 108, 273–278.

Fischer, H. B., List, E. J., Koh, R. C. Y., Imberrger, J. and N. H. Brooks (1979): Mixing in Inland and Coastal Waters, Academic Press, New York.

Gardner, W. R. (1960): Dynamic aspects of water availability to plants. Soil Sci. 89, 63–73.

Germann, P. F. and B. Levy (1986): Rapid response of shallow groundwater tables upon infiltration. EOS 91, 2–25.

Germann, P. F. (1990): Macropores and hydrological hillslope processes. In: Anderson, M. B. and T. P. Burt (Ed.): Process Studies in Hillslope Hydrology. John Wiley & Sons Ltd., 327–363.

Green, J. L., D. J. Durbon, G. H. Wolf and C. A. Angell (1990): Water and solutions at negative pressures: Raman Spectroscopic Study to -80 megapascals. Science 249, 649–652.

Hargrave, K. R., Kolb, K. J., Ewens, F. W. and D. Davis (1994): Conduit diameter and drought-induced embolism in salvia mellifera Greene (Labiatae). New Phytologist 126, 695–705.

Herkelrath, W. N., E. E. Miller and W. R. Garner (1977): Water uptake by plants, 2, The root contact model. Soil Sci. Am. J. 41, 1039–1043.

Huang, B. and P. S. Nobel (1994): Root hydraulic conductivity and its components, with emphasis on desert succulents. Agron. J. 86, 767–774.

Merz, B. (1996) Modellierung des Niederschlag-Abfluß-Vorgangs in kleinen Einzugsgebieten unter Berücksichtigung der natürlichen Variabilität, IHW, Heft 56, Institut für Hydrologie und Wasserwirtschaft Universität Karlsruhe.

Mohr, H. and P. Schopfer (1992): Pflanzenphysiologie, Springer, Berlin.

Molz, F. J.(1981): Models of water transport in the soil-plant system: A review. Water Resour. Res., 17, 1245–1260.

Moreshet, S., Huang, B. and M. G. Huck (1996): Water permeability of roots. In: Plant roots, the hidden half. Marcel Dekker, Inc.

North, G. B. and P. S. Nobel (1996): Radial hydraulic conductivity of individual root tissues of opuntia ficus-indica L. Miller as soil moisture varies. Annals of Botany 77, 133–142.

Oertli, J.J. (1996): Transport of water in the rhizosphere and in roots. In: Waisel, Y., Eshel, A. and U. Kafkafi: Plant roots — the hidden half. Marcel Dekker, Inc. New York.

Pockman, W. T., J. S. Sperry and J. W. O'Leary (1995): Sustained and significant negative water pressure in xylem. Nature 378, 715–716.

Renner, O. (1915): Theoretisches und Experimentelles zur Kohäsionstheorie der Wasserbewegungen. Jahrbuch für Wissenschaftliche Botanik 56, 617–667.

Richards, L. A. (1931): Capillary conduction of liquid through porous medium. Physics 1, 318–333.

Roth, A., V. Mosbrugger and H. J. Neugebauer (1994): Efficiency and evolution of water transport systems in higher plants — a modelling approach, I. The earliest land plants. Phil. Trans., R. Soc. Lond. B., 137–152.

Steudle, E. and A. B. Meshcheryakov (1996): Hydraulic and osmotic properties of oak roots. J. Exp. Bot. 47, 387–401.

Steudle, E. and C. A. Peterson (1998): How does water get through roots? J. Exp. Bot. 49, 775–788.

Strauss, W. A. (1995): Partielle Differentialgleichungen. Vieweg & Strauss. Braunschweig/Wiesbaden.

van Genuchten, M. Th. (1980): A closed-form equation for predicting the hydraulic conductivity of unsaturated soils. Soil Sci. Soc. Am. J. 44, 892–896.

Tyree, M. T. (1997): The cohesion-tension theory of sap ascent: current controversies. Journal of Experimental Botany 48, 1753–1765.

Tyree, M. T., V. Velez and J. W. Dalling (1998): Growth dynamics of root and shoot hydraulic conductance in seedlings of five neotropical tree species: scaling to show possible adaptation to differing light regimes. Oecologia 114, 293–298.

Zimmerman, M. H. (1983): Xylem Structure and the Ascent of Sap. Springer-Verlag, Berlin.

Part IV

Scale-Related Approaches to Geo-Processes

Scale-Related Approaches to Geo-Processes

The last chapter of our book is devoted to problems of geoscience, which we are yet and possibly will never be able to model in physical process models. We finally try to tackle the true Earth-system — and we have to admit, that the processes have now become much too complex and intertwined to model adequately with first principles using our beloved differential equations. We start with observations. With the full complexity of the geo-system having its trace marks everywhere in the data, we have to step back and resort to completely different kind of models. While we have shown in the previous chapter, that the multi-scale concept must be an integral part in physical process models of geo-systems, scales — their determination and interpretation — are literally the starting point of the problems which come into focus now. In several contributions of this chapter we venture to deduct the relevant geo-processes from the multi-scale structure of the observations at hand.

We start with a contribution by Hergarten, which elucidates — following a brief general introduction to fractals —, what we have learnt from the evidence of fractals in almost any geo-data. They tell us from either scale-invariant processes at work (when fractals are found as a static appearance in the data), or from systems in the state of self organised criticality (SOC, when fractals develop in a dynamic way). SOC is best known from the famous sand heap fed from a point source of sand grains. The grains maintain the critical slope of the heap not by finding a position on the slope individually, but by eventually creating small-scale landslides once the critical lope is exceeded. Analogous processes might be at work in the atmosphere in a convectively unstable state, when precipitation is generated in organised convective events, which re-stabilise the atmosphere. In another contribution Hergarten analyses measured land slides and fluvial erosion patterns in the SOC-framework and finds, that the SOC-concept seems to be a guiding mechanism for the prior ones, while the role of SOC in the latter is still unclear. Another application of the fractal concept is demonstrated by Leonardi, who finds inherent fractal structures in drilling profiles, which hint to self-similarities ranging from smallest scales to the complete record. This questions the applicability of homogeneous models in interpreting geological records on any scale.

Specific geological objects like a part of the Earth crust are conceived as three-dimensional spatial models combining e. g. stratigraphic surfaces with the knowledge about the physical processes (faulting, erosion etc.) leading to the object. Siehl and Thomson explain in detail, how multi-scale considerations and the knowledge of the trained scientist, which is not included in the data to model, enter the computer-aided re-construction process by means of object-oriented modelling. The contribution by Schäfer elucidates even more on the vast amount of information hidden in any geological survey in a general way. He elaborates especially on the different scales in the geological records and how geologists believe they have been created. Of special interest in this contribution are the processes, which are at work in a coastal environment, the typical setting for lower Rhine valley in geological time scales. Unravelling the series of processes in sediment development, when the coastline is approaching or receding, and deducting the appropriate times scales from the highly structured multi-scale geological records, and connecting the findings at different sites still resembles detective work by creating models in the mind of the scientist. Process models to aid in the reconstruction seem to lie still far ahead of us.

Gebhardt et al. reconstruct climate zones from remnants of the living environment of the past, namely terrestrial botanical data. They show that in using this data as proxies of climate (existence or non-existence of botanical species in certain time slices) one must be extremely careful not to mix the scales involved in creating the proxies. Many conditions must be met prior to conserving and also finding the witnesses of the past climate which all together form their own spatial scales in the observations. So, detecting scales and effectively avoiding them in the interpretation is a mayor task . In our final contribution by Steinhorst, Simmer and Schilling the interpretation of scales in terms of the underlying processes is the mayor issue. High resolution precipitation records at individual measuring stations are used to construct data models, which are able to reconstruct artificial time records from only a few statistical measures. The authors show that when extending current models to high-resolution records which contain contributions of the internal variability found within convective events, rainfall records do not exhibit complete self-similarity anymore. Obviously different models must be applied for different temporal scales.

Fractals – Pretty Pictures or a Key to Understanding Earth Processes?

Stefan Hergarten[*]

Section Geodynamics, Geological Institute, University of Bonn, Germany

Abstract. We illustrate two basically different definitions of fractals and their relevance to nature, especially to earth sciences. After this, two concepts that are able to explain the occurrence of fractals are discussed with respect to the use of fractals for elucidating the underlying processes: scale–invariant processes as a well–understood, hierarchical mechanism leading to static fractals, and self–organised criticality as a not completely understood phenomenon resulting in dynamic fractals.

1 A Little History

A stone, when it is examined, will be found a mountain in miniature. The fineness of Nature's work is so great, that, into a single block, a foot or two in diameter, she can compress as many changes in form and structure, on a small scale, as she needs for her mountains on a large one; and, taking moss for forests, and grains of crystal for crags, the surface of a stone, in by far the plurality of instances, is more interesting than the surface of an ordinary hill; more fantastic in form, and incomparably richer in colour — the last quality being most noble in stones of good birth (that is to say, fallen from the crystalline mountain ranges).

J. Ruskin, *Modern Painters*, Vol. 5, Chapter 18 (1860)

This citation is to be found in Turcotte's (1992) book on fractals and chaos in geology and geophysics. Shall it be considered as the birth of the concept of scale invariance? However, it took more than 100 years to embed these impressions into a quantitative scientific concept; although the idea has, for instance, been present in geology for many years. Each photograph of a geological feature should include an object that determines the scale, for instance a coin, a rock hammer or a person. Without this, it might be impossible to recover whether the photograph covers 10 cm or 10 km.

The first quantitative analyses and the introduction of a formal definition of scale invariance go back to a work of Mandelbrot (1967) who showed how the measured length of a coast line depends on the length of the ruler used for measuring. The mathematical Hausdorff dimension can be used for assigning a

---

[*] hergarten@geo.uni-bonn.de

non-integer dimension to such objects; for a coast line it is between one and two. The fractional dimension leads to the term *fractals*.

Since the late 1970's, Mandelbrot's ideas have attracted many scientists from various disciplines. Although fractals themselves may be somehow beautiful, it is more their apparent relevance to nature that makes them attractive. Simple algorithms led to fractal surfaces that were not just pretty, but looked very similar to the earth's surface (Voss 1985, Feder 1988, Turcotte 1992). Fig. 1 shows such a surface generated using a Fourier transform algorithm as illustrated by Turcotte (1992). However, the fact that such surfaces look realistic does not imply at all that the land surface is fractal in the sense discussed in the following.

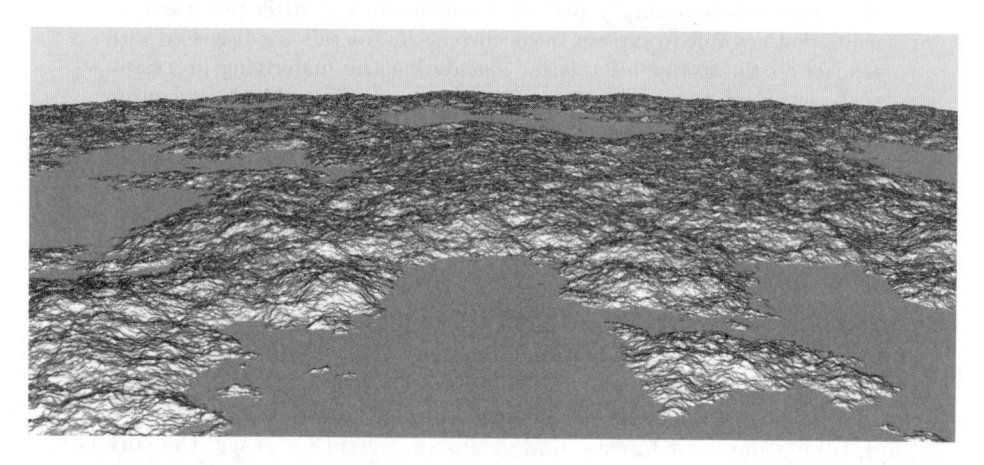

Fig. 1. A computer-generated fractal surface. For a realistic impression, the relief was filled with water up to a certain level.

Still more than the pattern itself, the imagination of its formation makes the idea of scale invariance fascinating to scientists. How does nature generate scale-invariant objects? What distinguishes their evolution from that of the majority of scale-dependent patterns? Do fractals guide us to any fundamental mechanism that unifies processes which seem to be completely different from our limited point of view?

2 What is a Fractal?

There is not a universal, straightforward definition of fractals; there are several definitions that are not equivalent. All these definitions represent the heuristic definition:

A fractal is an object that looks the same on all scales

or better:

Fractals – Pretty Pictures or a Key to Understanding Earth Processes?

A fractal is an object that is statistically similar on all scales.

Let us begin with an example. Fig. 2 shows two computer-generated patterns. We can imagine them as a very simple model of a fractured rock. Which one is fractal?

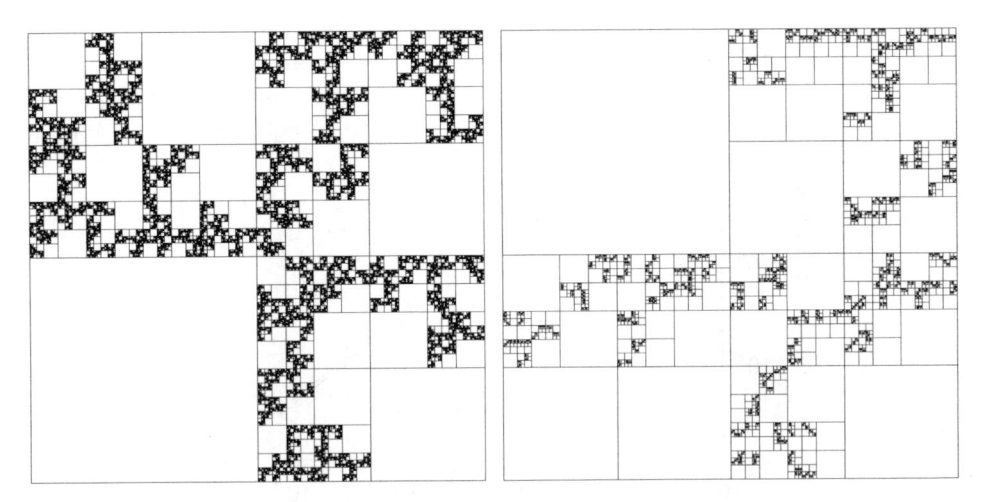

Fig. 2. Two computer-generated multi-scale patterns. Which one is fractal?

Obviously, both cover a wide range of scales, but it is not obvious that only one of them is fractal. Our eyes are able to distinguish whether a line is straight or whether two lines are parallel, but we are not practised in looking at different scales simultaneously. Thus, we must focus on different scales explicitly — as if we look trough different magnification glasses — if we want to see scale invariance or scale dependence.

Let us now switch to a more precise definition that includes a reasonable definition of the term *statistically similar*:

A fractal consists of objects of all sizes, and the number N of objects with a linear size (e. g., length, square root of area or cubic root of volume) greater than r decreases as r^{-D}.

The *power law distribution*

$$N(r) \sim r^{-D} \tag{1}$$

is called *fractal distribution*. D is the *fractal dimension* of the distribution. Strictly speaking, $N(r)$ should be replaced by a probability $P(r)$ that an arbitrarily chosen object has a linear size greater than r. But still then, the power law distribution (Eq. 1) requires that there is a minimum object size. However, from the view of a physicist or a geo-scientist this limitation is not crucial because scale invariance of natural patterns should end at least at the atomic scale.

Fig. 3 shows the number of fragments $N(r)$ with linear sizes (length of the edges) greater than or equal to r for the examples from Fig. 2. The data from the left hand part of Fig. 2 can be fitted with a straight line; the fractal dimension is $D = 1.585$. In contrast, the number of small fragments is too low in the right hand image, so that it cannot be fractal according to the definition given above.

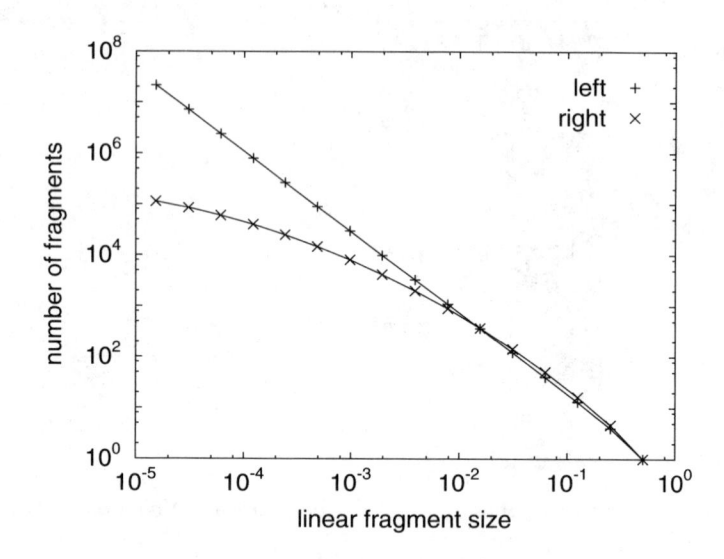

Fig. 3. Number of fragments $N(r)$ with linear sizes greater than or equal to r for the examples from Fig. 2.

The example shows that even the presence of objects at all scales is not sufficient for scale invariance. There is a large class of patterns between fractals and those which are governed by a single scale; the relation between objects at different scales distinguishes fractals from other patterns. In this sense, the necessity of including an object that determines the scale into a photograph of a geological feature might be closer to the idea of scale invariance than the impressions of J. Ruskin.

There are several examples for power law statistics in earth science. In this context it should be mentioned that any data set obtained from nature consists of a finite number of objects with a minimum and a maximum object size. Thus, a power law distribution can only be recognised over a limited range of scales; this distinguishes pure mathematical fractals from those found in the real world. Moreover, observed distributions will mostly exhibit power law behaviour only approximately. Both the range of scales and the deviations from the power law are criteria for the quality of a fractal in the real world.

Rock fragments do — in fact — often exhibit fractal statistics. Fig. 4 shows data from granite broken by a nuclear explosion (Schoutens 1979), broken coal (Bennett 1936), and projectile impact on basalt (Fujiwara al. 1977). The fragment sizes obey power law statistics over two to four orders of magnitude; the fractal dimension is close to 2.5. However, this value is not universal for rock fragments, (Turcotte 1992), and there are several examples where the distribution is not fractal at all.

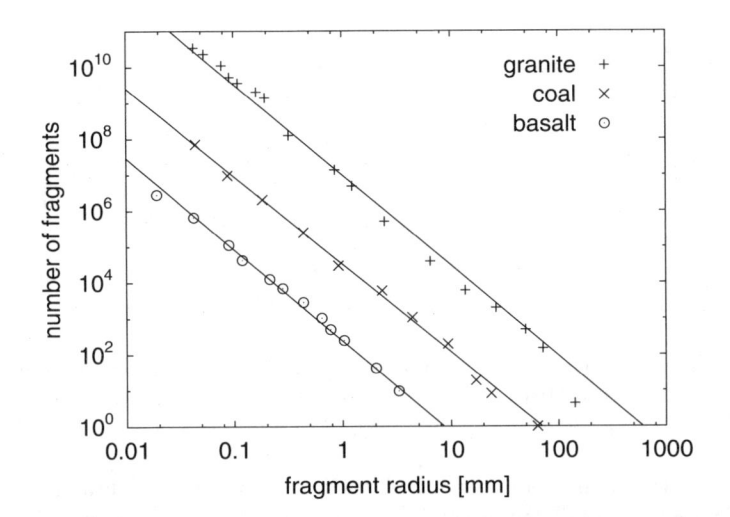

Fig. 4. Size distribution of rock fragments from granite broken by a nuclear explosion (Schoutens 1979), broken coal (Bennett 1936), and projectile impact on basalt (Fujiwara al. 1977). The data were taken from Turcotte (1992).

Earthquakes are the most famous example of scale invariance in geo-science. Gutenberg and Richter (1954) were the first to state the statistical frequency-size relation of earthquakes:

$$\log_{10} N(m) = a - bm, \qquad (2)$$

where m is the magnitude, $N(m)$ is the number of earthquakes (per fixed time interval) in a certain region with a magnitude larger than m, and a and b are parameters. While a describes the regional (or global) level of seismic activity and thus varies strongly, b is found to be between 0.8 and 1.2 in general. The Gutenberg-Richter law states that an earthquake of magnitude larger than $m+1$ is by a certain factor (about 10) less probable than an earthquake of magnitude larger than m, independently of the absolute value of m. The Gutenberg-Richter law turned out to be valid over a wide range of magnitudes; as an example, Fig. 5 shows the earthquake statistics from Southern California between 1932 and 1972 (Main and Burton 1986).

However, the magnitude of an earthquake is related to the energy of the seismic waves released by the earthquake, but does not define a linear object size immediately. So we need a little theory from seismology. During an earthquake, a displacement of material along a fault occurs. The rupture area is

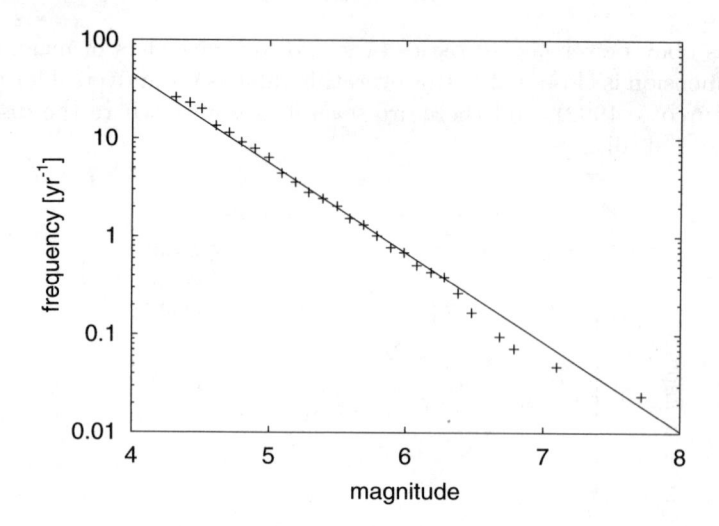

[Figure: frequency vs magnitude log-linear plot]

Fig. 5. Frequency of earthquake occurrence in Southern California between 1932 and 1972 (Main and Burton 1986). The data were taken from Turcotte (1992).

that part of the fault which is displaced; its size A is roughly proportional to 10^m (Kanamori and Anderson 1975, Turcotte 1992, Lay and Wallace 1995). With this, the Gutenberg-Richter law can be rewritten in the form

$$N(A) \sim A^{-b}. \tag{3}$$

If we assume that the linear extension is proportional to the square root of the rupture area, the Gutenberg-Richter law implies a power law distribution of the rupture areas with a fractal dimension that is twice the exponent b.

In our examples, defining the objects was straightforward. Our eyes distinguish the fragments in Fig. 2 immediately, real rock fragments can be weighted, and the rupture areas of earthquakes are related to the measured magnitude.

However, we could have taken the fractures (straight lines) in Fig. 2 instead of the fragments (squares), and it is not clear whether both obey similar statistics. As soon as the data set consists of a map, a photograph, a digital elevation model or something similar, recognising objects and measuring their sizes becomes a difficult task. Mostly, the recognition has to be done manually. Moreover, we have to keep in mind what objects we refer to. For instance, if we want to use the statistical definition for investigating fractal properties of the earth surface, we have to define and recognise objects such as mountains, slopes, plains, and rivers. We will end up with the specific result that some of these objects are fractally distributed and some are not, but should not ask whether the earth surface itself is fractal.

For this reason, the statistical definition of fractals is straightforward at first sight, but complicated. There are several other definitions of fractals that do not require any definitions of objects. The most common one is based on the *box-counting* method and can be applied to any set:

Fractals – Pretty Pictures or a Key to Understanding Earth Processes?

The set is covered by a regular grid of squares (in two dimensions) or cubes (in three dimensions) of linear size r. Let $N(r)$ be the number of boxes needed for covering the set. If $N(r)$ follows the power law relation (Eq. 1)

$$N(r) \sim r^{-D}$$

with a non-integer exponent D, the set is called fractal; D is the fractal dimension.

If a set is fractal in the sense of the box-counting definition, its fractal dimension D coincides with its Hausdorff dimension. Fig. 6 illustrates how a set is covered by boxes of different sizes.

If we apply the box-counting method to the patterns from Fig. 2, we first have to decide whether the set consists of the fragments or of the fractures. If we take the fragments, the whole area will be covered with boxes, no matter how small they are. Thus, the pattern is not fractal in this sense.

Fig. 7 shows the result of applying the box-counting method to the fractures from Fig. 2. The results are qualitatively the same as we obtained using the statistical definition: Only the left hand part is fractal. It may be surprising that the fractal dimension of $D = 1.585$ coincides with that obtained from the size statistics; but we shall not forget that we once considered fragments and once fractures. Thus, the coincidence should not be over-interpreted.

Although the power law relations in the box-counting definition and in the statistical definition are formally the same, the meaning of the quantities is essentially different. Thus, the two definitions are not equivalent in general.

The fundamental difference between both definitions can be seen in different terms of scales. In the statistical definition, the scale is defined by the sizes of the objects, while the scale in the box-counting definition (i.e., the linear box size) can be interpreted as a resolution. Because our eyes are used to focus on objects rather than on a resolution, the statistical definition seems to be more straightforward than the box-counting definition, although more complicated in detail.

If we compare the results from analysing the non-fractal right hand part of Fig. 2 in Figs. 3 and 7, we see that the deviations from the power law are more significant if the statistical definition applied than in box counting. Thus, the statistical definition apparently yields a more reliable criterion for elucidating fractal properties than box counting does if the available range of scales is limited.

Fig. 8 shows that box counting may lead to strange results; the data were obtained from applying the box-counting algorithm to the artificial example from Fig. 6 which is surely non-fractal. A straight line with a slope of of $D = 2$ fits the data fairly well for small box sizes up to 8; which means that the set looks like a two-dimensional area at small scales. This is not really surprising because the edges of letters are quite smooth. Between $r = 16$ and $r = 512$, a straight line with $D = 1.3$ looks reasonable. This illustrates the major problem with applying the box-counting method to the real world: Due to limited observing facilities, the available range of scales may be quite small; unfortunately, it often covers

$$N(r) \sim r^{-D}$$

$$N(r) \sim r^{-D}$$

$$N(r) \sim r^{-D}$$

$$N(r) \sim r^{-D}$$

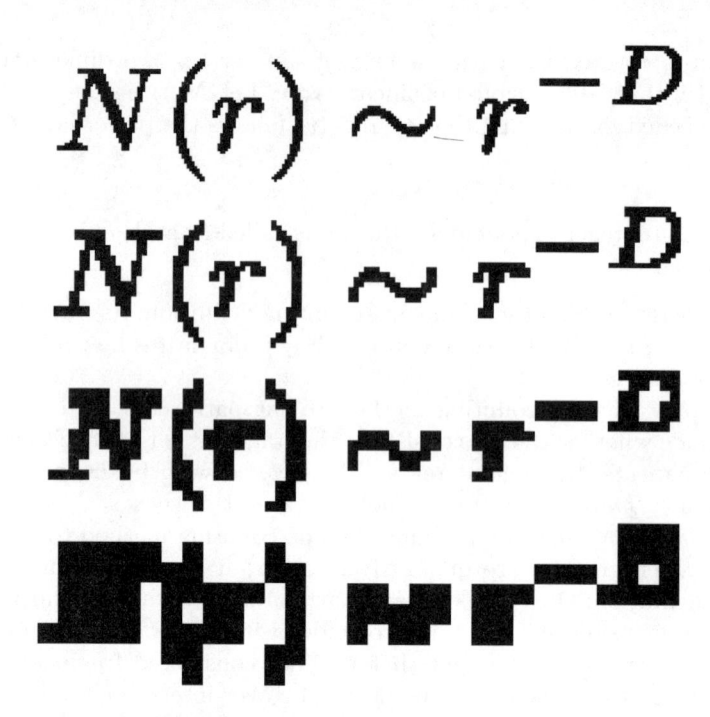

Fig. 6. Illustration of the box-counting algorithm. The box size increases from 16 (top) to 128 (bottom).

only one or two orders of magnitude. Our example shows that a fair power law behaviour obtained from box counting over such a small range is a poor criterion and does not comply with the concept of scale invariance in a reasonable way. Unfortunately, many authors looking for fractals in nature seem not to be aware of this problem, so that apparent fractals are often a result of over-interpretation.

Let us conclude this section with some statements:

- Not everything that covers a wide range of scales is fractal.
- The term *fractal* should not be used in an unspecific way; the definition referred to and — in case of the statistical definition — the considered objects should be mentioned.
- The statistical definition requires a definition of objects and their sizes, while the box-counting definition can be applied to arbitrary sets. In this sense, the statistical definition represents our intuitive ideas of scales better than the box-counting definition, but it is more complicated in detail.
- In the real world, fractals can only cover a limited range of scales; in many cases, limited observation facilities reduce this range further.
- A fair power law obtained from box counting over one or two orders of magnitude is a poor criterion for scale invariance.
- The statistical definition only considers object sizes, while box counting combines sizes, distances and alignment.

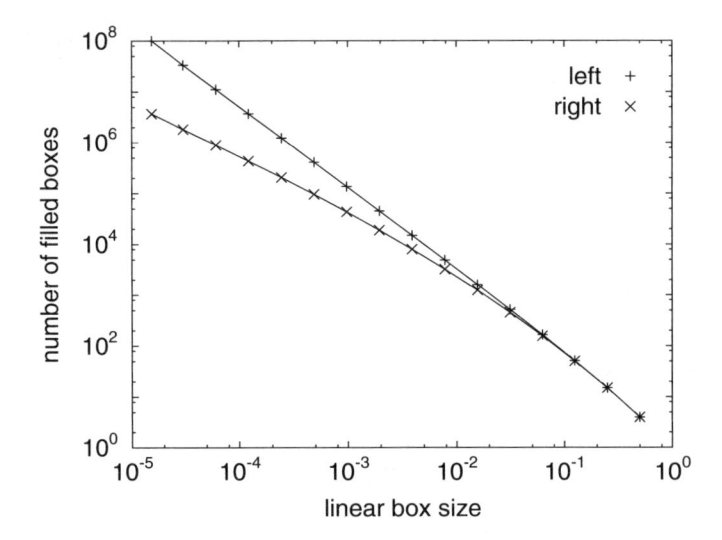

Fig. 7. Results of applying the box-counting method to the fractures in Fig. 2.

- The box-counting dimension is between zero and the dimension of the space the set is embedded, while the fractal dimension of the size distribution may exceed the Euclidian dimension of the embedding space.
- In contrast to box counting, the statistical definition allows overlapping objects as a result of spatio-temporal processes, such as the rupture areas of earthquakes.

This short introduction into the basics of scale invariance should be enough for a first insight, although it cannot cover this wide field completely. We have left out some further definitions of fractals, the difference between (isotropic) self-similar fractals and self-affine fractals (which have anisotropic scaling properties), and multifractals as a more general type of fractals. Self-affine scaling may occur in the analysis of one-dimensional data series, which is discussed later in this volume (Leonardi 2002). For further reading on these topics, the books of Mandelbrot (1982), Feder (1988) and Turcotte (1992) are recommended.

3 Application of Fractals

We should now know what fractals are, but the question what they are good for is still open.

In some cases, direct implications can be drawn from fractal properties. For instance, the Gutenberg-Richter law tells us something about the role of small and large earthquakes for crustal deformation. Obviously, the few large earthquakes cause much more damage than the large number of small earthquakes. But apart from the damages, a large part of crustal deformation is performed by earthquakes rather than by solid state flow of rocks.

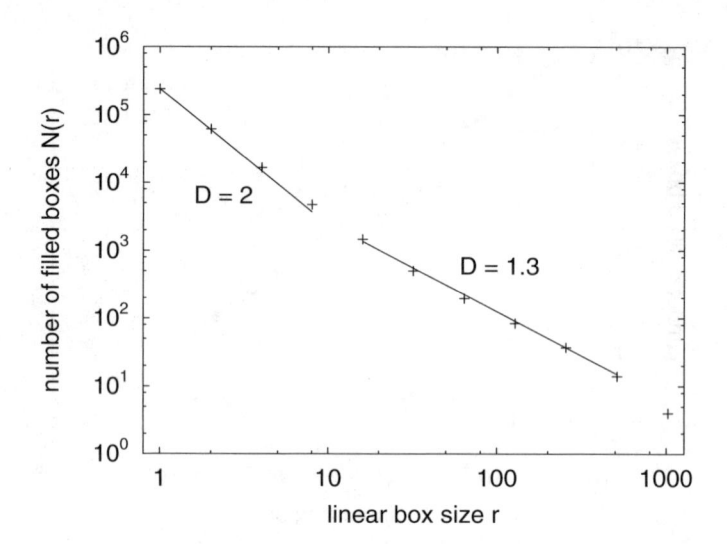

[Figure: plot of number of filled boxes N(r) vs linear box size r, showing two regimes with D = 2 and D = 1.3]

Fig. 8. Result of the box-counting algorithm applied to the image from Fig. 6.

With little more theory, the fractal size distribution of the rupture areas (Eq. 3) enables us to assess whether the deformation is governed by the majority of small earthquakes or by a few large events. The contribution of an earthquake to crustal deformation does not only depend on the size A of its rupture area, but on the displacement occurring along the rupture area, too. This result can be quantified using the seismic moment M that is proportional to the product of A and to the average displacement measured over the rupture area. Theory states that the displacement increases with the square root of the rupture area, so that $M \sim A^{\frac{3}{2}}$ (Kanamori and Anderson 1975, Turcotte 1992, Lay and Wallace 1995). Thus, Eq. 3 can be transformed into a distribution of the seismic moments:

$$N(M) \sim M^{-\frac{2}{3}b}. \tag{4}$$

When we introduced the power law distribution (Eq. 1), we already stated that such a distribution can only be valid above a minimum object size. On the other hand, the sizes of the rupture areas must at least be limited by the thickness of the earth's crust, so that there must be an upper limit for M, too. Let us assume that the Gutenberg-Richter law (and thus Eq. 4) is valid between a minimum moment M_1 and a maximum moment M_2. The total deformation induced by all earthquakes between M_1 and M_2 is then proportional to

$$\int_{M_1}^{M_2} M \frac{\partial N(M)}{\partial M} dM \sim \left| M_2^{1-\frac{2}{3}b} - M_1^{1-\frac{2}{3}b} \right|.$$

If we remember that b is between 0.8 and 1.2 in general, we see that the total deformation is governed by the small number of large earthquakes because the deformation tends towards infinity in the limit $M_2 \to \infty$, while it is bounded in the limit $M_1 \to \infty$. On the other hand, the majority of small earthquakes would govern the total deformation if b was greater than $\frac{3}{2}$.

This was a quite simple example for conclusions that can immediately be drawn from fractal properties. However, there are more sophisticated applications such as the computation of hydraulic parameters for porous media with fractal properties (Pape et al. 1999).

Extrapolation is another obvious, although sometimes misleading application of fractals. If we know that a distribution is fractal, we can use this property for extrapolating to object sizes that cannot be observed because they rarely occur. Let us take a look again on the earthquake statistics of California (Fig. 5). Assuming that the Gutenberg-Richter law holds for large magnitudes, we can extrapolate the risk of an earthquake of magnitude 8 or greater to 1 per 100 years, and the risk of an earthquake of magnitude 9 or greater to 1 per 800 years. However, the problem of extrapolating a fractal distribution is the same as for any kind of extrapolation because we assume that the relation holds outside the range of the data. Thus, we do in principle need an independent estimate of the validity of the Gutenberg-Richter law at large magnitudes. It is known that the Gutenberg-Richter law breaks down for large events; the upper limit appears to correspond to a limitation in rupture area due to the transition from the brittle crust to ductile behaviour. This limit seems to be somewhere between magnitude 8 and 9, so that we can conclude that the estimate of the risk of a magnitude 8 earthquake may be quite realistic, while the risk of a magnitude 9 earthquake should be lower than estimated from the fractal distribution if it occurs at all.

There are several other scientific applications of fractals. For instance, fractal distributions can be used for generating models of fracture networks in rocks with similar statistical properties as natural fracture networks. Such data sets can serve as a basis for modelling flow of water in fractured rocks (Kosakowski et al. 1997).

However, beside such rather direct applications, the question about the origin of fractals in nature is the major challenge of research on fractals. In the middle of it, there is the idea that there might be a fundamental type of processes behind the variety of fractal patterns, and that this type of process may open a view that unifies several processes which seem to be different from our limited point of view yet.

4 Fractals as a Result of Scale-Invariant Processes

Mostly, small things are made with small tools, and large things with large tools. From this experience, we tend to identify the scales of the patterns we see with the scales of the processes that generated the patterns. This leads us to a first idea on the origin of fractals: scale-invariant processes. There may be processes that are able to act on different scales and do not depend on the scale in any way. However, most physical processes do not fall into this category, but have their own characteristic scales.

Let us begin with a simple example — the growth of a fractal tree. It starts with a stem of unit length; we assume that n branches of length $\lambda \in (0,1)$ grow out of the stem. This procedure is continued on the smaller scales, so that each branch of length r has n smaller branches of length λr. Fig. 9 shows such a tree, built with $n = 4$ and $\lambda = 0.417$.

Fig. 9. Fractal tree.

It can easily be seen that the lengths of the kth order branches are $r_k = \lambda^k$, where $k = 0, 1, 2, \ldots$, and that the number of branches of length r_k is n^k. Thus, the number of branches of length r_k or greater can be computed using a geometric series:

$$N(r_k) = \sum_{i=0}^{k} n^i = \frac{n^{k+1} - 1}{n - 1} = \frac{n\,n^k - 1}{n - 1} = \frac{n\,n^{\frac{\log r_k}{\log \lambda}} - 1}{n - 1} = \frac{n\,r_k^{\frac{\log n}{\log \lambda}} - 1}{n - 1}.$$

Apart from the term -1 in the dominator, the distribution exhibits power law behaviour with a fractal dimension

$$D = -\frac{\log n}{\log \lambda}.$$

For large values k, i.e., for small object sizes, it approaches a power law distribution because the dominator is governed by the first term then. The deviations

Fractals – Pretty Pictures or a Key to Understanding Earth Processes?

from the power law at large object sizes are a result of the finite size of the tree; They would vanish if we assumed that there are $\frac{1}{n-1}$ objects that are larger than the stem.

The algorithm is straightforward and coincides with our imagination of growing trees, although the stem and the branches grow simultaneously in nature. Despite this, we must not claim that trees are fractal in nature because our algorithm used the scale-invariance of the growth process, namely the scale-invariant relation of the numbers and sizes of branches. So this is a nice model that is able to explain *why* trees are fractal if they are, but not to explain *whether* they are.

For the parameter values used in our example — $n = 4$ and $\lambda = 0.417$ — we obtain $D = 1.585$. This is exactly the same fractal dimension as we obtained for the left hand example in Fig. 2, although the pictures look completely different.

It is not surprising that two images obeying the same object size statistics may look different; it only reminds us on the necessity of keeping in mind what the objects are. We may not expect that a picture made of squares (our simplified geometry of fragments in Fig. 2) looks the same as a picture made of lines (the branches in Fig. 9).

The standard fragmentation algorithm leading to the scale-invariant statistics of fragments is quite similar to that of the fractal tree. It works as follows: We start with one block of unit volume $V_0 = 1$ and split it into f pieces of equal volume $V_1 = \frac{V_0}{f}$. We then assume than any of this first order fragments may be fragmented into f pieces at a given probability p; then we apply the same procedure to these second order fragments of volume $V_2 = \frac{V_1}{f}$, and so on. We assume that the parameters of the fragmentation process — f and p — are arbitrary, but the same at all scales. For fragmentation of rocks, $f = 2$ seems to be reasonable; however, the examples shown in Fig. 2 were constructed with $f = 4$.

It can easily be seen that the number of first order fragments is $n_1 = f(1-p)$ because $1 - p$ is the probability that a fragment does not break up again. The number of second order fragments is then $n_2 = f^2 p(1-p)$; this can be generalised to the order k:

$$n_k = f^k p^{k-1}(1-p).$$

The number of fragments of size $V_k = \frac{V_0}{f^k}$ or greater can be computed in analogy to the fractal tree; it obeys a proper power law for the small fragments:

$$N(V) \sim V^{\frac{\log(fp)}{\log f}}.$$

For computing the fractal dimension D of the distribution, we have to transform the fragment volumes V into linear sizes r using the relation $V \sim r^d$, where d is the spatial dimension. This leads to a fractal dimension

$$D = d \frac{\log(fp)}{\log f}.$$

As already mentioned, the size distribution of rock fragments often follows a power law with a fractal dimension $D \approx 2.5$. In the simple fragmentation model, this can be achieved assuming $p = 0.89$ if $f = 2$.

The left hand side of Fig. 2 results from a two-dimensional application of the fragmentation algorithm with $f = 4$ and $p = 0.75$, yielding the fractal dimension $D = 1.585$.

But what about the right hand size that is not fractal? It was generated using basically the same algorithm, but assuming that the fragmentation process is scale-dependent. The probability of being split is $p = 0.75$ for the first order fragments and decreases with decreasing fragment size. Here, we have used a power law relation between fragment size and probability: $p \sim r^e$ where $e = 0.1$. Although this dependence is not very strong, it leads to a significant deviation from the fractal distribution (Fig. 3) that can easily be recognised with box counting, too (Fig. 7).

Does this hypothetic scale-dependence of the fragmentation process have any relevance to nature? It should have because it is not obvious why natural fragmentation processes should be scale-invariant. A scale-invariant fragmentation may be realistic for explosions or impact of projectiles as shown in Fig. 4, but as soon as a macroscopic stress field becomes important, there should be a scale dependence. Each step of fragmentation should release stress, and thus should lead to a reduced probability of further fragmentation processes. Hence patterns resulting from thermal cracking due to cooling, desiccation or faulting due to tectonic forces should have a size distribution like that of the right hand pattern in Fig. 2 as shown in Fig. 3 rather than a fractal one; the deviation from the fractal distribution should reveal valuable information about the underlying processes.

5 Self-Organised Criticality

The explanation of fractals discussed in the previous section was based on the intuitive identification of process and pattern scales. In their two seminal papers (Bak et al. 1987, Bak et al. 1988), Per Bak and his coworkers introduced a completely different mechanism for generating fractals — *self-organised criticality* (SOC). The concept was developed introducing a simple cellular automaton model, often referred to as Per Bak's sandpile or Bak-Tang-Wiesenfeld (BTW) model.

The lattice is a regular, quadratic grid in two dimensions. We start with picking one of the sites randomly and put something at this site. In this rather abstract model it does not matter whether it is a grain of sand, a portion of energy, a sheet of paper on the desk of an bureaucrat or something else; let us for simplicity speak of grains. In the following, further grains are added without consequences until there are four grains at any site. The site with four grains becomes unstable, and the four grains are distributed homogeneously among the four neighbours of the unstable site. After this, one or more of the neighbouring sites may become unstable, too, because their number of grains is four. If so, they relax simultaneously in the same way; then their neighbours are checked, and so on. The procedure is repeated until all sites have become stable again. If a site at the boundary becomes unstable, the grains passing the boundary are lost.

Thus, the BTW automaton is based on two rules. The first rule — adding individual grains — represents some gentle driving of the model. The second rule is the threshold relaxation rule that may induce avalanches under certain circumstances. It can easily be seen that the effect of the threshold relaxation rule strongly depends on the actual state of the system, i.e., on the number of grains and on their distribution. If there are very few grains, there will be no large avalanches; in contrast, if there are three grains at each site, one additional grain will immediately cause a large avalanche. Fig. 10 shows the course an avalanche of medium size taken from a long simulation on a 128×128 grid. Only a 10×10 region is plotted. During the avalanche, 28 sites became unstable; and it took 11 relaxation cycles until all sites became stable again. 27 sites participated in the avalanche, which means that one cell became unstable twice. We will refer to the number of participating sites, which can be seen as a measure of the area affected by the avalanche, as the *cluster size*. The third stage shown in Fig. 10 shows that there may be more than four grains at some sites temporarily. In this case, four grains are redistributed, while the rest remains at the site.

Fig. 11 shows the number of grains that are present on the lattice versus the total number of grains supplied. During the first phase of the simulation, both numbers nearly coincide, which means that very few grains are lost through the boundary. The reason for this behaviour is that the number of grains per site is small, so that only few relaxations take place. After about 35,000 grains have been added, a long term equilibrium between the number of supplied grains and those lost through the boundaries is achieved. In this state, the average number of grains per site is about 2.1. The lower part of Fig. 11 shows that this state is not a steady state; the number of grains is not constant but fluctuates in an irregular way. This indicates the occurrence of avalanches of various sizes: While only single grains are added, a large number of grains (several hundreds) can be lost during one avalanche.

A quantitative analysis of this behaviour led to a surprising result: As shown in Fig. 12, the distribution of the cluster sizes A follows a power law — the model generates fractals. It should be mentioned that the plot presents the probability density $p(A)$ and not the number of avalanches $N(A)$ with a cluster size larger than A as required for the statistical definition of fractals. However, the latter can be obtained from integrating the probability density.

The exponent of the power law is close to 1.05:

$$p(A) \sim A^{-1.05}.$$

Integration leads to

$$N(A) \sim A^{-0.05},$$

and thus to a fractal dimension $D \approx 0.1$. If the number of avalanches $N(A)$ with a cluster size larger than A is plotted directly, it does not exhibit a clean law behaviour. The reason is to be seen in the low fractal dimension in combination with the finite size of the lattice: Cutting off the probability density at the model

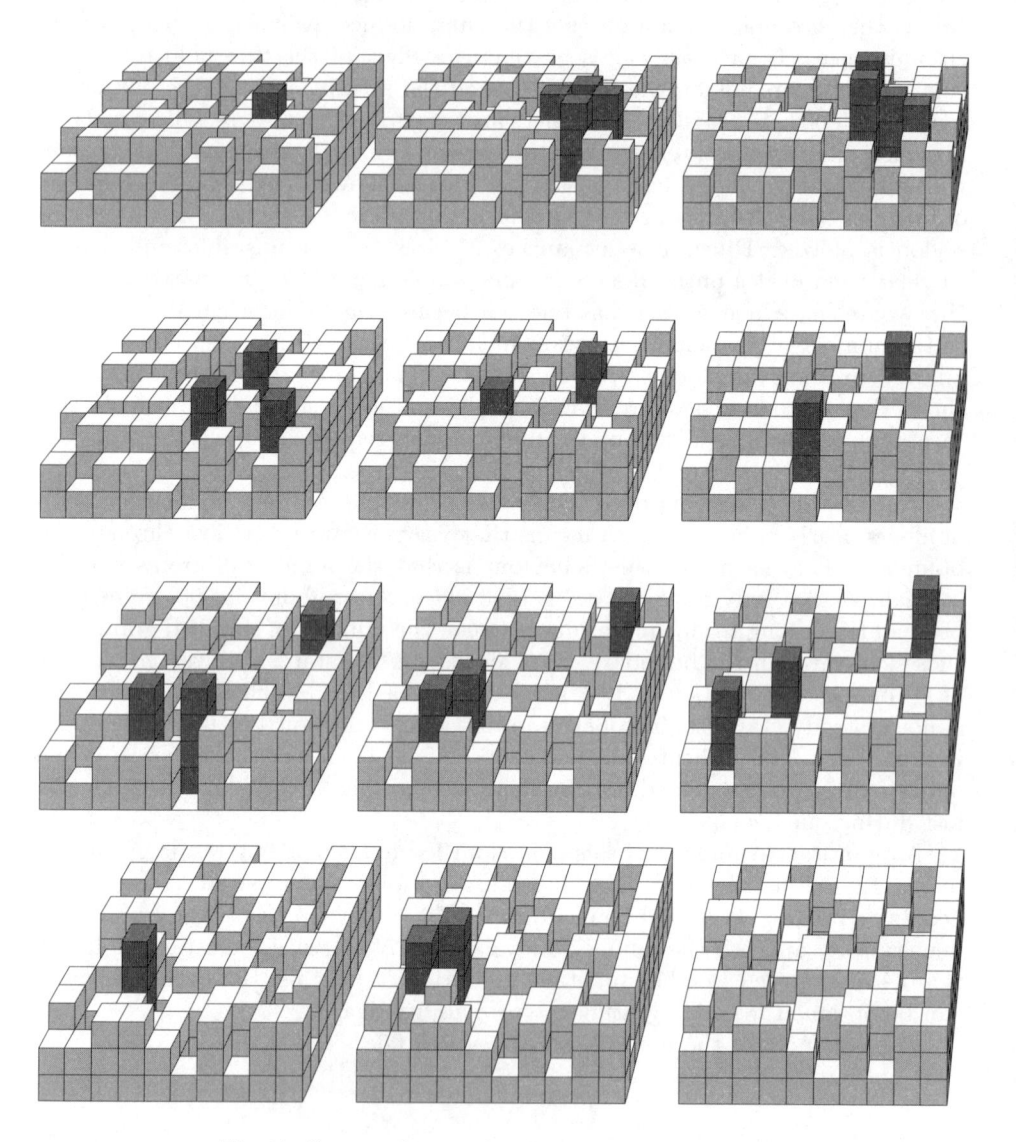

Fig. 10. Course of an avalanche in the BTW model.

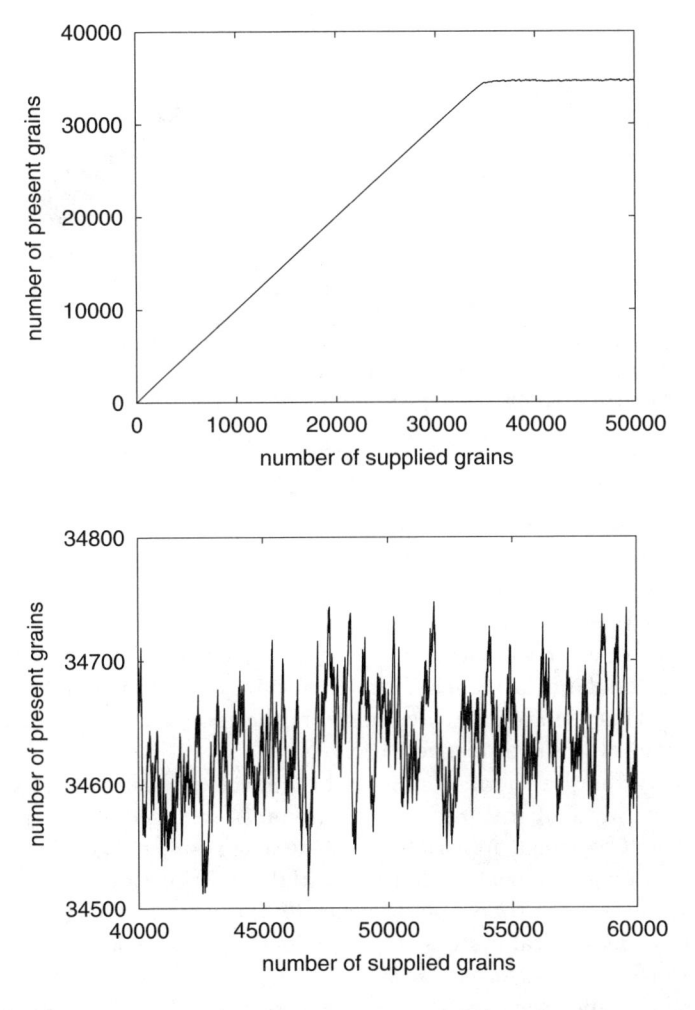

Fig. 11. Number of grains that are present on the lattice versus the total number of grains supplied on a 128 × 128 grid in the BTW model.

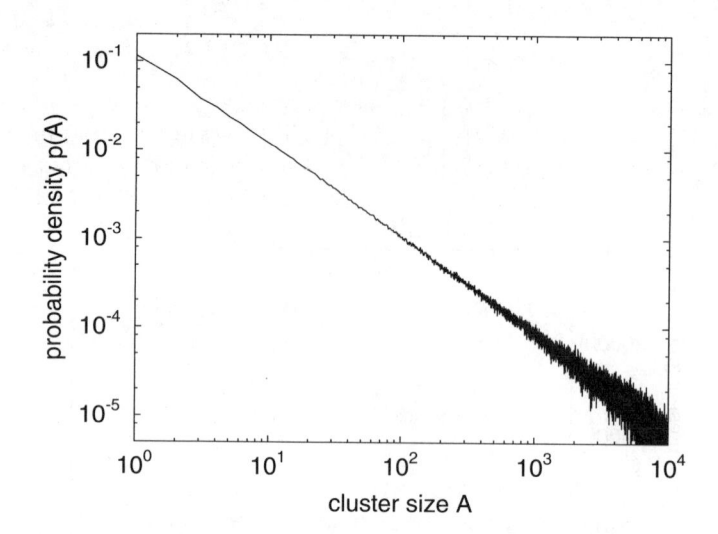

Fig. 12. Probability density of the cluster size distribution in the BTW model on a 128×128 grid. The statistics include 10^6 avalanches. The first 10^5 avalanches were skipped in order to avoid effects of the initial state.

size leads to a shift of the cumulative distribution in downward direction. This is the same effect which restrained the size distribution of the tree's branches from being a clean power law function. The smaller the fractal dimension is, the more important this effect becomes, so that the power law could only be recognised on very large grids.

But where does the name *self-organised criticality* come from? The state where small and large avalanches occur obeying a fractal statistics is called *critical state*. This term originally comes from thermodynamics where the critical state has scale-invariant properties, too.

Classical systems in thermodynamics become critical only under special conditions; in general, temperature and pressure must be tuned to the critical point. In contrast, SOC systems approach this state as a result of their own dynamics, without any tuning. No matter what the initial conditions are or how the system is disturbed, after some time it approaches its critical state. In terms of nonlinear dynamics, the critical state is an *attractor* in phase space. This property is expressed in the term *self-organised*.

Strictly speaking, we have left out an aspect of the definition of SOC. In the critical state, the temporal signal of the system should be a *pink noise*, also called *1/f noise* or *flicker noise*. This means that the power spectrum (obtained by a Fourier transformation) should be a power law with an exponent close to one. However, it is in general not clear what the temporal signal of the system is; in the BTW model it might be the number of present grains, the number of relaxations or the number of grains lost at the boundaries (as functions of time, i.e., of the number of supplied grains). Because the power spectra depend on this choice, this aspect of SOC is less clear than the fractal distribution of the cluster sizes. Moreover, power spectra tend to be noisy, so that it may be different to distinguish whether they show power law behaviour or not.

After discussing the basic ideas of SOC, let us revisit the example of earthquakes. The straightforward explanation of the fractal distribution of their rupture areas is that the distribution of fault sizes is fractal, and that each fault has its characteristic earthquake. In this model, the fractal distribution of the fault sizes is transferred to the distribution of the rupture areas. Although there is evidence for a fractal distribution of fault sizes (Turcotte 1992), the argument is not conclusive: The SOC concept shows that each fault could be in principle be able to produce a fractal earthquake statistics, and that the fault size might play a minor part. One could think of transferring the BTW model to the rupture dynamics at a fault, but we should we aware that the fractal dimension is predicted wrongly ($D = 0.1$ instead of $D \approx 2$) then. However, there are extensions of the BTW model (Olami et al. 1992) which are able to predict the fractal size distribution of earthquakes correctly in the framework of SOC.

Since the 1990's, much research on the SOC concept has been done. Several models for phenomena in physics, biology, economics, and geo-science turned out to exhibit SOC behaviour at least roughly. For an overview, the books of Bak (1996) and Jensen (1998) are recommended. It seems that the common property of SOC systems is that they are *slowly driven, interaction dominated threshold systems* (Jensen 1998). However, apart from such classifications, the nature of SOC is not well understood yet.

The SOC concept provides an explanation for the generation of *dynamic fractals*, which means that the distribution of the event sizes is scale-invariant. In contrast, the scale-invariant processes discussed in the previous section led to *static fractals*, i.e., to a final, steady state with fractal properties. However, we will see later in the volume (Hergarten 2002) that there are, too, systems such as drainage networks which seem to be somewhere between static and dynamic fractals.

6 Conclusions

Fractals are among the most fascinating patterns. Unfortunately, they cannot be defined in a unique way; there are several definitions which are not equivalent. In contradiction to mathematical fractals, fractals in the real world only cover a limited range of scales. Depending on the definition referred to, an apparent scale-invariant behaviour in images or data sets from nature should be interpreted carefully.

Beside some direct applications, the question for the origin of fractals in nature makes them attractive to scientists. There is not a unique mechanism than can be assumed to be responsible for the occurrence of fractals; there are at least two basically different ideas: scale-invariant processes as a well-understood, hierarchical mechanism leading to static fractals, and self-organised criticality as a not completely understood phenomenon resulting in dynamic fractals. The occurrence of different types of fractals and the deviations from scale-invariant behaviour in some cases provides a valuable tool for elucidating the underlying processes.

References

Bak. P., C. Tang, and K. Wiesenfeld (1987): Self-organized criticality. An explanation of 1/f noise. Phys. Rev. Lett., 59, 381–384.
Bak. P., C. Tang, and K. Wiesenfeld (1988): Self-organized criticality. Phys. Rev. A, 38, 364–374.
Bak. P. (1996): How Nature Works: the Science of Self-Organized Criticality. Copernicus, Springer, New York.
Bennett, J. G. (1988): Broken coal. J. Inst. Fuel, 10, 22–39.
Feder, J. (1988): Fractals. Plenum Press, New York.
Fujiwara, A., G. Kamimoto, and A. Tsukamoto (1977): Destruction of basaltic bodies by high-velocity impact. Icarus, 32, 277–288.
Gutenberg, B., and C. F. Richter (1954): Seismicity of the Earth and Associated Phenomenon. 2nd edition, Princeton University Press, Princeton.
Hergarten, S. (2002): Is the Earth's surface critical? The role of fluvial erosion and landslides. In: Dynamics of Multi-Scale Earth Systems, edited by H. J. Neugebauer and Clemens Simmer, this volume.
Jensen, H. J. (1998): Self-Organized Criticality. Emergent Complex Behaviour in Physical and Biological Systems. Cambridge Lecture Notes in Physics, 10, Cambridge University Press, Cambridge.
Kanamori, H., and D. L. Anderson (1975): Theoretical basis of some empirical relations in seismology. Seis. Soc. Am. Bull., 65, 1072–1096.
Kosakowski, G., H. Kasper, T. Taniguchi, O. Kolditz, and W. Zielke (1997): Analysis of groundwater flow and transport in fractured rock — geometric complexity of numerical modelling. Z. Angew. Geol., 43(2): 81–84.
Lay, T., and T. C. Wallace (1995): Modern Global Seismology. International Geophysics Series, 58, Academic Press, San Diego.
Leonardi, S. (2002): Fractal variablity in the drilling profiles of the KTB — implications for the signature of crustal heterogeneities. In: Dynamics of Multiscale Earth Systems, edited by H. J. Neugebauer and Clemens Simmer, this volume.
Main, I. G., and P. W. Burton (1986): Long-term earthquake recurrence constrained by tectonic seismic moment release rate. Seis. Soc. Am. Bull., 76, 297–304.
Mandelbrot, B. B. (1967): How long is the coast of Britain? Statistical self-similarity and fractional dimension. Science, 156, 636–638.
Mandelbrot, B. B. (1982): The Fractal Geometry of Nature. Freeman, San Francisco.
Olami, Z., H. J. Feder, and K. Christensen (1992): Self-organized criticality in a continuous, nonconservative cellular automaton modeling earthquakes. Phys. Rev. Lett., 68, 1244–1247.
Pape, H., C. Clauser, and J. Iffland (1999): Permeability prediction for reservoir sandstones based on fractal pore space geometry, Geophysics, 64(5), 1447–1460.
Schoutens, J. E. (1979): Empirical analysis of nuclear and high-explosive cratering and ejecta. In: Nuclear Geophysics Sourcebook, vol. 55, part 2, section 4, Rep. DNA OIH–4–2 (Def. Nuclear Agency, Bethesda, MD).
Turcotte, D. L. (1992): Fractals and Chaos in Geology and Geophysics. Cambridge Univ. Press, Cambridge.
Voss, R. F. (1985): Random fractal forgeries. In: Fundamental Algorithms in Computer Graphics, edited by R. A. Earnshaw. Springer, Berlin, 805–835.

Fractal Variability of Drilling Profiles in the Upper Crystalline Crust

Sabrina Leonardi*

Section Applied Geophysics, Geological Institute, University of Bonn, Germany

Abstract. A composite statistical analysis has been carried out on several petrophysical log data recorded in the deep and super-deep boreholes of the German Continental Deep Drilling Program (KTB). The results show that all the logs examined present similar statistics suggesting the fractal variability of magnetic, electric, acoustic and bulk properties of rocks down to 9 km depth. The inferred scale invariance may help modelling the complexity of small scale heterogeneities in crystalline crust.

1 Introduction

The history of the geodynamical processes of rock formations has been recorded in the structures of the Earth's crust, which are distributed over many orders of scales. On a macroscopic scale (kilometres and tens of kilometres), the crust is characterised by volumes of different lithological types, which were fractured by mainly tectonic events. On a microscopic scale (centimetres and even millimetres), rocks are composed of a framework and a pore space, the latter resulting from interstitial voids and micro-fissures.

Direct observations are thus in open contrast with the majority of geophysical models, which presuppose the homogeneity of the Earth's crust on large scales (Fig. 1a). A homogeneous medium is characterised by the unvarying propagation of physical properties, like e.g. seismic waves or electric pulses. The assumed statistical homogeneity of the Earth's crust with depth should be thus counterchecked by the uniform scattering of borehole physical parameters, which can be, in turn, approximated by white noise (Fig. 1b). White noise sequences are characterised by completely uncorrelated data values and flat power spectra (e.g. Priestley, 1981). From a statistical point of view, white noise distributions are described in terms of average values μ and standard deviations σ.

However, several geophysical variables recorded in boreholes at different geological sites (e.g. Hewett, 1986; Pilkington and Todoeschuck, 1990; Pilkington et al., 1994; Wu et al., 1994; Gritto et al., 1995; Holliger, 1996; Leonardi and Kümpel, 1996; Shiomi et al., 1997; Dolan et al., 1998; Leonardi and Kümpel, 1998; 1999), cannot be interpreted with white noise models and exhibit, on the contrary, a *scaling* or *1/f-noise* behaviour, meaning that their power spectra $P(k)$ are proportional to the wave number k according to a power law of type

$$P(k) \propto \frac{1}{k^b} \tag{1}$$

---

* sabrina@geo.uni-bonn.de

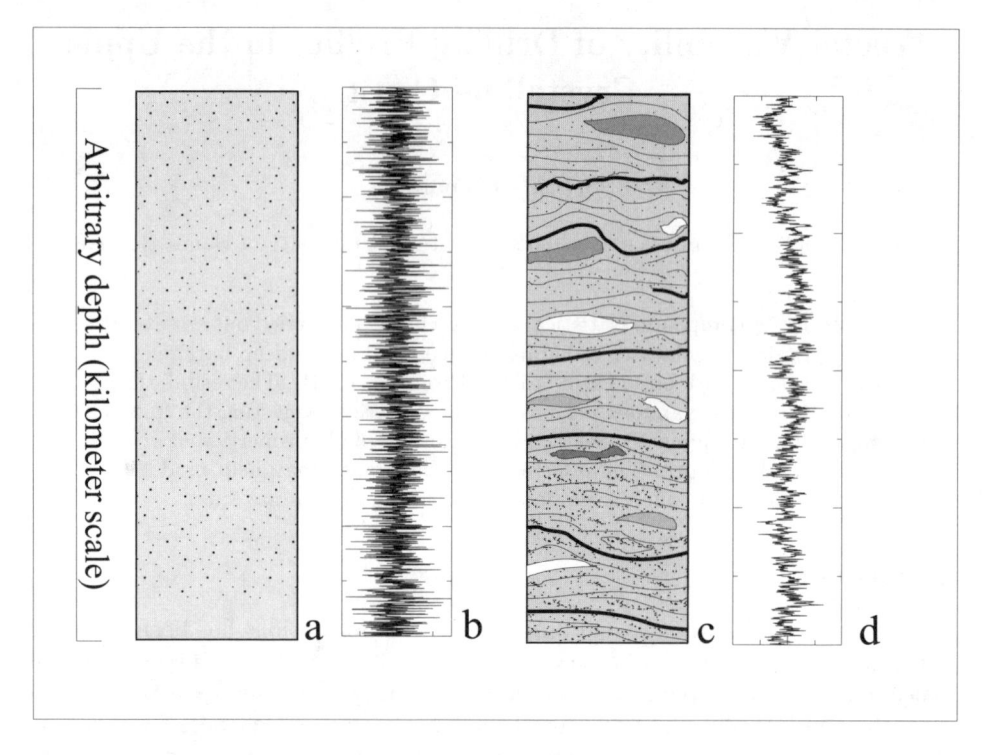

Fig. 1. Alternative types of rock variability in the crystalline crust: Macroscopically homogeneous model (a) and white noise scattering of physical properties (b); heterogeneous model (c) and scaling behaviour of physical properties approximated by $\frac{1}{f}$-noise with $b = 1$ (d).

where the spectral exponent b typically ranges between 0.5 and 1.5 (Fig. 1d). In case of scaling behaviour, μ and σ values alone are insufficient to describe the statistics of petrophysical properties; further parameters, that quantify the degree of scaling are needed. Crustal inhomogeneities, in turn, occur on a wide range of scales and cannot be levelled out by means of scale reduction (Fig. 1c). The concept of scale invariance is strictly related to fractal geometry (Mandelbrot, 1982), that branch of mathematics dealing with objects and variables preserving their statistical similarity on different orders of scales. The degree of scale invariance of a given object can be quantified by means of fractal dimension (Hergarten, this volume).

In this study, a composite statistical analysis is carried out on several log data from the German Continental Deep Drilling Program (KTB). The parameters recorded range from magnetic to electric to acoustic to bulk properties of rocks. Such a wide spectrum of petrophysical quantities reflect the lithological and structural character of the geological formations encountered. The aim is to relate the dominant statistical features of the KTB log data to the signature of crustal heterogeneities.

Fractal Variability of Drilling Profiles 259

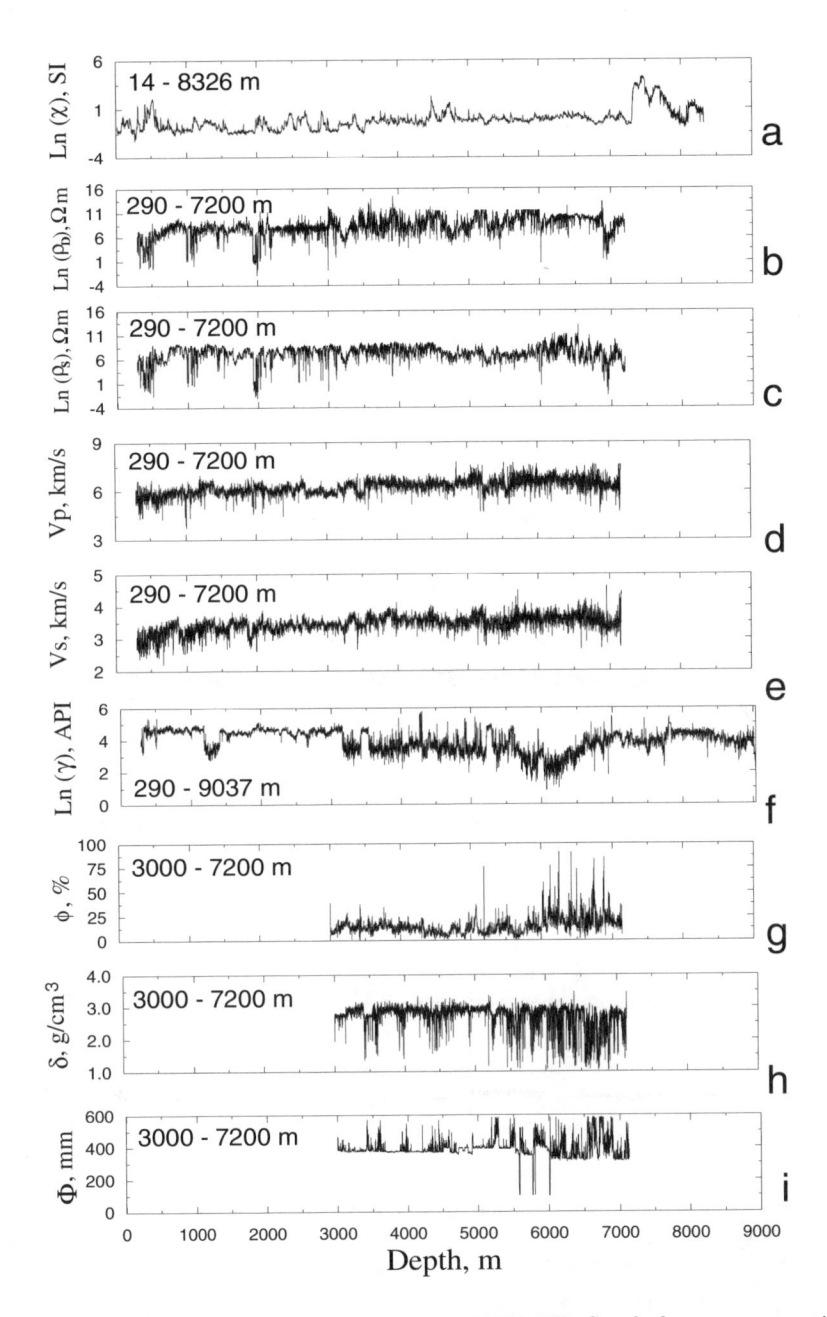

Fig. 2. Petrophysical properties recorded in the KTB-HB. Symbols: χ = magnetic susceptibility; ρ_D = electric resistivity, deep mode; ρ_S = electric resistivity, shallow mode; Vp = seismic velocity, P-waves; Vs = seismic velocity, S-waves; γ = natural radioactivity; δ = density; ϕ = porosity; Φ = caliper of the borehole (reprinted from Geophy. J. Int., 135, S. Leonardi and H.-J. Kümpel: *Variability of geophysical log data and the signature of crustal heterogeneities at the KTB*, pp. 964–974, copyright 1998, with kind permission from Blackwell Science).

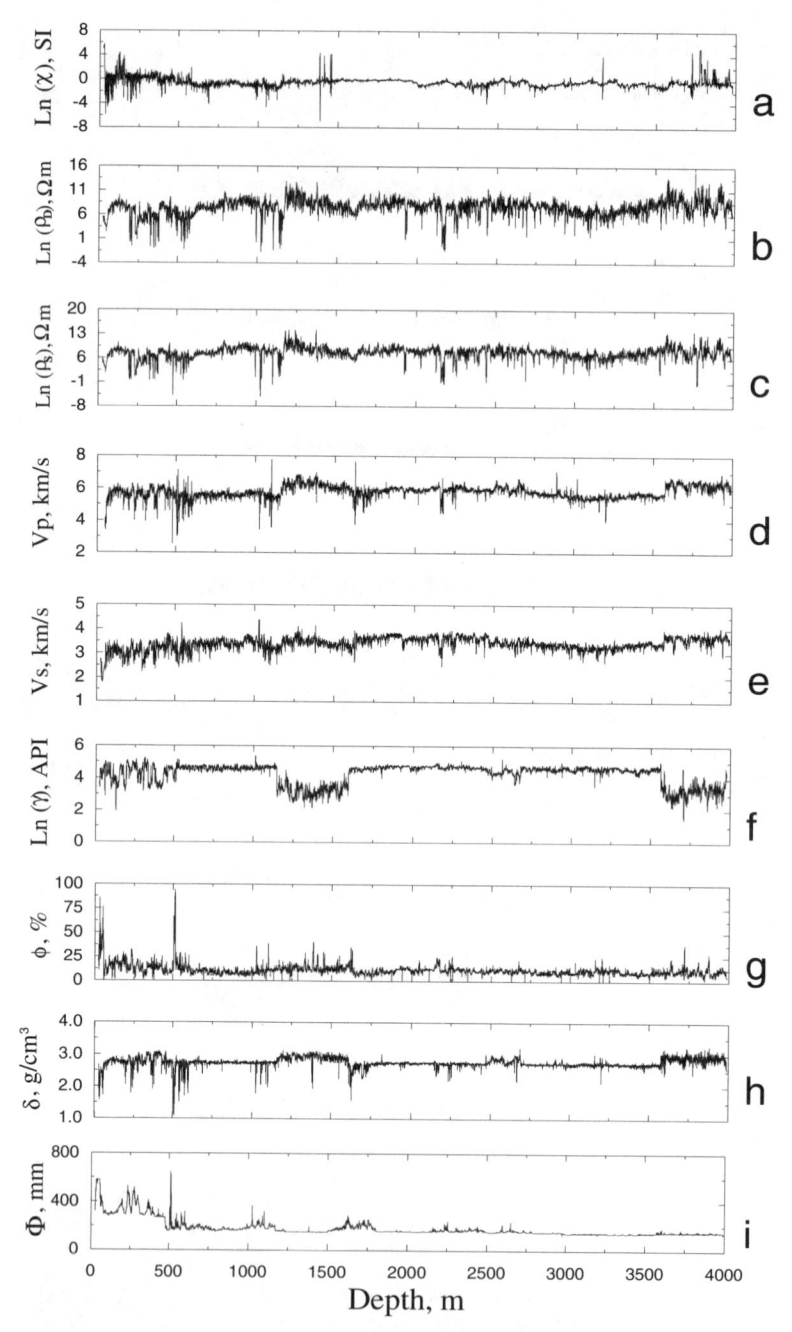

Fig. 3. Petrophysical properties recorded in the KTB-VB. Symbols as in Fig. 2 (reprinted from Geophy. J. Int., 135, S. Leonardi and H.-J. Kümpel: *Variability of geophysical log data and the signature of crustal heterogeneities at the KTB*, pp. 964–974, copyright 1998, with kind permission from Blackwell Science).

2 The KTB petrophysical logs

The KTB project has been carried out in the crystalline basement of the western border of the Bohemian Massif, geographically located in the Bavarian Oberpfalz, Germany. Two boreholes were drilled: a 4,000 m deep pilot hole ('Vorbohrung', VB hereafter) and a 9101 m super-deep main hole ('Hauptbohrung', HB hereafter), 200 m far from each other. Lithologically, this area is characterised by three main types of rocks: paragneisses — derived from former greywacke sediments —, metabasites — mainly consisting of amphibolites and metagabbros —, and alternations of gneiss-metabasites of probably volcano-sedimentary origin (Hirschmann et al., 1997). The protolith age of metabasites and paragneisses is estimated to be pre- to early Ordovician (about 500 Ma ago); the former rock successions were metamorphosed during the Variscan orogeny (from 480 to 380 Ma; Hirschmann et al., 1997) at temperature estimated to be 650–750°C and pressure of 8–14 kbar.

With one exception, the petrophysical properties that are analysed in this work were recorded *in situ* in both boreholes with a sampling interval $z_\epsilon = 0.1524$ m (Figs. 2–3). The magnetic susceptibility data from the HB was obtained by laboratory measurements on rock cuttings at $z_\epsilon = 2$ m. Details relative to the recording of log data are summerised by Leonardi and Kümpel (1998). The great depths of the boreholes and the high sampling rate imply a large amount of data values (20,000 to 60,000 for each log). Since the vertical resolution of most recording tools is estimated to be 0.5 to 1.0 m (Leonardi and Kümpel, 1998), the effective number of independent data values decreases to 3,000 to 18,000 for each log. Such a high number of independent data points is seldom available from boreholes, but is of basic importance for the significance of statistical studies.

3 Composite statistical analysis

The fluctuations within the log data recorded in the KTB boreholes (Figs. 2–3) may be mainly explained with lithological and structural variations within the drilled geological formations (Leonardi and Kümpel, 1998; 1999). Major fluctuations are related to main lithological types and faults; in particular some petrophysical properties — like γ radiation — match with rock composition, while others — like electric resistivity — are more sensitive to tectonic structures (Fig. 4). Note that some fluctuations within the logs are related to the geometry of the borehole caliper and are thus caused by drilling effects (Figs. 2i, 3i). Consistently, small-scale fluctuations (Fig. 5) should be related to petromineralogical changes as like small-scale cracks and fissures.

A deterministic analysis of all the small-scale variations would require a virtually infinite amount of information about the petro-physical-chemical and textural-structural features of the crust at the KTB site down to 9 km; alternatively an empirical statistical approach can be used. In this case, it is general rule to approximate the data to some known statistical distribution. In order to analyse the small-scale fluctuations of the KTB log data (Figs. 2–3), their deterministic linear trends were first subtracted. The latter are usually identified

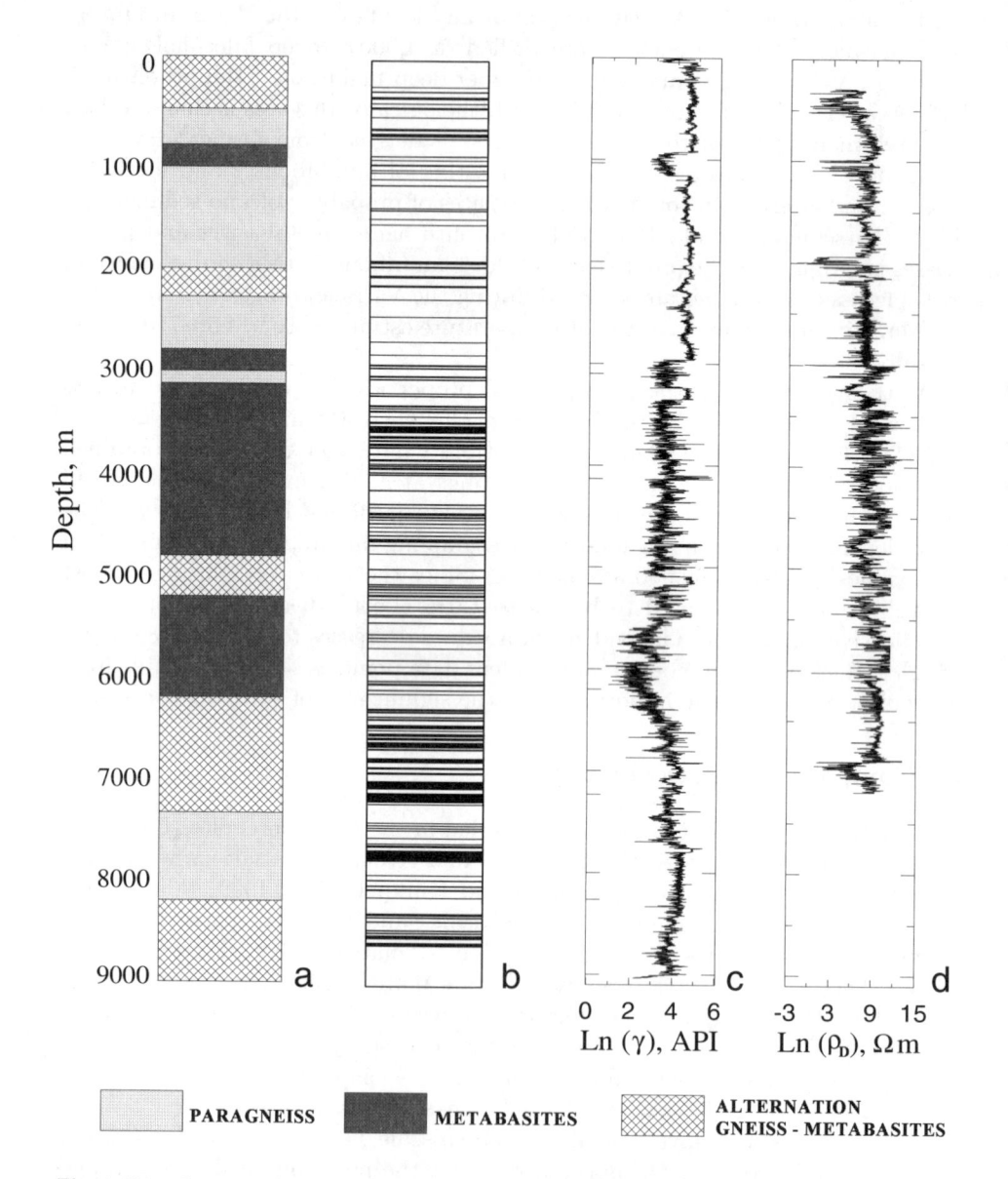

Fig. 4. Distribution of main lithological types (a) and faults (b) in the KTB-HB, compared with γ-ray (c) and electric resistivity (d) logs.

Fractal Variability of Drilling Profiles 263

with the steady increasing in the values of some physical parameters — like temperature and pressure — with depth (Wu et al., 1994; Holliger, 1996; Leonardi and Kümpel, 1998).

The classical methods of data series analysis assume stationary sequences and Gaussian probability distribution of the data values (Priestley, 1981); in other words the statistical properties of a given log should not change with depth. This condition is, of course, difficult to be fulfilled by real signals. The logs from the KTB boreholes, in particular, presented an evident bimodal probability distribution, caused by the lithological alternation of paragneisses and metabasites (Fig. 4; Leonardi and Kümpel, 1998). However, it has been proved that the statistical properties of the KTB logs do not significantly vary within the different rock formations (Leonardi, 1997). The lack of Gaussian distribution is thus restricted to the few main lithological 'jumps' and does not affect the results of our analysis.

Methods like autocorrelation functions, spectral and rescaled range (or R/S-) analyses are suitable tools to quantify the degree of correlation and of scale invariance within the log data.

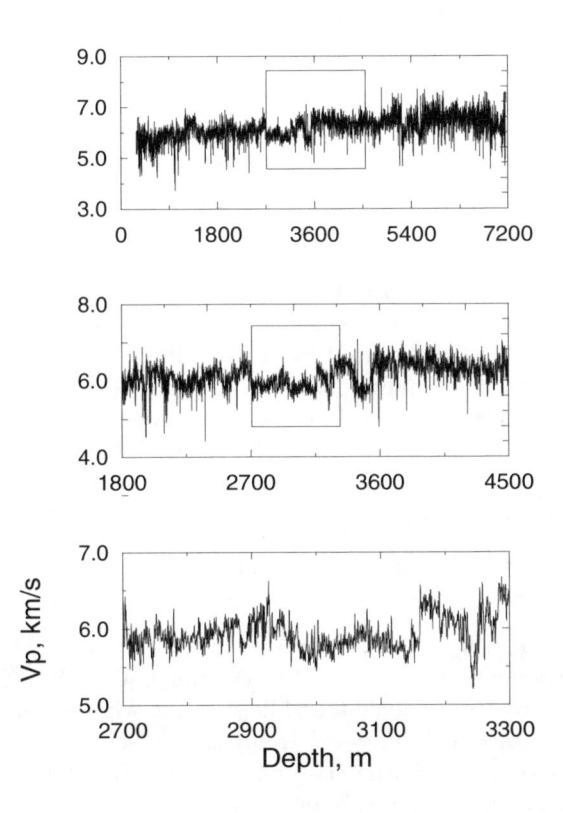

Fig. 5. Seismic velocity (P-wave) recorded in the KTB-HB at different depth scales.

Autocorrelation functions estimate the standardised covariance between the observed value $y(z_i)$ and that $y(z_i + \Delta z)$, where Δz is the depth shift. White noise sequences are completely uncorrelated and their autocovariances as like their correlation lengths a are equal to zero, showing that every two adjacent values are independent. Correlated data series, on the other side, present positive autocovariance values and therefore $a > 0$.

Spectral analysis estimates the distribution of the amplitudes of a given data series over an interval of wave numbers k. White noise sequences are characterised by the homogeneous distribution of amplitudes over the whole k range and their power spectra $P(k)$ are fitted by flat lines with spectral exponents $b = 0$. Correlated noise sequences, on the contrary, exhibit power law spectra of type eq. (1); in this case one speaks of *scaling noise* and it is possible to determine a fractal relation over a certain range of wave numbers k. Scaling noise sequences with $0.5 \leq b \leq 1.5$ preserve indeed their statistical properties under a change of resolution (Mandelbrot, 1982). Given the log $y(z)$ with scaling noise features, if the depth z is rescaled by rz, then $y(z)$ should be replaced by $r^H y(z)$, where r is a constant and H depends on the value of b. Such a transformation, that scales the two co-ordinates z and $y(z)$ by different factors is called *affine* and curves that reproduce themselves under an affine transformation are called *self-affine* (Turcotte, 1992). Voss (1988) established a direct mathematical relation between the exponent H and the fractal dimension D_F for scaling noise, which in one dimension gives:

$$D_F = D_E - H \qquad (2)$$

and D_E is the Euclidean dimension. D_F is also related to the spectral exponent b through

$$D_F = \frac{5 - b^*}{2} \qquad (3)$$

for one dimensional scaling noise, where $b^* = b + 2$ (Voss, 1988).

A method to calculate the exponent H is given by the rescaled range (or R/S-) analysis. This was first introduced by Hurst (1957), therefore H is also named Hurst exponent. The R/S-analysis allows to quantify the *persistence* of a signal, i.e. how a local trend is continued in the whole sequence. For practical purposes, the data series $y(z)$ is integrated over the depth interval z and then divided in subintervals of N data values. The range $R(N)$, or the difference between maximum and minimum amplitude value is estimated and H is calculated from the relation

$$H \propto \frac{\log R^*}{\log N} \qquad (4)$$

and R^* corresponds to R/S, i.e. it represents the standardised value of the range R. The Hurst exponent assumes values in the range $[0,1]$. $H > 0.5$ are typical of persistent signals, where the local trend depicts the whole sequence on a smaller scale; whereas for $H < 0.5$ the local trend is reversed compared to the whole sequence. Values $H = 0.5$ characterise uncorrelated or white noise, where each value is independent from its neighbours.

4 Results and discussion

Correlation lengths of the KTB petrophysical logs are illustrated in Fig. 6; spectral and Hurst exponents are shown in Fig. 7 and Fig. 8 respectively. The composite statistical analysis ascertain and quantify positive correlation and signal persistence in all the examined logs. Details about the methods of analyses are reported by Leonardi and Kümpel (1998).

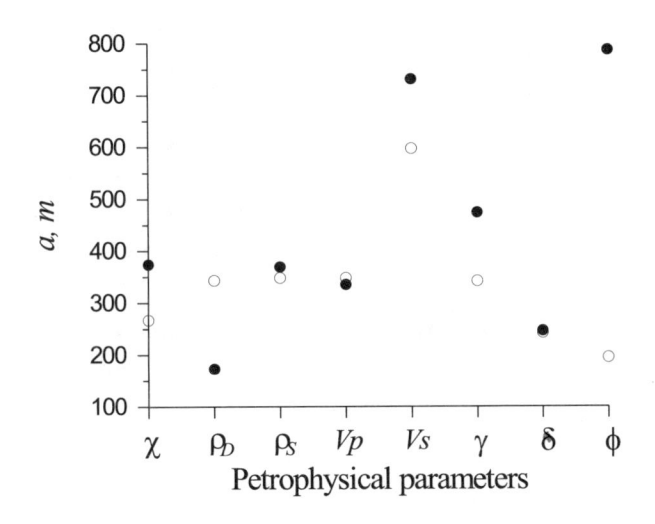

Fig. 6. Correlation lengths a in m of the KTB petrophysical properties from HB (black circles) and VB (white circles). One standard variation uncertainties are of the order of 10 %. Symbols as in Fig. 2.

Study of autocorrelation functions shows that the KTB log data behave as correlated noise with correlation lengths of the order of hundreds of metres (Fig. 6). Although the resolution of the recording tools may be in part responsible for this positive correlation, there is no doubt that, within each log, the recorded values are related to the upper and deeper ones. In other words, the variations of the petrophysical properties within the two boreholes are not sudden, as in case of white noise distributions, but gradual, like for correlated noise.

The spectral exponents for depth spacing from $10z_\epsilon$ to $1{,}000z_\epsilon$, i.e. in the range from metres to hundreds of metres, show values $0.6 \leq b \leq 1.9$ (Fig. 7). In this scale range the KTB log data can be statistically described as flicker or $\frac{1}{f}$-noise (Mandelbrot, 1982; Voss, 1988). The results of this study confirm previous analyses carried out on log data from different boreholes (Pilkington and Todoeschuck, 1990; Leary, 1991; Pilkington et al., 1994; Wu et al., 1994; Kneib, 1995; Holliger, 1996; Leonardi and Kümpel, 1996; Shiomi et al., 1997; Dolan et al., 1998). For smaller scales (less than metres), the influence of the recording tools, i.e. of their vertical resolutions, becomes predominant and the

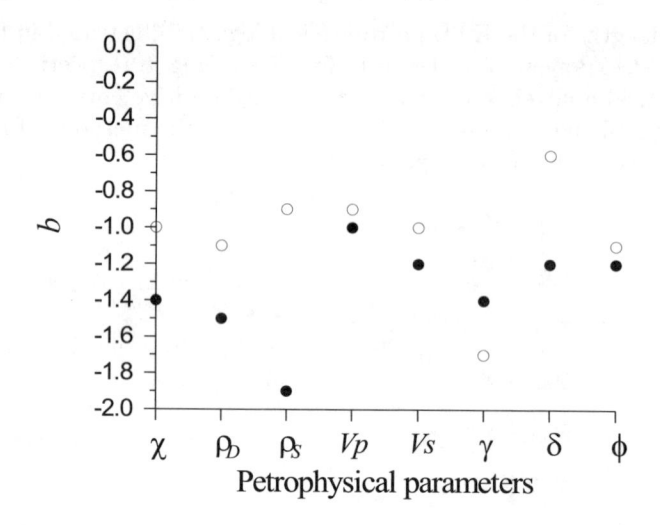

Fig. 7. Spectral exponents b of the KTB petrophysical properties from HB (black circles) and VB (white circles). One standard variation uncertainties are of the order of 10 %. Symbols as in Fig. 2.

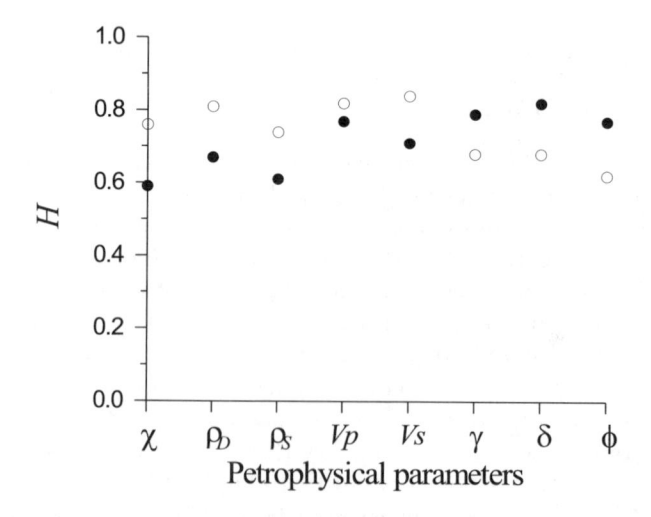

Fig. 8. Hurst exponents H of the KTB petrophysical properties from HB (black circles) and VB (white circles). One standard variation uncertainties are less than 5 %. Symbols as in Fig. 2.

Fractal Variability of Drilling Profiles   267

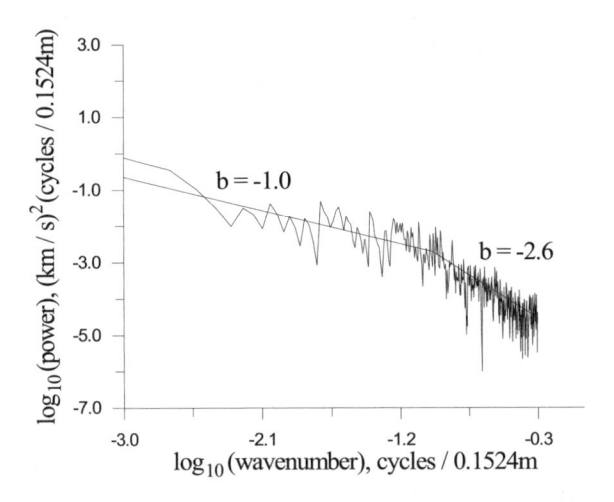

Fig. 9. Spectral density of the velocity log from the HB. A straight line with $b = 1.0$ is the best linear fit for a range of metres to hundreds of metres. At lower scale, larger b values are needed.

correlation can be better fitted by Brownian motion models (Fig. 9). This result also suggests that different b values should be taken for different scale ranges (Goff and Holliger, 1999; Marsan and Bean, 1999).

The R/S-analysis shows that the Hurst exponents, for depth scaling from $10z_\epsilon$ to $1{,}000z_\epsilon$, i.e. in a scale range from metres to kilometres, take values from 0.59 to 0.84 (Fig. 8), reflecting persistence within the corresponding signals. In other words, what we see on a smaller scale can be statistically found on a larger one.

Fractal properties characterise all the logs recorded in the KTB boreholes. Fractal dimensions D_F may be calculated by means of Hurst exponents (eq. 2) or recurring to spectral exponents (eq. 3). Following eq. 2, $1.16 < D_F < 1.41$; whereas according to eq. 3, $0.6 < D_F < 1.2$. The discrepancy between the two results are mainly related to the lack of Gaussian probability distributions and to the multiscaling properties of the KTB logs (Fig. 9; Leonardi and Kümpel, 1998; 1999; Goff and Holliger, 1999).

5 Conclusions

Several analysis techniques, like the study of autocorrelation functions, spectral methods and rescaling put into evidence that a wide range of petrophysical parameters from boreholes down to 9 km depth preserve their statistical properties at different scales, possibly from decimetres to kilometres. Scale invariance features can be quantified by means of spectral and Hurst exponents, or alternatively — following eqs. 2 and 3 — by fractal dimensions.

Scaling properties of rock parameters seem to be a universal feature of the upper crust, since results similar to those presented here were found not only for crystalline basements but also for logs recorded in volcanic and sedimentary formations (Pilkginton and Todoeschuck, 1990; Shiomi et al., 1997). One dimensional model for vertical log data may be extended to the two horizontal dimensions (Holliger et al., 1994; Wu et al., 1994) to get a 3D picture of the upper crustal petrophysical properties.

Since at least three independent parameters, such as average values μ, standard deviations σ and fractal dimensions D_F are required to describe the variability of log data, all of them should be used for modelling the geodynamical processes of the Earth's upper crust in a more realistic way.

References

Dolan, S. S., Bean, C. J. and Riollet, B., 1998. The broad-band fractal nature of heterogeneity in the upper crust from petrophysical logs. Geophys. J. Int., 132, 489–507.

Goff, A. J., Holliger, K., 1999. Nature and origin of upper crustal seismic velocity fluctuations and associated scaling properties: Combined stochastic analyses of KTB velocity and lithology logs J. Geophys. Res., 104, 13, 169-13, 182.

Gritto, R., Kaelin, B. and Johnson, L. R., 1995. Small scale crustal structure: consequences for elastic wave propagation at the KTB site. In: KTB Report 94-1 (K. Bram and J. K. Draxler eds.), 351–365.

Hergarten S., 2002. Fractals — Pretty pictures or a key to understanding geoprocesses?, this volume.

Hewett, T. A., 1986. Fractal distributions of reservoir heterogeneity and their influence on fluid transport. SPE Paper No. 15386, New Orleans.

Hirschmann, G., Duyster, J., Harms, U., Konntny, A., Lapp, M., de Wall, H., Zulauf, G., 1997. The KTB superdeep borehole: petrography and structure of a 9-km-deep crustal section. Geol. Rundsch., 86, Suppl., 3–14.

Holliger, K., 1996. Upper-crustal seismic velocity heterogeneity as derived from a variety of P-wave sonic logs. Geophys. J. Int., 125, 813–829.

Holliger, K., Carbonell, R., Levander, A., and Hobbs, R., 1994. Some attributes of wavefields scattered from Ivrea-type lower crust, Tectonophysics, 232, 267–279.

Hurst, H. E., 1957. A suggested statistical model of some time series which occur in nature. Nature, 180, 494.

Kneib, G., 1995. The statistical nature of the upper continental crystalline crust derived from in-situ seismic measurements. Geophys. J. Int., 122, 594–616.

Leary, P., 1991. Deep borehole evidence for fractal distribution of fractures in crystalline rock. Geophys. J. Int., 107, 615–627.

Leonardi, S., 1997. Variability of geophysical log data and the signature of crustal heterogeneities. An example from the German Continental Deep Drilling Program (KTB). Galunder Verlag, Wiehl.

Leonardi, S. and Kümpel, H.-J., 1996. Scaling behaviour of vertical magnetic susceptibility and its fractal characterisation: an example from the German Continental Deep Drilling Project (KTB). Geol. Rundsch., 85, 50–57.

Leonardi, S. and Kümpel, H.-J., 1998. Variability of geophysical log data and the signature of crustal heterogeneities at the KTB. Geophys. J. Int., 135, 964–974.

Leonardi, S. and Kümpel, H.-J., 1999. Fractal variability in superdeep borehole — Implications for the signature of crustal heterogeneities. Tectonophysics, 301, 173–181.

Mandelbrot, B. B. (1982): The Fractal Geometry of Nature. Freeman, San Francisco.

Marsan, D. and Bean, C. J., 1999. Multiscaling nature of sonic velocities and lithology in the upper crystalline crust: Evidence from the KTB main borehole. Geophys. Res. Lett., 26, 275–278.

Pilkington, M. and Todoeschuck, J. P., 1990. Stochastic inversion for scaling geology. Geophys. J. Int., 102, 205–217.

Pilkington, M., Gregotski, M. E., and Todoeschuck, J. P., 1994. Using fractal crustal magnetization models in magnetic interpretation. Geophys. Prosp., 42, 677–692.

Priestley, M. B., 1981. Spectral Analysis and Time Series. Academic Press, London.

Shiomi, K., Sato, H. and Ohtake, M., 1997. Broad-band power-law spectra of well-log data in Japan. Geophys. J. Int., 130, 57–64.

Turcotte, D. L. (1992): Fractals and Chaos in Geology and Geophysics. Cambridge Univ. Press, Cambridge.

Voss, R. F., 1988. Fractals in nature: from characterization to simulation. In: The Science of Fractal Image (H.-O. Peitgen and D. Saupe eds.). Springer, Berlin, 21–70.

Wu, R. S., Xu, Z. and Li, X.-P., 1994. Heterogeneity spectrum and scale-anisotropy in the upper crust revealed by the German continental deep-drilling (KTB) holes. Geophys. Res. Lett., 21, 911–914.

Is the Earth's Surface Critical? The Role of Fluvial Erosion and Landslides

Stefan Hergarten*

Section Geodynamics, Geological Institute, University of Bonn, Germany

Abstract. Fluvial erosion and landsliding are among the most important landform evolution processes; both result in fractal structures at the earth's surface. We discuss these two phenomena with respect to the framework of self–organised criticality (SOC) and point out some basic differences between both. While SOC provides an explanation of fractal landslide statistics, the role of SOC in fluvial erosion is still unclear.

1 Fractal Properties of the Earth's Surface

The enthusiasm about scale invariance in the 1980's did not stop at the earth's surface. Increasing computer power allowed the generation and visualisation of artificial fractal surfaces that looked very much like the earth's surface at first sight (Voss 1985, Feder 1988, Turcotte 1992, Hergarten 2002). Apparently, the richness of details contained in fractal surfaces and the hope of finding a unifying concept that is able to explain the shape of the earth's surface may have made scientists insensitive to the question whether the earth's surface is really fractal. Fractal analyses of the real surface often covered only a small range of scales and were focused on determining fractal dimensions rather than on justifying the hypothesis of scale invariance.

Fractal models of the earth's surface mostly compose objects of different sizes explicitly or are based on spectral methods (Feder 1988, Turcotte 1992). They are not able to give reasons why objects of different sizes occur, and the resulting surfaces cannot be interpreted as a result of a physical evolution process. Thus, they cannot deepen our understanding of Earth surface processes in any way. There have been attempts to generate fractal surfaces with a process-based background, such as a nonlinear Langevin equation suggested by Sornette and Zhang (1993) which was successful in generating at least fractal profiles in a physically reasonable way. However, it is well-known that the earth's relief is shaped by a variety of processes; much research has been done on these processes and on their dependence on climate and topography. From this point of view, describing the earth's surface with the help of just one simple equation or algorithm seems to be oversimplified. Thus, all these fractal approaches were never accepted as comprehensive models of the earth's surface.

Detailed analyses of fractal properties of the earth's surface confirmed these doubts. Evans and McClean (1995) showed that the earth's relief is in fact not

* hergarten@geo.uni-bonn.de

as fractal as it seemed to be, and that the deviations mirror the fact that earth surface processes depend on climate and topography. There is no doubt that there are some fractal elements in the earth's surface; so recent research focuses on the analysis of these components and on the question whether they tell us something about the underlying processes.

As already discussed in in this volume (Hergarten 2002), self-organised criticality (SOC) has become the seminal concept in this discussion. However, recognising SOC in earth surface processes is still difficult, and there are still many open questions. In the following, we try to shed some light on these aspects in the examples of fluvial erosion and landsliding, which are among the most important processes contributing to pattern formation at the earth's surface.

2 Fluvial Erosion

2.1 Fractal Properties of Drainage Networks

Scale-invariant properties of drainage networks were discovered long before the framework of fractals was established. In the 1940's, Horton found relations between the numbers of rivers of different orders that can — afterwards — be interpreted as some kind of scale invariance. However, this interpretation is not straightforward; so we focus on other fractal properties of drainage networks.

Hack's (1957) law is the most established fractal relation for drainage networks; it relates the sizes of drainage areas with river lengths. As long as braided rivers are not considered, river networks have a tree structure. This means that rivers join, but do not branch; from any point there is a unique way to the outlet of the drainage basin. For this reason, a unique drainage area can be assigned to each point of the river network; it consists of that part of the basin which delivers its discharge to the considered point. Let A be the size of the drainage area of an arbitrary point, and L be the length of the longest river within this area. Hack's law relates these quantities for different locations in a basin:

$$L \sim A^h. \qquad (1)$$

It should be mentioned that this is only a statistical relation; neither natural drainage basins nor the artificial networks obtained by the models discussed later obey Hack's law exactly. Observational values of the exponent h are between 0.52 and 0.6 (Maritan et al. 1996b, Rigon et al. 1996).

Hack's law does not immediately refer to a fractal in the sense discussed earlier (Hergarten 2002); especially, it does not provide a fractal distribution of drainage areas or river lengths, but only states a relation between both. But what makes Hack's law fractal? It is the deviation of h from 0.5, which is the expected exponent of non-fractal length-area relations.

Hack's law allows two basically different interpretations:

1. Rivers are convoluted at all scales, so that their lengths grow faster than the distances between source and end point.

2. Small and large drainage areas differ in shape, their length grows faster than their width. This means that drainage areas are not self-similar, but may be self-affine (Mandelbrot 1982, Feder 1988, Turcotte 1992).

The first explanation is quite straightforward and fits into our imagination of rivers quite well. However, detailed studies (Rigon et al. 1996, Maritan et al. 1996b) have shown that both effects are present and are comparable in their importance.

The power law distribution of drainage areas was discovered later (Rodriguez-Iturbe et al. 1992):

$$P(A) \sim A^{-\beta} \qquad (2)$$

where $P(A)$ is the probability of finding a drainage area larger than A. β is between 0.41 and 0.46 for real basins (Maritan et al. 1996b). This means that the drainage areas obey a fractal distribution with a fractal dimension $D = 2\beta$ lower than one.

2.2 Chance-Dominated Network Evolution Algorithms

Leopold and Langbein (1962) were the first to reproduce fractal properties of drainage networks with a model. In their approach, a rectangular region is divided regularly into quadratic tiles; precipitation acts uniformly on all tiles. In the first step, the source of the first river is randomly chosen. This river then performs some kind of random walk: The drainage direction from the source is randomly chosen among the connections to the four nearest neighbours; diagonal flow directions are not allowed. Then a random flow direction is chosen for this site under the restriction that backward flow is not allowed. The process ends when the river leaves the model area. After this, a second river is generated by the same procedure; it ends if the river leaves the area or if it joins another river. Rivers are generated until the whole area is drained; some more rules have to be imposed in order to avoid loops.

The model reproduces Hack's law (Eq. 1) with an exponent $h \approx 0.57$, which is in perfect coincidence with the mean value for real basins. However, the rivers are more convoluted than natural rivers (Rinaldo et al. 1998); and in return the elongation of the drainage areas (i.e., their self-affinity) is underestimated; thus, the good value of h is misleading. The model leads to fractal distributions of the drainage areas (Eq. 2), but the obtained exponent $\beta \approx 0.52$ is significantly too high.

Several scientists have followed the random walk idea, e.g., Scheidegger (1967). His model aims at simulating the drainage in alpine valleys, so he introduced preferential flow directions. This restriction immediately avoids the problem of loops, but the quantitative results are further away from nature than those of Leopold and Langbein; in the limit of infinite model size, $h = \frac{2}{3}$ and $\beta = \frac{1}{3}$ are obtained (Huber 1991, Maritan et al. 1996a). A review on further chance-dominated network evolution algorithms is given by Rinaldo et al. (1998).

2.3 Drainage Networks Emerging from Landform Evolution Models

Although chance-dominated algorithms that describe the growth of a network may provide qualitatively reasonable results, they cannot really deepen our understanding of network organisation. The major problem is that their statistical rules do not represent any knowledge about the flow of water and about sediment transport. Intuitively, we tend to prefer models where the drainage network emerges as a result of landform evolution. Especially in the 1990's, a large number of landform evolution models was developed. Among them are rather comprehensive models that distinguish between fluvial and hillslope processes (e. g., Howard 1994, Kooi and Beaumont 1996, Tucker and Bras 1998), but also simplified models that focus on fractal properties of the drainage pattern (Takayasu and Inaoka 1992, Inaoka and Takayasu 1993, Kramer and Marder 1992).

The question of scales in sediment transport processes, especially on the gap between the grain sizes and the — in general much larger — model scale that is determined by the macroscopic geometry of the system, was discussed earlier in this volume (Hergarten et al. 2001). In a numerical model, the surface heights are mostly defined on a discrete grid which determines the spatial resolution of the model. For instance, if we represent a basin of $100 \times 100 \, \text{km}^2$ by a grid of one million nodes, we obtain a spatial resolution of 10 m; this only allows an appropriate representation of rivers that are at least several ten meters wide. Hence there is a second gap between scales in this kind of model — the spatial resolution of the model does not even allow to resolve the structures that govern the macroscopic flow field. Thus, such a large scale landform evolution model must refer to a description of rivers as linear (one-dimensional) objects which may be characterised by additional parameters that describe the cross section and the properties of the river bed.

It is convenient to align the rivers to the grid on which the surface heights are defined. This means that only a discrete set of flow directions is allowed, i. e., flow from any nodes to their neighbours. The most common, regular grid topologies used in this context are quadratic (Kramer and Marder 1992, Rinaldo et al. 1993) and hexagonal (Takayasu and Inaoka 1992, Inaoka and Takayasu 1993). On a hexagonal lattice, each node in the bulk has six neighbours of equal distance, so that the flow directions are discretised in 60° steps. On a quadratic grid, each node in the bulk has four direct neighbours; in order to reduce grid anisotropy, the four diagonal neighbours are included into the neighbourhood in contrast to the random walk approach of Leopold and Langbein (1962) discussed in the previous section, so that there are eight discrete flow directions. Flow routing between nodes is then straightforward; each node delivers its discharge to that neighbour where the steepest downslope gradient occurs.

In order to generate any discharge, precipitation has to be regarded. Let us cover the model area with tiles of equal size in such a way that every tile contains exactly one node, and let us assume that each node collects the precipitation from its tile. Then, the discharge at any node — measured in units of the rainfall per tile — coincides with the drainage area of this node — measured in units of the tile area. Thus, we do not need to distinguish between discharge and drainage areas in the following.

Erosion and deposition in fluvial systems are governed by the their capability of detaching particles from the river bed and by their transport capacity. Depending on discharge, slope, and sediment properties, either or both may become the limiting factor to erosive action. Let us, for simplicity, focus on the detachment-limited case; this means that all particles that are once detached leave the basin through the outlet. In this case, balancing the downstream sediment transport can be replaced by a local erosion rule for the surface height H_i at the node i:

$$\frac{\partial H_i}{\partial t} = -E(q_i, \Delta_i), \qquad (3)$$

where q_i is the discharge at the node i, Δ_i the (steepest) downslope gradient, and E the rate of erosion.

Expressions for the erosion rate $E(q_i, \Delta_i)$ can be partly derived from hydrodynamic principles; a common approximation is (Rinaldo et al. 1998, Sinclair and Ball 1996):

$$\frac{\partial H_i}{\partial t} = -E(q_i, \Delta_i) \sim -q_i^\mu \Delta_i^\nu. \qquad (4)$$

The parameters μ and ν are positive, and $\nu \approx 2\mu$. Following Takayasu and Inaoka (1992) we choose $\mu = 1$ and $\nu = 2$. In this case, an implicit time discretisation of the slope gradients Δ_i in Eq. 4 can be performed if the discharges are once computed. This avoids instabilities that may propagate in upstream direction, so that we do not need to limit the erosion rate at large discharges and slopes as Takayasu and Inaoka did. Introducing non-dimensional variables, the proportionality in Eq. 4 can be replaced by an identity. It should be mentioned that Eq. 4 leads to a permanent decrease of surface heights; in order to obtain a more realistic behaviour, tectonic uplift should be regarded, too. We introduce this phenomenon by adding a given uplift rate R, so that Eq. 4 is enlarged to

$$\frac{\partial H_i}{\partial t} = R - q_i \Delta_i^2. \qquad (5)$$

In addition, the surface heights can be scaled in such a way that $R = 1$.

Eq. 5 requires additional boundary conditions at these nodes where water leaves the model area. In the following, we only consider domains with a single outlet which is located at a corner. However, the number of outlets and their location is not a crucial point of the model. We assume an equilibrium between erosion and tectonic uplift at the outlet, so that its height remains constant.

Local depressions of the surface, which mostly result in lakes in reality, require a special treatment, too. Obviously, the straightforward idea of flow routing in direction of the steepest descent is not applicable here. One could think of simply allowing a flow in uphill direction in this case, but this may result in loops in the drainage network. Thus, in a realistic model these lakes must be treated in a more sophisticated manner, e. g. by filling them up with water

Fig. 1. Evolution of a drainage network. Only sites with a drainage area larger than 10 tiles are assumed to be channelised, smaller flows are not plotted. For clarity, the line widths are proportional to the fourth root of the discharge, so that the differences between large and small rivers are less pronounced than in nature. The lower left network describes the stage where the network first drains the whole area, i.e., immediately after the last lake has vanished; the lower right one shows the final state $(t \to \infty)$.

Is the Earth's Surface Critical? 277

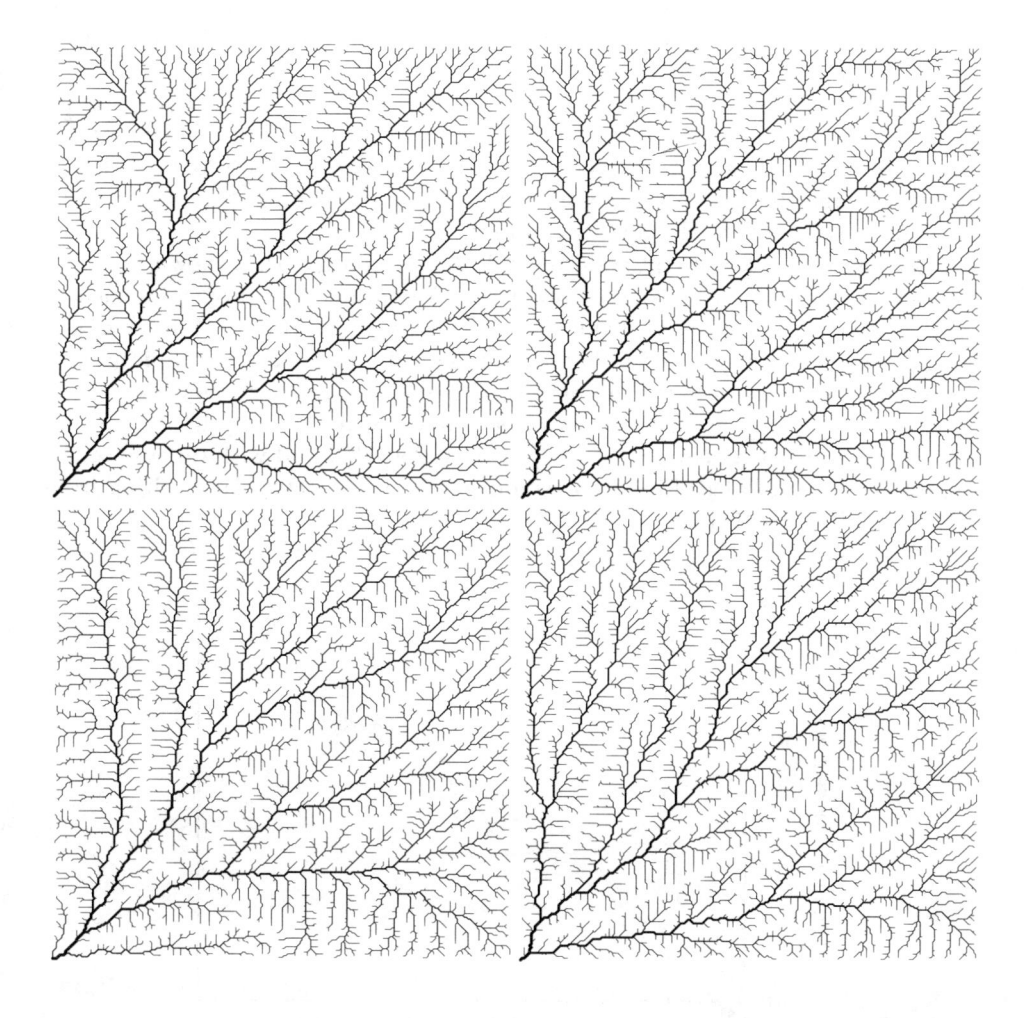

Fig. 2. Four examples of equilibrated drainage networks with random initial conditions. Only sites with drainage areas larger than 10 tiles are assumed to be channelised, smaller flows are not plotted. For clarity, the line widths are proportional to the fourth root of the discharge, so that the differences between large and small rivers are less pronounced than in nature.

(Takayasu and Inaoka 1992, Inaoka and Takayasu 1993). However, simulations show that all lakes vanish through time under the given boundary conditions; so we do not take care of lakes and simply assume that the water vanishes at local minima of the surface without any erosive action.

Fig. 1 shows a network evolution on a quadratic 256×256 grid. The initial surface was nearly flat with small, uncorrelated random values of H_i. The outlet is located at the lower left corner of the domain.

The drainage network grows like a tree, starting at the outlet. This upstream growth corresponds to the well-known fact that disturbances in fluvial systems preferably proceed in upstream direction. Nevertheless, the lower parts of Fig. 1 illustrate that there is still a significant reorganisation after the pure growth process ceases, until finally a steady state is achieved.

Because of the small statistics, a single simulation like that shown in Fig. 1 is not sufficient for recognising or predicting fractal network properties such as Hack's law (Eq. 1) or the distribution of drainage areas (Eq. 2). Larger model areas provide better statistics, but the numerical effort increases rapidly with increasing grid size because the high discharges require short time steps, and it takes a long time until the surface becomes stationary. For this reason, an ensemble average over several simulations on a 256×256 grid with random initial conditions is performed. Fig. 2 shows four equilibrated networks resulting from different initial conditions.

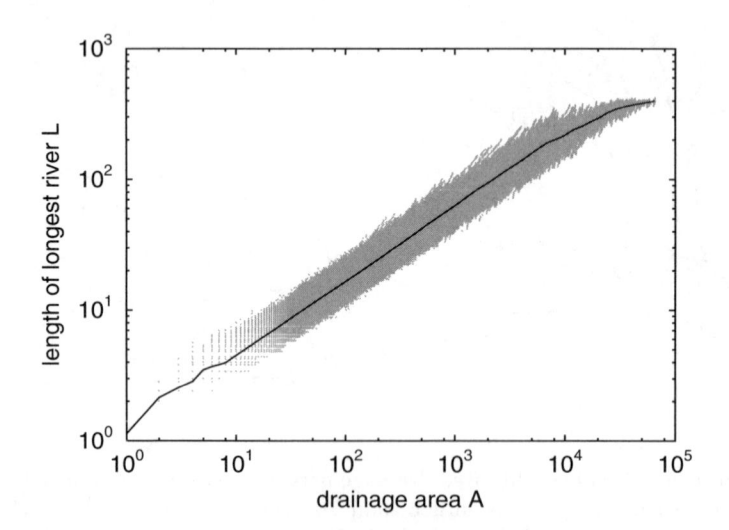

Fig. 3. Length of the longest rivers within a given drainage area versus drainage area, obtained from 100 realisations on a 256×256 grid. The dots represent the individual values obtained from all $100 \times 256 \times 256$ points, the solid line is an average.

Fig. 3 shows the length of the longest rivers versus the drainage areas, obtained from 100 realisations. The simulations reproduce Hack's law (Eq. 1) quite well with an exponent of 0.57 over three decades in area. The dots illustrate the

Is the Earth's Surface Critical?

statistical character of Hack's law; the river lengths scatter by half an order of magnitude for a given drainage area.

Fig. 4 shows the cumulative probability distribution of the drainage areas, again obtained from 100 realisations. The plots roughly show power law behaviour, but deviate at large and at small areas. The deviations at small areas are not surprising because the grid structure with only eight discrete flow directions certainly affects network organisation at small scales. On the other hand, comparing the curves for different model sizes suggests that the deviations at large areas result from the finite model size. Thus, one should not simply fit a power law to one of the curves, but extrapolate the straight part (that between 10^1 and 10^2 for the 256×256 grid) in order to get an estimate of the exponent in the limit of infinite model size. For the model discussed here, this leads to an exponent $\beta = 0.46$ in the discharge distribution (Eq. 2), which is at the upper limit of the range observed in nature.

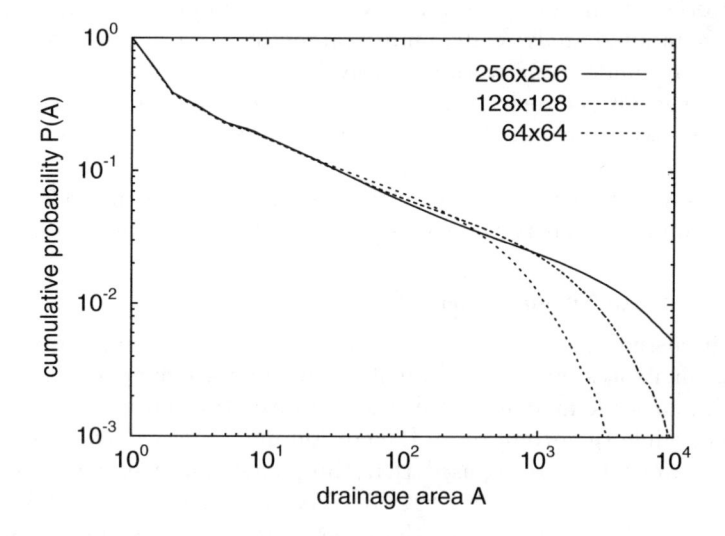

Fig. 4. Cumulative size distribution of the drainage areas, obtained from 100 realisations on grids of different sizes.

The results obtained from this landform evolution model are significantly better than those from the random walk approaches discussed before. This is not really surprising because local erosion rules are closer to our intuitive understanding of drainage network evolution than random walk is. However, if we keep in mind that the model is very simple and considers just one aspect of complex landform evolution, it is surprising that the fractal network properties are so close to those observed in nature.

2.4 Optimal Channel Networks

In the previous section we have learned that simple landform evolution models may result in drainage networks with quite realistic fractal properties. However, our understanding of the process would be deepened if the final, stationary network could be characterised without considering the whole evolution process.

A characterisation of steady states is possible in many physical systems; let us, for instance consider a ball being placed at an arbitrary position on a surface. No matter how complicated the motion of the ball is, we know that it will finally end at a local minimum of the surface, provided that gravity is the driving force.

Unfortunately, the steady state version of Eq. 5 (left hand side vanishes) is not sufficient for such a characterisation; it allows much more solutions than those with realistic fractal properties. Thus, an additional criterion is needed for characterising realistic networks.

Rinaldo et al. (1992, 1993) proposed the total energy dissipation of the water flow through the network as a criterion. They claimed that those networks are preferred where the energy expenditure is minimal and introduced the term "optimal channel networks" (OCNs) for networks obeying this condition. However, networks which are optimal in the sense of minimum energy expenditure cannot be computed analytically; numerical methods are required for optimising an initial (e. g. random) network successively.

Apparently, OCNs reproduce the fractal properties of drainage networks gradually better than landform evolution starting at a random surface (Rinaldo et al. 1998); introducing random components (Caldarelli et al. 1997) improves the results further. Nevertheless, theoretical basis and relevance of the principle of minimum energy expenditure in landform evolution are still unclear.

2.5 Are Drainage Networks Critical?

We have discussed three different model concepts for predicting fractal properties of drainage networks; all of them finally result in stationary drainage patterns. In the random walk models and in the optimisation approach, stationarity is pre-defined, while the evolution in the erosion models ceases over long times.

It was immediately recognised that this behaviour does not meet the criteria of self-organised criticality (SOC) (Bak et al. 1987, 1988, Bak 1996). As already discussed in this volume (Hergarten 2002), SOC requires that the system tends towards a state which is only steady in the long term average, and where fluctuations of all sizes occur. The fractal size distribution of these events is the spatial fingerprint of SOC systems, while the fractal properties of the drainage networks discussed here are static. Takayasu and Inaoka considered the tree-like growth observed in their erosion model (Takayasu and Inaoka 1992, Inaoka and Takayasu 1993) as a new type of SOC — "spatial SOC". However, discussion (Sapozhnikov and Foufoula-Georgiou 1996) has shown that this behaviour is far away from SOC in the original sense, and that considering it a new type of SOC does not provide any new insights.

From this point of view, drainage networks are not self-organised critical; but there are still two arguments that suggest that this is not the whole story:

1. The earth's surface is not in a steady state but undergoes permanent changes.
2. As mentioned above, the fractal properties of those networks obtained from optimising energy expenditure seem to be better than those obtained from landform evolution models with random initial conditions. Successive optimisation means that the actual network is disturbed, and the success of the disturbance is assessed with respect to minimum energy expenditure. This could mean that the landform evolution process starting at a random surface does not allow enough optimisation. The incision of the rivers during their growth from the outlet might be too strong to allow a sufficient reorganisation afterwards.

Hence the processes that lead to a permanent alteration of the earth's surface in nature may be a key to understanding network evolution within the framework of SOC. These processes may be

- meandering,
- variable tectonic uplift rates or
- variations in erodibility due to different layers of rock.

However, it is not clear at all whether these processes result in fluctuations that obey fractal statistics, as required for SOC, and whether these fluctuations have anything to do with the observed fractal properties of drainage networks.

3 Landslides

3.1 The Fractal Frequency-Magnitude Relation of Landslides

Landslides are another example of fractal properties at the earth's surface. Especially in the 1990's, several studies on the frequency-magnitude relation of landslides were carried out (Hovius et al. 1997, Pelletier et al. 1997, Sugai et al. 1994, Yokoi et al. 1995); most of them resulted in scale-invariant distributions of the affected areas, although the range of scales covered by a power law distribution is narrow. Fig. 5 shows results obtained by Hovius et al. (1997) from mapping about 5,000 landslides in New Zealand. The logarithmically binned probability density can be fit by a power law with an exponent $b \approx 1.16$. Because of the logarithmic bin widths, this exponent coincides with the exponent b of the cumulative size distribution:

$$P(A) \sim A^{-b}, \tag{6}$$

where $P(A)$ is the probability that the area affected by a landslide is larger than A. Most studies result in exponents that are slightly larger than one; but Pelletier et al. (1997) also report on distributions with exponents up to two.

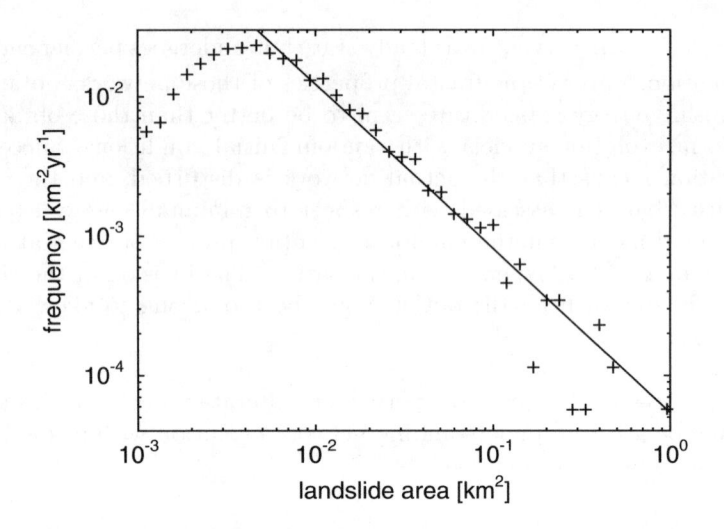

Fig. 5. Frequency-magnitude relation obtained from landslide mapping for a part of a mountain belt in New Zealand (Hovius et al. 1997). The distribution is non-cumulative, but represents a logarithmically binned probability density. The straight line shows a power law with an exponent of 1.16.

3.2 Are Landslides like Sandpile Avalanches?

Obviously, the size distributions of landslide areas resemble dynamic fractal properties as required for SOC. Thus, explaining the fractal frequency-magnitude relation of landslides as a SOC phenomenon is tempting.

Landsliding is a dissipative phenomenon because material moves downslope, except for the runout of rapid flows. Thus, SOC behaviour can only be expected if the long term driving forces that steepen the surface are considered, too; otherwise a landslide model must result in a quite flat equilibrium surface in the end. Depending on the scales under consideration, these driving processes may be tectonic uplift or erosion, for instance the incision of a river at the foot of a slope. In the following, we focus on the scale of individual slopes and assume fluvial erosion to be the major driving process (Fig. 6).

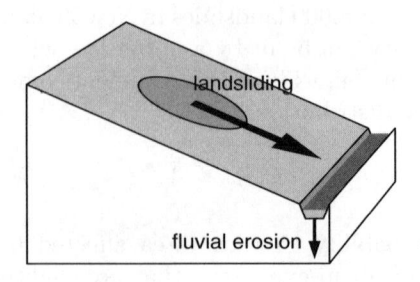

Fig. 6. Sketch of fluvial erosion as the long term driving force to landsliding.

Is the Earth's Surface Critical?

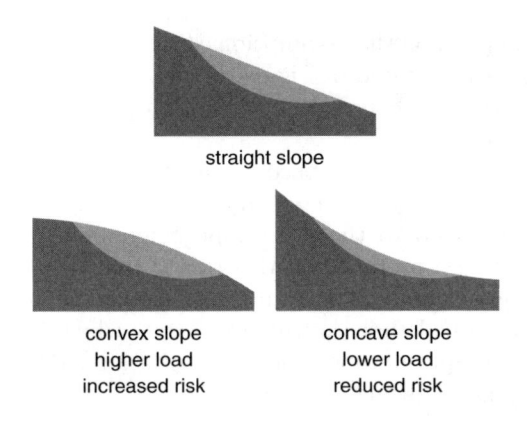

Fig. 7. Effect of profile curvature on landslide probability.

Obviously, the probability of landslide occurrence depends strongly on the surface gradient; but as Fig. 7 illustrates, the curvature may play an important part, too, because it affects the load acting on the slip surface. The effect of plan curvature (in direction perpendicular to the slope gradient) may differ significantly from that of profile curvature (taken in direction of the slope gradient) because plan concavity focuses surface runoff and subsurface flow. A straightforward idea is combining gradient (in main slope direction) and curvature linearly to an expression for the landslide probability u:

$$u(x_1, x_2) = \alpha \frac{\partial}{\partial x_1} H(x_1, x_2) - \beta_1 \frac{\partial^2}{\partial x_1^2} H(x_1, x_2) - \beta_2 \frac{\partial^2}{\partial x_2^2} H(x_1, x_2). \quad (7)$$

$H(x_1, x_2)$ denotes the surface height; we assume that x_1 is the main upslope direction; α, β_1, and β_2 are parameters. While α and β_1 are positive in general, β_2 may be negative according to the effects of drainage discussed above. Let us — for simplicity — skip the difference between β_1 and β_2 and replace both by a single parameter β.

If we switch to a discrete formulation on a quadratic grid with unit spacing, Eq. 7 turns into:

$$u_i = \alpha(H_i - H_k) + \beta \sum_{j \in N(i)} (H_i - H_j), \quad (8)$$

where u_i and H_i are landslide probability and surface height at the node i, $N(i)$ is the nearest neighbourhood of the node i (consisting of four nodes in the bulk), and k is the downslope neighbour of the node i.

As mentioned above, the long term driving forces such as fluvial erosion, which lead to continuous, slow changes in the surface heights, must be regarded. Deriving Eq. 8 with respect to the time leads to an expression for the slow changes in landslide probability at the node i through time:

$$\frac{\partial}{\partial t} u_i = r_i, \quad (9)$$

where the rates r_i depend on the assumptions on the driving forces. If only fluvial erosion is assumed as illustrated in Fig. 6, r_i vanishes everywhere, except at the lower boundary. If we — in contrast — consider a plane that is slowly tilted, r_i is the same at all nodes.

Let us assume that the node i becomes unstable as soon as $u_i \geq 1$. In this case, material is removed from this node until $u_i = 0$. For simplicity, the unstable material is not transferred to the downslope neighbours, but simply removed from the system. An interpretation of this rule is that unstable material moves to the foot of the slope rapidly and is immediately carried away by the river.

It should be mentioned that the rule $u_i \to 0$ — which is somehow arbitrary — introduces an assumption about the depth of a landslide; inserting this rule into Eq. 8 leads to the following rule for H_i:

$$H_i \to H_i - \frac{u_i}{\alpha + 4\beta}. \tag{10}$$

Thus, the depth of the slide is pre-defined to be $\frac{u_i}{\alpha+4\beta}$. However, assuming that u_i does not decrease to zero, but to another (positive or negative) value, alters the depth without any further effects on the model.

The relaxation of a node i ($u_i \to 0$) does not only affect the node itself, but also alters the landslide probability at its neighbours. Writing Eq. 8 for the neighbours of the node i and inserting Eq. 10 leads to the following relaxation rules:

$$u_i \to 0, \quad u_j \to u_j + a_{ij} u_i \quad \text{for} \quad j \in N(i), \tag{11}$$

where

$$a_{ij} = \begin{cases} \frac{\alpha+\beta}{\alpha+4\beta} & \text{if } j \text{ is the upslope neighbour of } i \\ \frac{\beta}{\alpha+4\beta} & \text{else} \end{cases}. \tag{12}$$

Obviously, $\sum_{j \in N(i)} a_{ij} = 1$ in the bulk; this means that the landslide probability of the unstable node i is entirely transferred to its neighbourhood. Such a relaxation without any loss in sum is called "conservative relaxation".

Except for the lower boundary that is defined by a river at the slope foot, posing reasonable long term boundary conditions is difficult. Periodic boundary conditions provide an easy way out of this problem; Fig. 8 shows two variants. The black lines illustrates which sides are linked with each other through the periodic boundary conditions; both variants are equivalent.

This cellular automaton is quite similar to the seminal SOC model — the Bak-Tang-Wiesenfeld (BTW) model (Bak et al. 1987, 1988). It can be interpreted as some kind of idealised sandpile (e. g. Jensen 1998); its properties were discussed earlier in this volume (Hergarten 2002). Apart form the anisotropic relaxation rule (Eq. 11), the model coincides with the conservative limiting case of the Olami-Feder-Christensen (OFC) model (Olami et al. 1992) which is used for predicting the fractal frequency-magnitude relation of earthquakes.

Thus, it is not surprising that our landslide model exhibits SOC behaviour as the BTW model and the OFC model do; as long as the anisotropy is not too strong, it does not affect the results. So this simple model is able to explain

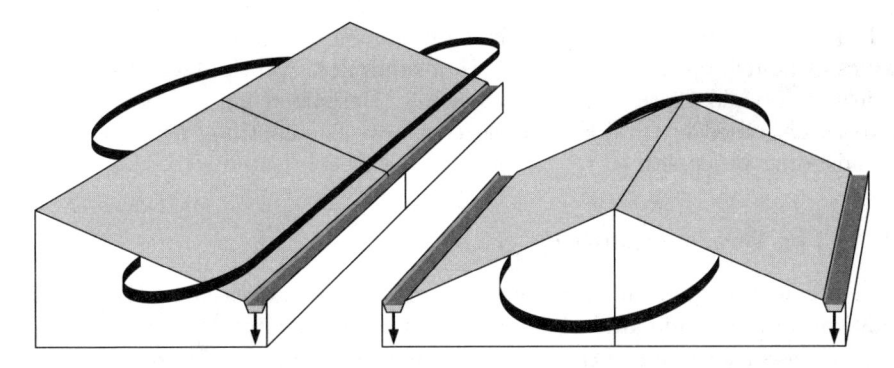

Fig. 8. Boundary conditions for a slope with a river at the slope foot.

the fractal frequency-magnitude relation of landslides qualitatively in analogy to sandpiles.

It is not surprising, too, that the exponent of the fractal frequency-magnitude relation is close to that of the BTW and conservative OFC models. Both yield probability densities with an exponent of about one, which means that the exponent of the cumulative size distribution b is close to zero. Compared with the observed values of about one or greater, the exponent from the model is much too low, so that the predicting capabilities of this model are poor in a quantitative sense.

One could think about modifying the model in order to obtain higher exponents, for instance in the following ways:

– Combining slope and curvature (Eq. 7 respectively. Eq. 8) nonlinearly.
– Assuming that u_i does not decrease to zero in Eq. 11.
– Moving unstable material towards the downslope neighbours instead of removing it immediately.

The second and third ideas do not impose any significant change; the model remains similar to the conservative OFC model. Apparently, a nonlinear combination of slope and curvature makes the problem still worse.

Although this discrepancy in the exponents might be considered a minor problem from a qualitative point of view, it should be emphasised that b determines the relation between small and large events (Hovius et al. 1997). Does the small number of large events contribute more than the large number of small events or vice versa? Let us consider the contribution of the landslides with areas between A_1 and A_2 to the total area of all landslides. As shown earlier for the example of earthquakes (Hergarten 2002), this contribution is proportional to $|A_2^{1-b} - A_1^{1-b}|$. Thus, the minority of large events governs the distribution for $b < 1$, while the majority of small events is more important if $b > 1$. Finally, $b \approx 1$ is an interesting situation where the contributions of large and small events are of comparable importance. Although the exact value of b may be of minor relevance, the question whether b is larger, smaller or about one is even qualitatively important.

From this point of view, there is a fundamental difference between the dynamics of landslides and that of idealised sandpiles, and the fact that there are models that yield more realistic exponents (Hergarten and Neugebauer 1998, 2000) as discussed in the following section tells us something about the nature of landsliding processes.

3.3 A Two-Variable Approach

In the previous section we have seen that landslide dynamics cannot be explained quantitatively by some kind of sandpile dynamics based on the geometric properties of the surface. Thus, there must be another component beside the surface geometry that contributes to landslide probability considerably.

It is well-known that the mechanical properties of soil and bedrock and the water pressure are important factors in assessing landslide risk; so integrating them into the model is a quite straightforward idea. Water pressure varies rapidly at the time scales of weeks and months; hence we suspect that changes in water pressure rather act as a trigger than impose a long term increase of landslide probability. For this reason we focus on time-dependent weakening, for instance as a result of weathering.

In order to introduce a time-dependent weakening in a lumped way, we introduce a second variable v_i and assume that the landslide probability now depends on both the geometric component u_i (as defined in Eq. 8) and v_i. The time dependence of weakening is incorporated by a linear increase of v_i through time. It seems to be reasonable that the rate of increase is the same at all nodes, so that the time scale can be chosen in such a way that the rate is one. Thus, the long term driving rule of the model (Eq. 9) is enlarged to:

$$\frac{\partial}{\partial t} u_i = r_i, \quad \frac{\partial}{\partial t} v_i = 1. \tag{13}$$

We assume that the site i becomes unstable as soon as $u_i v_i \geq 1$. For some reasons, this product approach is better than, for instance, adding u_i and v_i (Hergarten and Neugebauer 2000).

As already mentioned, unstable material is immediately removed from the system. Thus, v_i is reset to zero in case of instability without transferring it to the neighbourhood, so that the relaxation rule (Eq. 11) is enlarged to:

$$u_i \to 0, \quad v_i \to 0, \quad u_j \to u_j + a_{ij} u_i \quad \text{for} \quad j \in N(i). \tag{14}$$

General properties of this model in the isotropic case ($a_{ij} = \frac{1}{4}$ for $j \in N(i)$) with open boundaries were discussed by Hergarten and Neugebauer (2000). After some transient period, the system approaches a quasi-steady state where the size distribution of the avalanches shows power law behaviour (Eq. 6), i.e. a critical state. Thus, the model exhibits SOC behaviour; the exponent b is found to be close to one.

Anisotropy and boundary conditions affect the results slightly. Fig. 9 shows the cumulative size distribution of the events for $\alpha = 1$ and $\beta = 2$ with the boundary conditions from Fig. 8. Comparing the results of simulations for different grid sizes shows a significant finite-size effect that makes fitting a power law distribution difficult. However, plotting the probability density instead of the distribution and extrapolating to infinite model sizes shows that the size distribution tends towards a power law with an exponent $b = 0.85$.

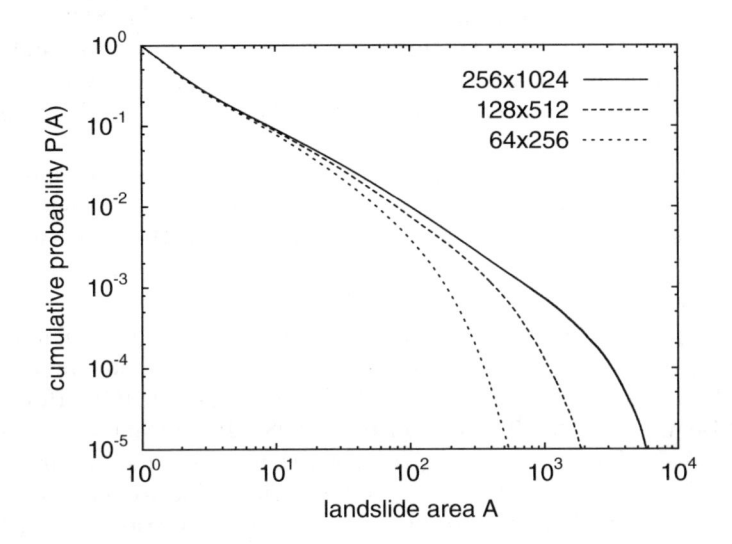

Fig. 9. Cumulative size distribution of the landslide areas, obtained from 2×10^7 events each (after the critical state has been reached) for different model sizes.

Compared to the first approach that only considers surface geometry, this two-variable approach leads to a quite realistic size distribution of the landslides. From this we conclude that the fractal frequency-magnitude relation of landslides can be explained within the framework of SOC by the dynamics of at least two system components which may be interpreted in terms of surface geometry and soil or bedrock properties.

4 Conclusions and Open Questions

After discussing two landform evolution processes — fluvial erosion and landsliding — with respect to SOC, we should now come back to the question about criticality of the earth's surface. It should be clear that actually nobody is able to answer this question: Landform evolution consists of a variety of processes, and we are far away from understanding all the complex interactions between them.

Thus, we can only elucidate the relevance of the SOC concept to individual earth surface processes, perhaps coupled with one more process in order to regard long term driving forces. Fluvial erosion and landsliding are just two examples, but were not chosen without reasons: In both, fractal distributions of object sizes are observed in nature, but the results are non-unique: While the concept of SOC provides an explanation for scale-invariant landslide statistics as dynamic fractals, the fractal nature of drainage patterns seems to be of static origin, although we may hope to recognise SOC in drainage network evolution, too, quite soon.

In order to obtain a more comprehensive view of the role SOC in landform evolution, soil erosion caused by surface runoff of water and by wind must be considered, too. Apparently, erosion rills have fractal properties over a limited range of scales, but these scales are smaller than that of drainage basins or landslides.

Finally, the question remains whether the process or only the model is self-organised critical or not. It is easy to say that only the model is critical, and that the model has nothing to do with the real world. However, one can also argue the opposite way: If the model in not critical, but we are convinced that nature is, we can say that the model is not appropriate. But when is a model appropriate? There are in fact many landform evolution models that capture more processes and describe them in more detail than the two models discussed here (e.g., Coulthard et al. 1998, Densmore et al. 1998, Howard 1994, Kooi and Beaumont 1996, Tucker and Bras 1998). If we want to predict quantities such as erosion rates under given conditions, these models are all better than the simple ones discussed here. But are they able to generate the large statistics required for investigating a phenomenon with respect to SOC? Does their accuracy not depend crucially on the availability and accuracy of parameter values? And do we really understand where a certain effect comes from if such a variety of processes is considered? Thus, the question whether a model is appropriate cannot simply be answered by checking if all the processes which seem to be important are regarded, but depends crucially on the purpose.

Regardless of all these open questions, at least on the landscape scale models seem to be the only link between observations and rather abstract concepts such as SOC. SOC cannot be observed directly because most processes cannot be resolved with respect to time with adequate accuracy. Laboratory experiments, e.g., of braided rivers (Sapozhnikov and Foufoula-Georgiou 1997), may become a step between nature and theoretical models. However, observing system behaviour may still be difficult in laboratory; and we must not forget that, strictly speaking, a laboratory experiment is a model, too.

In summary, there are still many open questions concerning the role of SOC in earth surface processes, but this seminal concept seems to be on its way to occupy its field in geomorphology.

References

Bak. P., C. Tang, and K. Wiesenfeld (1987): Self-organized criticality. An explanation of 1/f noise. Phys. Rev. Lett., 59, 381–384.

Bak. P., C. Tang, and K. Wiesenfeld (1988): Self-organized criticality. Phys. Rev. A, 38, 364–374.

Bak. P. (1996): How Nature Works: the Science of Self-Organized Criticality. Copernicus, Springer, New York.

Caldarelli, G., A. Giacometti, A. Maritan, I. Rodriguez-Iturbe and A. Rinaldo (1997): Randomly pinned landform evolution. Phys. Rev. E, 55, R4865–4868".

Densmore, T. J., M. J. Kirkby, and M. G. Macklin (1998): Non-linearity and spatial resolution in a cellular automaton model of a small upland basin. Hydrology and Earth System Sciences, 2(2–3), 257–264.

Densmore, A. L., M. A. Ellis, and R. S. Anderson (1998): Landsliding and the evolution of normal-fault-bounded mountains. J. Geophys. Res. 103(B7), 15203–15219.

Evans, I. S., and C. J. McClean (1995): The land surface is not unifractal: variograms, cirque scale and allometry. Z. Geomorph. N. F., Suppl. 101, 127–147.

Feder, J. (1988): Fractals. Plenum Press, New York.

Hack, J. T. (1957): Studies of longitudinal profiles in Virginia and Maryland. US Geol. Survey Prof. Paper 294–B. Washington D. C.

Hergarten, S. (2002): Fractals — just pretty pictures or a key to understanding earth processes? In: Dynamics of Multiscale Earth Systems, edited by Horst J. Neugebauer and Clemens Simmer, this volume.

Hergarten, S., and H. J. Neugebauer (1998): Self-organized criticality in a landslide model. Geophys. Res. Lett., 25(4), 801–804.

Hergarten, S., and H. J. Neugebauer (2000): Self-organized criticality in two-variable models. Phys. Rev. E, 61(3), 2382–2395.

Hergarten, S., M. Hinterkausen, and M. Küpper (2001): Sediment transport — from grains to partial differential equations. In: Dynamics of Multiscale Earth Systems, edited by Horst J. Neugebauer and Clemens Simmer, this volume.

Hovius, N., C. P. Stark, and P. A. Allen (1997): Sediment flux from a mountain belt derived by landslide mapping. Geology, 25(3), 231–234.

Howard, A. D. (1994): A detachment-limited model for drainage basin evolution. Water Resour. Res., 30(7), 2261–2285.

Huber, G. (1991): Scheidegger's rivers, Takayasu's aggregates, and continued fractions. Physica A, 209, 463–470.

Inaoka, H., and H. Takayasu (1993): Water erosion as a fractal growth process. Phys. Rev. E, 47, 899–910.

Jensen, H. J. (1998): Self-Organized Criticality. Emergent Complex Behaviour in Physical and Biological Systems. Cambridge Lecture Notes in Physics, 10, Cambridge University Press, Cambridge.

Kooi, H., and C. Beaumont (1996): Large-scale geomorphology: Classical concepts reconciled and integrated with contemporary ideas via a surface process model. J. Geophys. Res., 101(B2), 3361–3386.

Kramer, S., and M. Marder (1992): Evolution of river networks. Phys. Rev. Lett., 68, 381–384.

Leopold, L. B., and W. B. Langbein (1962): The concept of entropy in landscape evolution. US Geol. Survey Prof. Paper 500–A. Washington D. C.

Mandelbrot, B. B. (1982): The Fractal Geometry of Nature. Freeman, San Francisco.

Maritan, A., F. Colaiori, A: Flammini, M. Cieplak, J. R. Banavar (1996a): Universality classes of optimal channel networks. Science, 272, 984–986.

Maritan, A., A. Rinaldo, R. Rigon, A. Giacometti, and I. Rodriguez-Iturbe (1996b): Scaling laws for river networks. Phys. Rev. E, 53(2), 1510–1515.

Olami, Z., H. J. Feder, and K. Christensen (1992): Self-organized criticality in a continuous, nonconservative cellular automaton modeling earthquakes. Phys. Rev. Lett., 68, 1244–1247.

Pelletier, J. D., B. D. Malamud, T. Blodgett, and D. L. Turcotte (1997): Scale-invariance of soil moisture variability and its implications for the frequency-size distribution of landslides. Engin. Geol., 48, 255–268.

Rigon, R., I. Rodriguez-Iturbe, A. Maritan, A. Giacometti, D. G. Tarboton, and A. Rinaldo (1996): On Hack's law. Water Resour. Res., 32(11), 3367–3374.

Rinaldo, A., I. Rodriguez-Iturbe, R. L. Bras, E. J. Ijjasz-Vasquez, and A. Marani (1992): Minimum energy and fractal structures of drainage networks. Water Resour. Res., 28(9), 2182–2195.

Rinaldo, A., I. Rodriguez-Iturbe, R. Rigon, and R. L. Bras (1993): Self-organized fractal river networks. Phys. Rev. Lett., 70, 822–825.

Rinaldo, A., I. Rodriguez-Iturbe, and R. Rigon (1998): Channel networks. Annu. Rev. Earth Planet. Sci., 26, 289–327.

Rodriguez-Iturbe, I., E. J. Ijjasz-Vasquez, and R. L. Bras (1992): Power law distributions of discharge mass and energy in river basins. Water Resour. Res., 28(4), 1089–1093.

Sapozhnikov, V. B., and E. Foufoula-Georgiou (1996): Do the current landscape evolution models show self-organized criticality? Water Resour. Res., 32(4), 1109–1112.

Sapozhnikov, V. B., and E. Foufoula-Georgiou (1997): Experimental evidence of dynamic scaling and indications of self-organized criticality in braided rivers. Water Resour. Res., 33(8), 1983–1991.

Scheidegger, A. E. (1967): A stochastic model for drainage patterns into an intramontane trench. Bull. Assoc. Sci. Hydrol., 12, 15–20.

Sinclair, K. C., and R. C. Ball (1996): A mechanism for global optimization of river networks from local erosion rules. Phys. Rev. Lett., 76, 3359–3363.

Sornette, D., and Y.-C. Zhang (1993): Non-linear Langevin model of geomorphic erosion processes. Geophys. J. Int., 113, 382–386.

Sugai, T., H. Ohmori, and M. Hirano (1994): Rock control on the magnitude-frequency distribution of landslides. Trans. Japan. Geomorphol. Union, 15, 233–251.

Takayasu, H., and H. Inaoka (1992): New type of self-organized criticality in a model of erosion. Phys. Rev. Lett., 68, 966–969.

Tucker, G. E., and R. L. Bras (1998): Hillslope processes, drainage density, and landscape morphology. Water Resour. Res., 34(10), 2751–2764.

Turcotte, D. L. (1992): Fractals and Chaos in Geology and Geophysics. Cambridge Univ. Press, Cambridge.

Voss, R. F. (1985): Random fractal forgeries. In: Fundamental Algorithms in Computer Graphics, edited by R. A. Earnshaw. Springer, Berlin, 805–835.

Yokoi, Y., J. R. Carr, and R. J. Watters (1995): Fractal character of landslides. Envir. & Engin. Geoscience, 1, 75–81.

Scale Problems in Geometric-Kinematic Modelling of Geological Objects

Agemar Siehl* and Andreas Thomsen

Geological Institute, University of Bonn, Germany

Abstract. To reveal, to render and to handle complex geological objects and their history of structural development, appropriate geometric models have to be designed. Geological maps, sections, sketches of strain and stress patterns are such well-known analogous two-dimensional models. Normally, the set of observations and measurements supporting them is small in relation to the complexity of the real objects they derive from. Therefore, modelling needs guidance by additional expert knowledge to bridge empty spaces which are not supported by data. Generating digital models of geological objects has some substantial advantages compared to conventional methods, especially if they are supported by an efficient database management system. Consistent 3D models of some complexity can be created, and experiments with time-dependent geological geometries may help to restore coherent sequences of paleogeological states. In order to cope with the problems arising from the combined usage of 3D-geometry models of different scale and resolution within an information system on subsurface geology, geometrical objects need to be annotated with information on the context, within which the geometry model has been established and within which it is valid, and methods supporting storage and retrieval as well as manipulation of geometry at different scales must also take into account and handle such context information to achieve meaningful results. An example is given of a detailed structural study of an open pit lignite mine in the Lower Rhine Basin.

1 Introduction

Three-dimensional geometric models of specific geological objects are based on conceptual process models of generic classes of geological objects. The observed supporting data normally are sparse, compared to the complexity of the objects and their geological history, and the patterns of geological facts can only be interpreted if the geologist experiments with alternative process-oriented hypotheses in mind how the structures could have developed. It is a typical procedure of inverse reasoning, yielding normally a set of possible solutions. This is the primary motivation why we make geometric models in geology using computer aided design methods: to ensure spatial consistency of the assembly of geological surfaces and bodies while finding the most likely assumption how they may

---

* siehl@uni-bonn.de

have developed establishing the best fit of ideas and data. Such consistent three-dimensional models of geological objects may define the present day boundary conditions for subsequent backward reconstructions to access paleogeological states, so time is coming in as fourth dimension. The final objective is to decipher the interaction of uplift and subsidence, erosion and sedimentation, burial history and subsequent structural development, by modelling a time series of balanced 3D restorations of paleostructural and paleogeographical stages.

In the course of modelling large geological objects, substantial amounts of raw and model data of some geometrical and topological complexity are generated, which have to be managed adequately. It is essential to provide efficient interoperability between modelling tools and a database supporting higher dimensional spatial access to stored geological objects. Object oriented modelling techniques appear to be a convenient communication basis for geoscientists and computer scientists, because a direct analogy of the conceptual hierarchy of real geo-objects can be established with the object oriented database representation. It is then possible to interactively select and combine different data sources and to provide a great variety of user-defined views and problem-oriented graphical outputs, which makes such numerical models superior to conventional printed maps, cross-sections and block drawings. But the geological model behind is always determined by the primary setting of its purpose, the size of the considered area, spatial distribution of accessible data and sampling resolution, which in turn defines a restricted range of scale to further processing and display. Moreover, when supervising the modelling process, the scientist incorporates specific geological background knowledge not included in the original data, either by applying certain rules and algorithms or by freeform adjustment of model geometry to data and geological plausibility, which again is scale-dependent. Different models with their specific domain size and resolution are thus causally related to their origin, and a meaningful integration of different models generated in different scales cannot be performed by a simple numerical procedure of up- and downscaling, but needs the formulation of integrity constraints to establish compatibility of different model versions of the same group of objects. This is a crucial point in the daily routine of geological mapping and 3D modelling, where the switching between different scales of representation is taking place continuously during the investigation and analysis of spatial relationships, but without the geologist always being fully conscious of the formal consequences thereof. The present example of a geometric/numerical modelling of parts of the Lower Rhine Basin provides us with the occasion to present some of the arising problems and to discuss first attempts to their solution.

2 The study area: Lower Rhine Basin

A joint project of geologists and computer scientists at Bonn University deals with the development of database-supported space-time models of the sedimentary and structural evolution of the Lower Rhine Basin. It is an intracratonic graben system and part of the Mid European earthquake belt, forming an active branch of the intraplate rift system which connects the North Sea Rift with the Upper Rhine Graben. The rather shallow earthquakes with hypocenters located at depths between 5 and 20 km reach a magnitude up to 6.

Target area for this study (Fig. 1) is the southern part where it is bordered to the East and Southwest by the Variscan Rhenish Massif and to the West by the Caledonian Brabant Massif. Wide spanned flexural subsidence commenced in Middle Oligocene about 35 Ma ago, and was followed from Lower Miocene onward (about 20 Ma) by dominantly fault controlled subsidence. In Oligocene and Miocene times the Southern North Sea repeatedly flooded the Lower Rhine Embayment and transgressed as far South as Bonn.

During the Miocene coastal swamps developed which today form enormous lignite deposits. Uplift of the Rhenish Massif and the regression of the North Sea

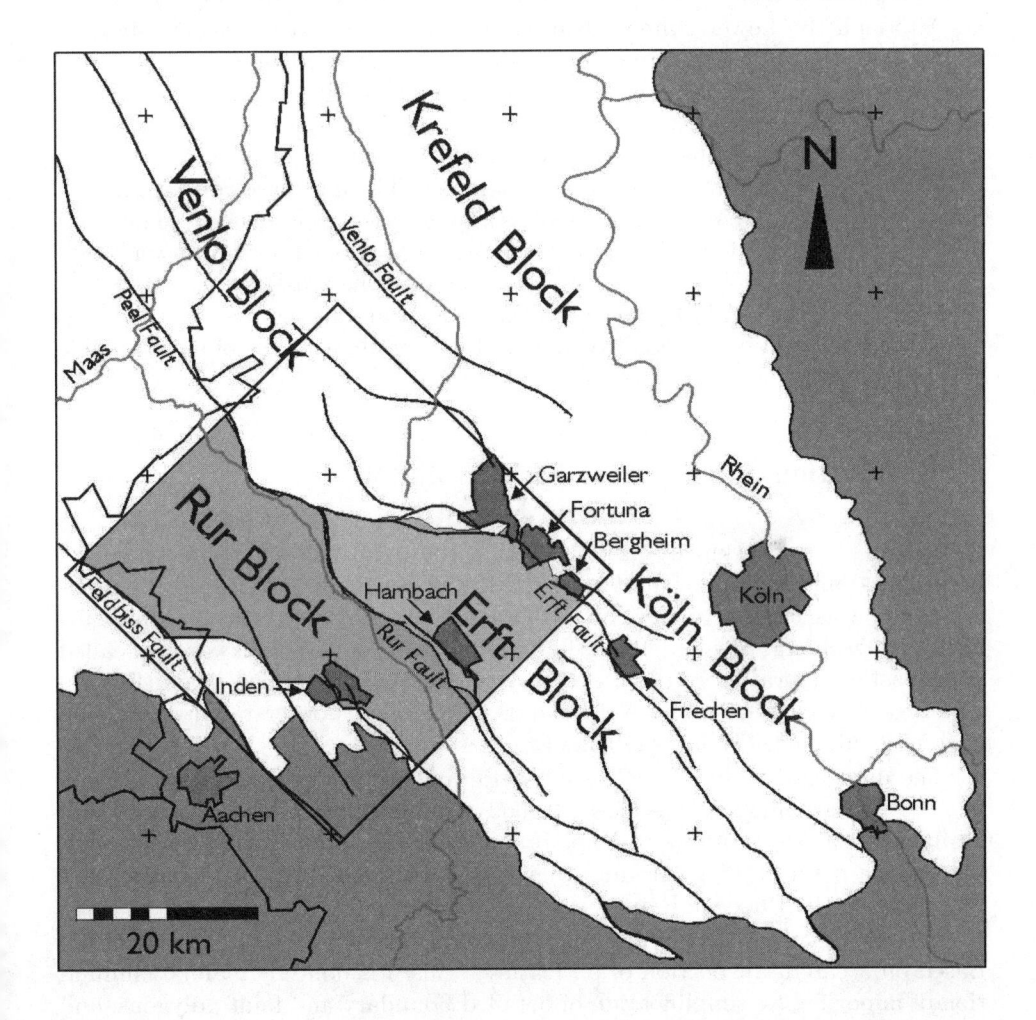

Fig. 1. Map of Lower Rhine Basin (LRB) with two work areas: Southern LRB (Rur and Erft blocks) and Bergheim lignite mine.

ended lignite formation in late Miocene time. Subsidence drastically accelerated in the last 3 Ma during the Pliocene and Quaternary. Fluvial and limnic deposits dominate the Pliocene sedimentation, while during the Quaternary the rivers Rhine and Maas deposited thick sand and gravel terraces.

The deepest parts of the basin have subsided more than 1,500 m below present sea level. Extensional tectonics with a rotational and a slight dextral wrench component created a half graben system of tilted blocks with a characteristic pattern of synthetic main normal faults (dipping SW) and antithetic conjugate faults (dipping NE).

The basin is still subsiding at a mean rate of 2 mm/a, according to GPS measurements (Campbell et al. 2000, Campbell et al. 1994). Maximum subsidence together with the anthropogenic component amounts up to 22 mm per year. There are indications for lateral dilatational movements up to 6 mm/a.

Although the Lower Rhine Basin has not been covered by seismic surveys, extensive exploration and mining of lignite at deep open pit mines has provided an excellent three-dimensional database for the reconstruction of basin development in the southern part. The deposits contain 55,000 million tons of lignite, about one third of all European resources, which are being exploited by Rheinbraun AG, Köln at several huge open pit mines. The database consists of drill hole and well log data and in mining areas also of series of closely spaced detailed lithological cross-sections. Selected stratigraphic horizons and faults were digitised from these sections and used as input to model the present state subsurface geological structure. Additional data from geological maps and survey sketches were evaluated and interpreted to support the overall geometry of stratigraphic and tectonic structures.

3 Restoration of meso-scale basin evolution

A study region of 42 km by 42 km covering the main tectonic units (Rur and Erft blocks) and the major fault zones (Erft and Rur fault systems, Peel boundary fault) was chosen to develop a 3D-model of the main stratigraphic surfaces (Jentzsch and Siehl 1999) as the starting position for a backward, so-called palinspastic restoration of basin evolution in the past 32 million years (Fig. 1). The area stretches from the Netherlands — German border to the Bergheim open cast mine near Köln (see below).

The data used to build the initial static model were taken from a series of 20 lithostratigraphic maps (scale 1 : 25,000) produced by the mining survey and available to us by courtesy of Rheinbraun AG. The dataset includes top and base of the main lithological units (lignite seams, major clay layers) from base Oligocene to the Pliocene Reuver clay.

In order to keep data volumes to a minimum, the maps had to be generalised. Based on a careful inspection of all features, this was done by manual elimination of minor faults, simplification of detailed boundary and fault polygons, and aggregation of complicated fault zones to single faults. The editing resulted in

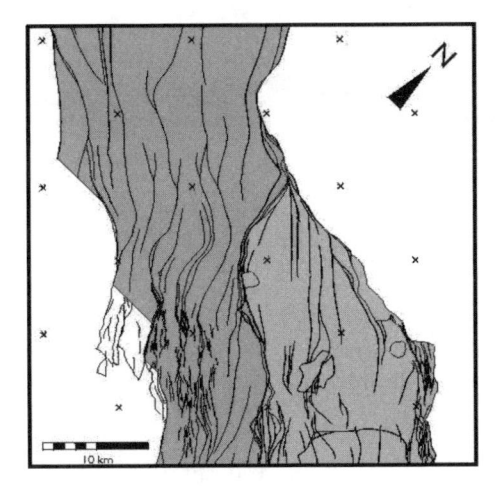

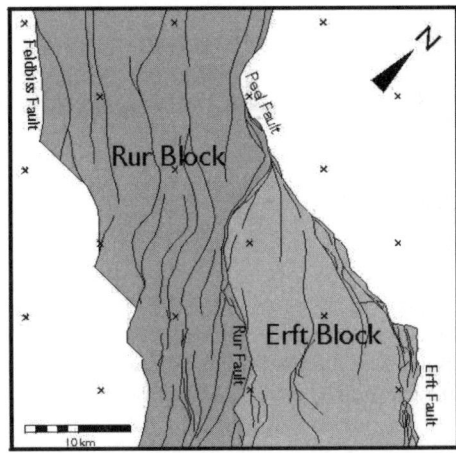

Scale Problems in Geometric-Kinematic Modelling of Geological Objects   295

Fig. 2. Generalisation of original detailed fault network by manual elimination, simplification and aggregation of features. Adapted from (Jentzsch, in prep.)

a simplified dataset of major structures only, corresponding to a generalisation scale of approx. 1 : 250,000 (Fig. 2).

A backstripping algorithm has been applied to a geometrical 3D-model consisting of prismatic volumes, constructed from an initial set of stacked triangulated surfaces. The result is a collection of palinspastically restored volumes for each timestep of basin evolution. The backstripped volumes of each layer are then arranged within a time scene, and the set of time scenes form an hierarchical time scene tree (Polthier and Rumpf 1995). By interpolating between successive key-frames, the subsidence history of the basin can be rendered as an interactive, continuous animation.

4 Detailed structural study of Bergheim open pit mine

A study of the Bergheim open pit lignite mine at the Ville Horst at the northeastern margin of the study area aims at the construction of a kinematic model of a typical small faulted domain with detailed spatial information on strata and fault geometry. The objective is to learn more about the characteristic mechanical behaviour of the graben infill which could be upscaled to the structural development of a larger area within the rift system. For this purpose a new method for balanced volume-controlled 3D-restoration of a system of independent fault-delimited volume blocks is developed, that permits relative displacements of blocks at faults, without affecting internal consistency.

The investigated crustal block comprises an area of 2 km ×2 km up to a depth of 0.5 km. Lateral and vertical extensions are of the same order of magnitude, so no vertical exaggeration has to be applied for modelling and visualisation. Whereas the fault geometry is quite intricate, stratification between faults is

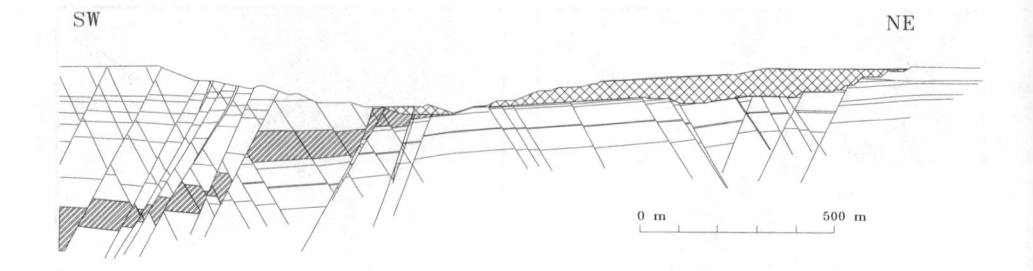

Fig. 3. A digitised cross-section of the Mio-Pliocene strata of the Bergheim open pit mine. The hatched area indicates the filled part of the mine, the lignite seam is marked grey (adapted from Thomsen et al., 1998).

rather regular, with the upper Miocene main lignite seam of about 100 m thickness in the middle part of the sections (Fig. 3). The initial data of this study consist of a set of about 20 SW-NE directed geological sections at a scale of 1 : 2,000, arranged parallel to each other at distances of about 200 m, designed by the mining survey. The geometry of a set of volume blocks delimited by more than hundred faults as sliding surfaces is reconstructed, and a kinematic model of the balanced displacements of such blocks is being developed, controlled only by the geometric configuration of the study area.

Techniques of the type "inclined shear" (White et al. 1986) are not applied to the restoration of this small area, because the listric form of the normal faults at greater depth cannot be observed on such a small near-surface section.

GOCAD software is used for the interactive modelling of the geometry (Mallet 1992), whereas storage and retrieval are managed by software developed in co-operation with the group of Breunig and Cremers at the Computer Science Department of Bonn University on the basis of the object-oriented 3D-geometry database software GeoToolKit and the commercial system ObjectStore (Balovnev et al. 1997, Breunig et al. 2002). Restoration of earlier states of development is done using Rouby's method of restoration (Rouby et al. 1996)(Fig. 4): The projection into the map plane of a horizon to be restored is divided into blocks separated by faults and additional artificial borders, if necessary. By a sequence of horizontal shifts and rotations of the blocks, a configuration is searched at which the total size of gaps and overlaps attains a minimum. This best fit configuration is interpreted as a possible restoration of the initial situation before extension. From the comparison of initial and actual configuration, a displacement vector field is derived. This automatic iterative optimisation procedure may lead to several distinct approximative solutions (local minima of the target function). It is controlled by the definition of neighbourhood relationships between blocks, by the choice of blocks to be kept fixed, and by the order in which blocks are chosen for displacement. Separate restoration of parts of the domain provides another means of guiding the method. The geologist may use these controls to introduce assumptions based on geological knowledge, e. g. on the order of deformation increments.

translation   rotation

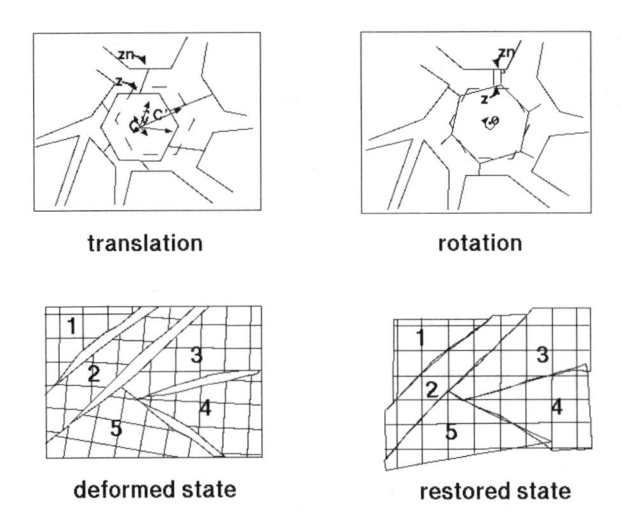

deformed state   restored state

Fig. 4. Rouby's method of restoration in the map plane. At each iteration step, one block is shifted and rotated in the map plane such as to minimise gaps and overlaps between neighbours (adapted from Rouby et al. 1996).

5 Working with different versions of the same geometric model

In some cases, the modelling of position, form and connectivity of geological surfaces can be done with statistical and numerical methods. A good example is Mallet's (1992) algorithm of Discrete Smooth Interpolation (DSI), that starts with a given set of points and constraints and constructs an approximation by a triangle mesh that minimises a combined target function measuring distance from the given points and overall curvature of the surface by an iterative nonlinear optimisation method. In general, however, there is no automatic method to turn data points into a complete geometric model, and such methods are only used as additional tools during the interactive design process. Even in the case of a numerical algorithm like DSI, the definition of the mesh, of constraints, and the calibration of parameters such as the weight of smoothness of the surface against distance from the given data points, the number of iterations etc. leave ample space for the design of several variants of a geometric model. The interactive design of a geometry model can be regarded as the building of successive versions of similar model objects, that are distinguished by variations of node co-ordinates, of connectivity, precision, and of detail. In general, however, it is assumed that the overall quality, as appreciated by the scientist on the basis of his geological knowledge, is increasing during the process of modelling, so that normally there is no need to conserve earlier versions. Therefore the last version is considered as the best, and retained. If, however, the available information contains questionable data or, otherwise, leaves enough freedom of interpretation that two or more different versions can be accepted as plausible, then the conservation of alternative models may be useful. A practical example would be, to have one of two alternative models including and excluding certain data points as outliers, and to submit both variants of the geometry models to a later decision process until additional data are available. Another example is the design of a time-dependent geometry model by constructing a series of different versions of essentially the same model structure with varying geometry and topology at different time instances.

6 Different versions of a time-dependent geometry model in kinematic reconstruction

The reconstruction of the movements and deformations of geological bodies in a kinematic model is an extensive inverse time-space problem, for which in most cases insufficient information leaves much freedom for various solutions and interpretations. In the present case, however, constraints of volume or mass balance and of physical plausibility are available to restrict the set of possible solutions. Instead, another question arises: given a sequence of reconstructed geometry models of the same region at different time steps by a method like Rouby's algorithm, is it principally possible to integrate the different states of the model in time into a consistent sequence of plausible movements? In case of insufficient available information from observation, the correct order of deformation increments can only be assumed, so different interpolation paths of the process trajectory may exist, which have to be discussed. More generally, it may be helpful to consider alternative reconstructions at given time steps, and judge the quality of the reconstruction by the resulting sequence of movements in total, rather than only by inspecting individual steps (Fig. 5).

Thus the reconstruction of the movements of a set of geological bodies naturally leads to the comparison of different versions of the same model, even when referring to the same time instance. Once a time-dependent sequence of geometry models has been accepted, it can be transformed into a kinematic model organised e. g. in a time scene tree (Polthier and Rumpf 1995).

7 Comparison of different, but related geometric models

In the previous examples, variants of the same geometric model of a set of geological bodies have been considered, and variations concerned mostly geometry (co-ordinates), topology (connectivity), scalar properties like density, qualitative properties like lithology, and dependent variables. When we wish to compare different models referring to the same geological problem, a comparison on the basis of the constituting variables alone is not sufficient and may lead to insignificant results. A geometric model of geological bodies cannot be separated from its specific *context*, which for the time being we loosely define as everything relevant to the use of the model, and not contained within its data structure. This

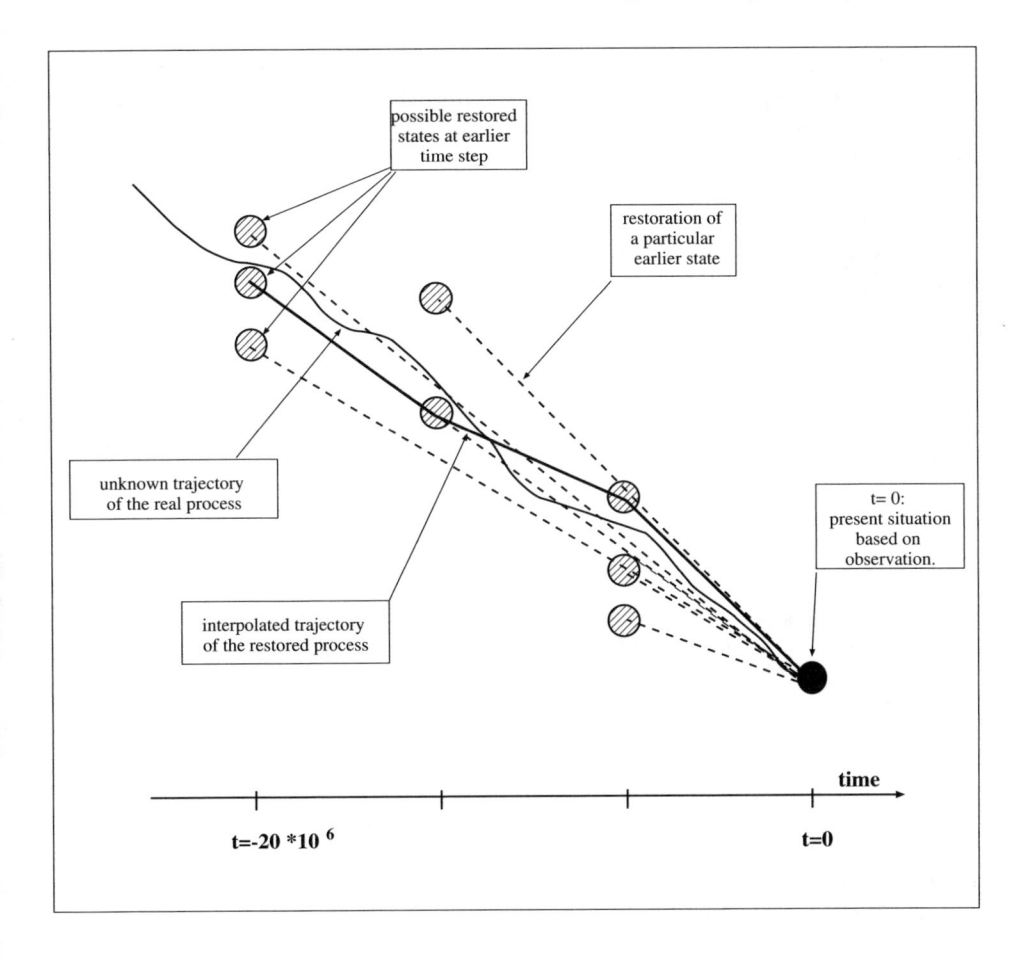

Fig. 5. Reconstruction of the development of a faulted domain from map plane restoration of individual stratum surfaces implies the problem of arranging possible restored states at different times into a coherent sequences of movements.

implies that the context initially is not accessible to the model software, and the question arises, what part of it has to be integrated into the numerical model. We therefore need to describe more explicitly what is necessary for two geometry models to be comparable in a meaningful way.

Some of the more important features of such a context are: initial purpose of the modelling, geological background knowledge of the domain that is being modelled and of the problem under consideration, evaluation of the input data as well as other data that exist but are excluded from the modelling, mathematical properties of the data structure and algorithms used, especially their limits, advantages and shortcomings, and last not least 3D-extension, scale and resolution, which is closely related to the level of detail of the model.

Often there exist other models with different properties that are relevant to the problem considered — in literature, in archives or in databases directly accessible to the computer. In the following, we suppose that the purposes of the models to be compared are closely related, and that the geometrical domains of the models partly overlap. We shall focus on questions concerning the handling of different scales and resolutions. As a comparison of two geometry models A and B may involve a great number of geometrical operations, let us assume that during this process a common third model C is emerging, even though this may be temporary, and never be displayed explicitly. The context of this model of course will comprise both initial contexts (Fig. 6).

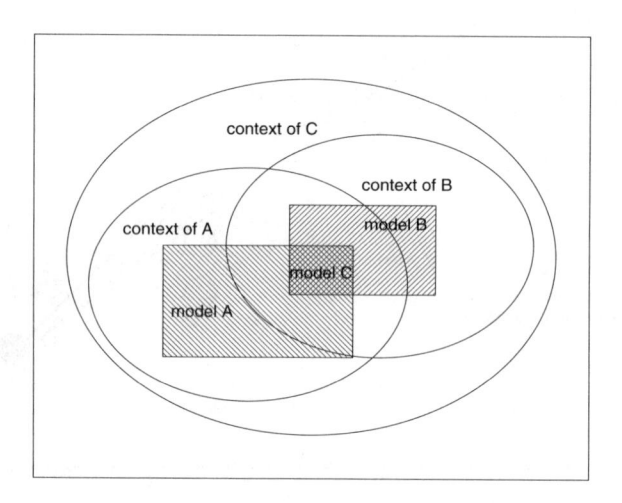

Fig. 6. Comparison of two geometry models can be considered as a geometrical operation and involves the contexts of both operands

As an example for the scales involved in such a common context, when comparing the Bergheim and the basin-scale model, we encounter scales ranging from 1 : 2,000 to 1 : 250,000, and extensions from e. g. 2 km to 100 km. Let us concentrate on the scale range of a common model and on the transformations necessary for establishing it (Fig. 7).

8 Complex geometry models involving different scales

When assigning a range of scales to a geological geometry model, besides the actual map scale, we must take into account the accuracy and numerical precision of input data, the resolution of the information contained in the model, defined by the smallest distinguished objects, the possible variation of interpretation, and the level of aggregation (cf. Kümpel 2002, and Förstner 2002 for a discussion of different aspects of scale in Geo-Data sampling and Geoscientific Modelling).

Thus any computer based geometry model not only is characterised by the restricted range of scales given by its level of detail and by the numerical precision of data and algorithms, but also by a larger range of scales covered by such

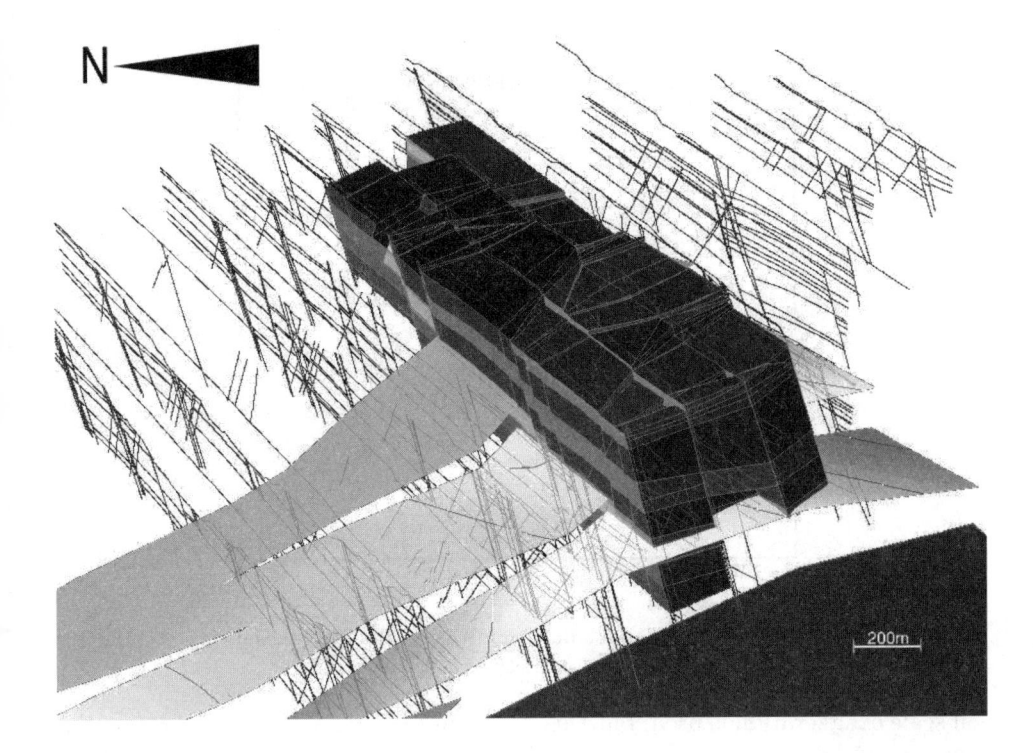

Scale Problems in Geometric-Kinematic Modelling of Geological Objects   301

Fig. 7. Embedding of part of the geometry model of the southern LRB into the geometry model "Tagebau Bergheim". Both volume blocks and profile cross-sections at scale 1 : 2,000 and a scaled down layer boundary surface (modelled at scale 1 : 250,000) are displayed. Near the centre the different amounts of detail of faults as well as differences of positions of layer surfaces are visible.

aspects as the origin of the input data, the geological problem posed, and the geologist's interpretation and background knowledge.

Input data for geometrical-geological models normally consist of a heterogeneous set of digitised maps and sections, additional geological field and laboratory sampling, lithological and geophysical borehole data tied together and more or less influenced by preceding interpretations. Each kind of data will have its own range of scale, which may substantially differ from that of the actual model, as input data originally may have been collected for purposes different than those of the actual modelling. Therefore transformations to the scale of the actual model may be necessary, in the course of which any lower scale details will be lost, averaged — or else have to be exaggerated, if especially important (Fig. 8).

In 3D-modelling, spatial anisotropy plays an important role. Extensions and scales of the range of geological processes, e. g. sedimentation, in vertical direction may have extensions of some centimetres or meters, whereas in horizontal direction they may cover several kilometres. For visualisation, normally a verti-

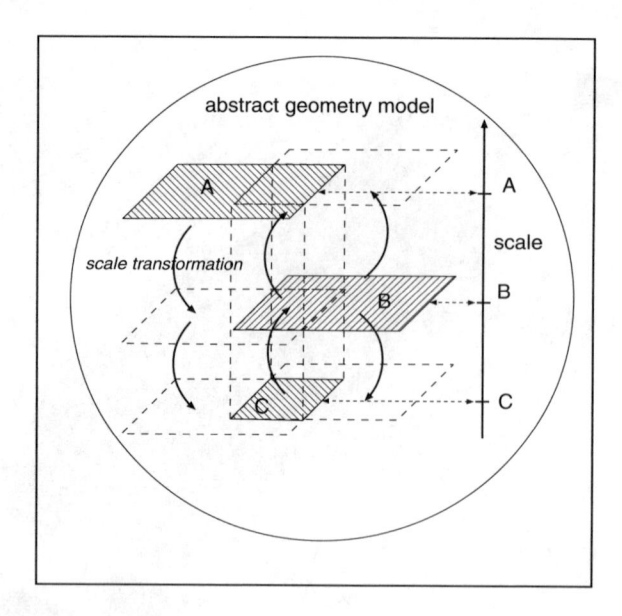

Fig. 8. An abstract geometry model may comprise different realisations at different scales, and include scale transformation methods.

cal scale exaggeration is used, but this causes unrealistic distortions of geometry, concerning the dip (vertical inclination) of strata and faults. Modelling tools like GOCAD do not handle such anisotropies in a consistent way. The principal reason for this shortcoming is that anisotropy cannot simply be linked to a special co-ordinate axis (z-axis) but must rather be seen as orthogonal to the bedding planes of the strata, which in a basin model are later deformed by compaction and tectonical movements, so that the present position and orientation have no direct relation to the primary sedimentation processes. The normal way to handle this problem is to model curved surfaces — the stratum boundaries — rather than 3D volumes, and to take care that the discretisation of a sequence of such surfaces is compatible, i.e. that artificial fluctuations in stratum thickness caused by inappropriate node positions and connectivity in subsequent surfaces are avoided.

Whereas such considerations apply to a 1 : 250,000 scale model of the southern Lower Rhine Basin, in the case of the Bergheim model, at a scale of 1 : 2,000, horizontal and vertical extensions are comparable, and therefore no vertical exaggeration is necessary. Rather, details of stratigraphy at a small scale are ignored to obtain volume bodies that have a length of some 100 m and a thickness of some 10 m. The purpose of this model is the balanced reconstruction of the horizontal and vertical movements, and the operational units of the computer model are volume cells rather than layer surfaces. A comparison of the Bergheim model with the larger scale Lower-Rhine Basin models (Jentzsch (in prep.), Alms et al. 1998, Klesper 1994) therefore does not only involve a change of scale and of detail, but also a change of the structure of the model.

The scope of processes the models are built to represent differ as well: large area movements and basin development are modelled with a limited number of normal faults, in one case using the inclined simple shear approach (Alms et al. 1998), in another a vertical subsidence model (Jentzsch (in prep.)) with a much greater number of simplified faults (Fig. 2). On the contrary, in the Bergheim model, a purely geometrical approach that resembles a 3D-puzzle is done to understand the formation of a great number of faults. Nevertheless, the global geological processes of extension, subsidence, sedimentation and compaction are identical, and the invariant, namely volume/mass balance, too.

9 Some basic principles and requirements

Summing up the above-mentioned examples and considerations, we suggest the following list of requirements and basic principles of a computer-based representation of a multi-scale geometric model of complex geological objects:

1. As the geological processes and the observations cover a range of scales rather than a single scale, the context of the model covers a scale interval.
2. The information contained in the model is only a limited extract of a much richer context, and input data may have different resolution.
3. Related geological data and models may have different scales as well.
4. The geological processes and the resulting geometry are highly anisotropic, but this anisotropy cannot simply be reduced to a difference in scale of the vertical vs. horizontal axes.
5. The geometry model itself has limited resolution, but it may be used in a context where a scale different from the original scale of the model is requested e. g. in a comparison of two models with different resolutions.
6. A change of scale may imply a different mathematical model of the geological process and different effects may be taken into account — e. g. inclined shear (White et al. 1986) vs. spatial puzzle solved using restoration in the map plane (Rouby et al. 1996).

10 Abstract geometry model and realisations at different scales

We conclude that a 3D-model should be considered at two levels of abstraction: On one hand, it is a geometry model of a certain geological problem, using a specific data structure and algorithms, at a given extension and level of detail or resolution. This is the normal approach when considering one model at a time with the exception that resolution and the assigned scale of such a model should be documented explicitly for each of the co-ordinates, when storing it in a data base, and that changes of resolution and scale of e. g. input data during the process of modelling must be documented together with the model.

On a higher level of abstraction, the geometry model does not have a specific scale, but is a complex object with different realisations depending on a numerical parameter, namely scale, which does not vary continuously, but in certain steps.

At each step, i.e. at each scale, a different set of details has to be modelled. The task arising is to determine rules and methods controlling a change in scale, or rather the generation of scale-dependent versions of the abstract geometry model. The scale-dependent realisations of the model may be considered as different versions of the abstract model depending on a numerical parameter. There is a certain similarity to the realisation of a time-dependent geometry model with different versions depending on the numerical parameter time. Even though the abstract model covers a range of scales, it is based on data at a certain scale and resolution, which limits the level of detail and precision at which the model can be realised.

We can arrange the rules and methods controlling such realisation in two groups:

- coarsening / generalisation with respect to the original data, i.e. suppression, averaging or exaggeration of details
- refining by embedding a more general geometry model in another model that requires more detail, which cannot be automatically provided.

An example would be to use part of a geometry build at a scale of 1 : 50,000 as starting point for a model at 1 : 10,000, followed if necessary by a refining of the mesh, and maybe a smoothing operation. After such a transformation, the resulting geometry may well fit into a model at 1 : 10,000, but still will not carry more information than at scale 1 : 50,000. To obtain a true 1 : 10,000 geometry, we need to supply additional information on detail visible at 1 : 10,000.

Both changes of realisation may not only imply a transformation of coordinates, but also a remeshing, or even a complete change of the modelling data structure. Normally, methods for generalisation will be more easy to define than methods to add information to embedded coarser geometrical objects. A special case is a model based on data with different resolution in different parts e.g. close to the surface, or near a mining area. In such a case, both generalisation and embedding may be applied at the same time in different regions or, in case of anisotropy, in different space directions.

The main difference from conventional methods is that original scales as well as rules and methods for changes of scale are to be documented explicitly and become part of the abstract data model.

From the point of view of data modelling, an important task is to reduce redundancy within such a multi-resolution geometry model by identifying a basic set of data and transformations from which versions at specific scales can be generated, rather than managing different versions in parallel, and to allow for a sort of top-down modelling by the incremental addition of detail.

11 Outlook: towards a complex geological multi-resolution geometry model

A geological model that is supported by geological interpretation of sparse subsurface data and has been designed interactively cannot be based only on numerical methods, and hence changes of resolution cannot be handled solely by numerical methods, such as have been presented by Gerstner (1999), Gerstner et al. 2002. Applying techniques of object-oriented modelling (cf. Breunig et al. 2002), it is desirable to define the geometry of geological objects in an abstract way, independent of a concrete representation by a certain geometrical structure (triangulated surfaces, boundary representation, voxels), and with the possibility of varying detail at different discrete levels of resolution.

A first approach is simply to store different versions of the same object at different resolutions, letting the scale of the representation be a stepwise varying parameter of the methods applied. With such a representation, however, the following problems arise:

- The construction of such an object with complete multiple versions requires an important amount of supplementary work.
- To ensure compatibility of different versions of the same object, integrity constraints must be defined. In fact, the formulation of such constraints would formally establish what we mean by saying that different representations at different scales are "compatible".
- A lot of redundancy will occur, with all the resulting problems caused by updates of the object.

If, however, we were able to define rules which control the derivation of generalisations from detailed models, different versions of the same object might be generated only when necessary, thus avoiding problems of redundancy.

The well known process of selection of a "window" into a 3D-model and of "zooming in", may be taken as a paradigm of how to work with a multi-resolution geometry model: in a top-down approach, a model first is constructed at a general level of resolution, leaving aside most of the details. Then, if a better understanding of the geology is gained, and if also more data are available, certain windows, i.e. parts of the modelled region are selected, where more detail can be added, e.g. based on information from a borehole database, or from seismic sections. The interesting question is to what extend this can be done incrementally. The resulting 3D-model will possess a different resolution in different parts, depending on the density of available information.

References

Alms, R., Balovnev, O., Breunig, M., Cremers, A.B., Jentzsch, T. and Siehl, A. (1998): Space-Time Modelling of the Lower Rhine Basin Supported by an Object-Oriented Database, EGS2001. — Physics and Chemistry of the Earth, Vol. 23, 251–260, Amsterdam (Elsevier).

Balovnev, O., Breunig, M. and Cremers, A.B. (1997): From GeoStore to GeoToolKit: The second Step. – In: M. Scholl, A. Voisard (eds.): Advances in Spatial Databases, Proceedings of the 5th Intern. Symposium on Spatial Databases (SSD 97) 223–237, Berlin, July 1997, LNCS 1262, Berlin (Springer).

Breunig, M., Cremers, A.B., Flören, J., Shumilov, S. and Siebeck, J.: Multi-scale Aspects in the Management of Geologically defined Geometries, this volume.

Campbell, J., Kümpel, H. J., Fabian, M., Görres,B., Keysers, Ch. J., Kotthoff, H. and Lehmann, K. (2000, submitted): Recent Movement Pattern of the Lower Rhine Basin from GPS Data, 9 p., 6 figs, 2 tables.

Campbell, J., Freund, P. and Görres, B. (1994): Höhen- und Schwereänderungen in der Rheinischen Bucht — Ein neues Projekt an der Universität Bonn. — Vermessungswesen und Raumordnung, 56/2, 109–120.

Förstner, W.: Notions of Scale in Geosciences, this volume.

Gerstner, T. (1999): Multiresolution Visualisation and Compression of Global Topographic Data, SFB 256 report, Universität Bonn, 1999, submitted to GeoInformatica.

Gerstner, T., Helfrich, H.-P. and Kunoth, A.: Wavelets, this volume.

Jentzsch, T. (in prep.)

Jentzsch, T. and Siehl, A. (1999): From Palinspastic Reconstruction to Kinematic Basin Models. — 19th GOCAD meeting, June 1999, Nancy, 6 p. [extended abstract].

Klesper, Ch. (1994): Die rechnergestützte Modellierung eines 3D-Flächenverbandes der Erftscholle (Niederrheinische Bucht). — Berliner Geowissenschaftliche Abhandlungen (B) 22, 117 p. 51 Abb., Berlin (FB Geowiss.).

Kümpel, H.-J.: Scale Aspects of Geo-Data Sampling, this volume.

Mallet, J. L. (1992): GOCAD: A Computer aided Design Program for Geological Applications. — In: A. K. Turner, Three–Dimensional Modeling with Geoscientific Information Systems, NATO ASI Series C, 123–141, (Kluwer Academic Publishers).

Polthier, K., and Rumpf, M. (1995): A Concept for Time-Dependent Processes. — In: Goebel, M., Müller, H., Urban, B. (eds.): Visualization in Scientific Computing, 137–153, Wien (Springer).

Rouby, D., Souriot, Th., Brun, J. P. and Cobbold, P. R. (1996): Displacements, Strains, and Rotations within the Afar Depression (Djibouti) from Restoration in Map View. — Tectonics, 15, 5, 952–965.

Thomsen, A., Jentzsch, T. and Siehl, A. (1998): Towards a Balanced Kinematic Model of a Faulted Domain in the Lower Rhine Basin. GOCAD ENSG Conference on 3D Modelling of Natural Objects: A Challenge for the 2000's, June 1998, Nancy, 12 p. [extended abstract].

White, N., Jackson, J. A. and McKenzie, D. P. (1986): The Relationship between the Geometry of Normal Faults and that of the Sedimentary Layers in their Hanging Walls. — J. Struct. Geol., 8., 897–909.

Depositional Systems and Missing Depositional Records

Andreas Schäfer*

Geological Institute, University of Bonn, Germany

Abstract. Depositional systems consist of components showing different scales. These are the scales of bedforms, environments, and sedimentary basins. The deposition of sediments within these scales provides strata, sedimentary sequences, and stratigraphic units. These altogether are geological documents. They contain stratigraphical gaps which can be evident as visible erosion surfaces or can be hidden within the sedimentary sequence. The missing thickness of strata and the time gaps are difficult to specify. Yet, depositional records and gaps both are necessary to describe depositional models, both a modern geomorphic setting and an ancient rock record.

1 Introduction

Sediments and sedimentary rocks document depositional processes. Amanz Gressley (1814–1865) demonstrated, that modern depositional environments on land or in the sea close to each other are able to follow on top of each other in an ancient stratigraphic record (Cross and Homewood 1997). Conterminous environments along an inclined depositional relief become vertically superposed as a consequence of the progradation of the environments — environments lower in the profile will be placed below environments higher in the profile (Walther 1894). As a consequence, paleogeographic settings of past landscapes can be reconstructed from measured sections, from wells, and from seismic lines with different scales and completeness. Ancient depositional systems can be deduced from observation and reconstruction, especially with the aid of sediment models under the scope of uniformitarian principles (Reineck and Singh 1980). The actual geomorphology of the earth's surface displays individual depositional systems. The ancient rock record superposes depositional systems forming facies tracts. These are of different scales and also contain erosion surfaces or structural discordances, where the geologic record is incomplete with respect to time and bed thickness. Yet, deposits and gaps both are depositional documents and both are necessary for the environmental interpretation of rocks.

At a *bedform scale*, the deposition of either fines, sands, and/or coarse clastics composing bedsforms and strata reflect the hydrodynamic variation of a depositional system (Petijohn et al. 1987). In aquatic environments — lakes, rivers, or the sea — the forces of waves and currents produce bedforms from non-cohesive

---
* schaefer@uni-bonn.de

sediments, reflecting the parent hydrodynamic conditions. Bedforms of sands at the sediment/water interface are formed as ripples and dunes. They build when sands are transported in suspension above unconsolidated ground. As a consequence, sedimentary structures may be used to infer, that the transport energy once was able to move and deposit the down-stream moving load (Allen 1982; Blum and Tornqvist 2000).

At an *environment scale*, outcrop and subsurface sections show successions of strata containing siliciclastic rocks, carbonates, or chemical precipitates together with the biotic record. Metres to decametres of clastic strata composed of layered, rippled, contorted strata, or smoothly alternating fines and sands may be part of and represent fluvial plains, coast-line clinoforms, shelves, or deep-sea fans. Channel lags of rivers, scours of cross-over bars of a braided alluvium, rapidly prograding alluvial systems, deeply eroding estuarine channels, high-energy coastlines reworked by the sea, long-time retrogradation of the bay-line of a delta — they all represent actual deposystems and may follow on precursor depositional systems, which lost a considerable part of their topmost stratigraphic record due to intense erosion. Therefore, facies interpretation of sedimentary rocks has to reconstruct depositional models of ancient paleogeographic settings from mostly incomplete strata (Reading 1996).

At a *basin scale*, depositional sequences of rocks demonstrate the balance between basin subsidence versus retrogradation or progradation of alluvial settings or coast lines of shelf seas. The depositional sequences will prograde toward the depocentre of the basin, if (1) the sediment supply to it from a source in the proximal part of the system, (2) the subsidence of the basin, and (3) the eustatic control of the sea-level are adequate to preserve a thick sedimentary fill. In case the accommodation space of the basin is small with respect to the sediment supplied to it, nothing will be deposited in its marginal positions. The sediments will pass by to deeper reaches of the basin (Einsele 2000).

The depositional systems, either small or large, are expected to be continuous under uniform hydrodynamic, structural and/or stratigraphic conditions. Nevertheless, natural environments at a bedform scale, at an environment scale, and at a basin scale mostly are not uniform. The sedimentary systems of a geologic record inhibit both structural discordances and erosional surfaces of various sizes. They also are time gaps in stratigraphy, as for instance was observed by Reineck (1960). And time gaps, missing depositional records are difficult to specify, yet can be deciphered by applying time related Wheeler diagrams (Wheeler 1964) to the stratigraphic records of ancient rock successions, where the time control is good.

The aim of this paper is to show that stratigraphic successions generally contain missing stratigraphic records.

2 Hierarchy of depositional processes

Facies document depositional processes in sedimentary basins. Sedimentary sequences are the basic units for the sedimentological interpretation of strata

(Leeder 1999; Walker and James 1992). Often, the sedimentary sequences are incomplete and contain stratigraphic gaps in their depositional record. Sedimentary sequences and their stratigraphic gaps may differ much with respect to time and space. Nevertheless, both are important for the sedimentological interpretation (Emery and Myers 1996).

2.1 Bedforms

Most processes of transport and deposition of sediments that are relevant to geology are made by turbulent waterflows. They provide vivid, non stable, and non reproducible vortices and discordances in the water body. With respect to the Reynolds number, the stream velocity of non-laminar, turbulent waters in a flume (or in rivers) with a hydraulic radius of 5 to 10 m has to be more than 0.2 to 0.5 mm/sec (Tanner 1978). Natural water flows usually have a higher velocity — the stream velocity will be about 18 cm/sec when fine sands of 0.2 mm in diameter are suspended in flowing water. Depending on the water depth with respect to the Froude number, tranquil (subcritical) stream velocities of about 10 to 120 cm/sec develop bedforms from non-cohesive sands of a grain-size between 30 to about 500 microns. Rapid (supercritical) stream velocities will destroy these bedforms again.

For the formation of bedforms, the sands have to be washed free from clay minerals and have to form a loose unconsolidated substrate, because the clay minerals tend to react with the salinity of the water body and form floccules (Sundborg 1967). Floccules from aggregates of clay build adhesive muddy substrates after their deposition. They are deposited from standing water bodies during stand-still phases of high-water and low-water tides in tidal environments, forming coherent mud layers on the tidal flat surface or delicate flasers in individual ripple troughs (Reineck and Singh 1980). These mud layers hardly can be reworked again to single mineral particles and therefore prevent a build-up of bedforms.

Tranquil waterflow in a water depth of roughly about 1 dm to 10 m is necessary to build geologically relevant bedforms. The sandy substrate forms eddies behind the relief of the ground. With accelerating stream velocity and increasing bed roughness, the sand is lifted to form suspensions. Transverse bedforms (ripples, dunes) originate, showing a progradation of their slip-faces downstream and a luvside erosion. The eddies are not continuous vortices, keep close to the leeside of the bedforms, are flattened at their upper side, and provide a backflow regime in the ripple troughs (Fig. 1). Depending on the dimension of the hydraulic active water way and the suspended sediment freight, bedforms strongly reflect the hydraulic conditions of the flume and natural water flows. Also waves of free water bodies and along shore lines show vortex trajectories in direct relationship to the height and length of the oscillating wave, from several millimetres to several metres in diameter (Allen 1979). This results in symmetrical bedforms, in wave ripples.

Bedforms from unidirectional waterflows range from cm-scale ripples to m-scale mega-ripples (dunes) in a two-dimensional and a three-dimensional shape

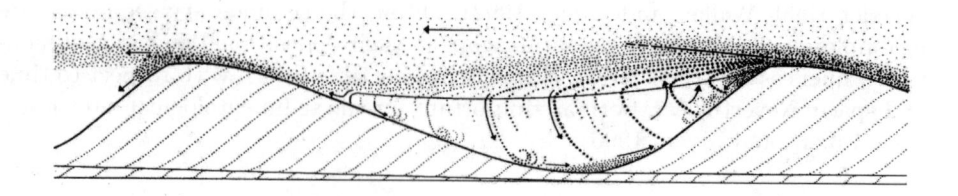

Fig. 1. Rollable fine-grained sands are suspended by running waters, forming eddies behind ripples. Backflow in their lee-side and avalanching of sands from the ripple crests develop downstream oriented concave-up slip-faces. The actual ripple generation provides basal unconformities preserving ripple slip-face sets from earlier generations (modified after Reineck and Singh 1980)

depending on grain size, on water velocity, and on the depth of the water way (Reineck and Singh 1980; Southard and Bogouchwal 1990). Their size generally is from centimetres to several tens of metres of ripple length and from centimetres to one or two metres of ripple height (Reineck et al. 1971). The height of current bedforms may also be of several metres (Singh and Kumar 1974). The resulting fabric is cross-bedding with trough-like basic scours and inclined leeside foreset faces, which are oriented downstream (Ashley et al. 1990). The downstream path of the water flow is unsteady and modified by the bed roughness. While the energy of the water flow is still low, fine grains form small scale ripples, whereas higher energy and larger grain sizes develop large scale mega-ripples (Rubin 1987). Ripples will form within minutes, mostly depending on the availability of sediments.

As the ripple crests provide obstacles to the water flow, deviation of the individual stream lines may occur. This results in unconformities formed at the lee-side slip-faces of the bedforms, as the position and orientation of the lee vortex behind their crests shifts with time. These unconformities are known as reactivation surfaces that form syngenetically during the ripple migration (McCabe and Jones 1977; De Mowbray and Visser 1984). They separate sets of lee-side slip-faces from each other with different orientation and form. Reactivation surfaces may be hidden in the fabric of small bedforms, because the grain population in small scale ripples is only small. On the other hand, in large-scaled dunes, different orientations of slip-face sets may cause marked unconformities. They form while the ripple is under construction, and thus demonstrate that they are as short-lived as are the individual slip-faces of the bedforms. They also represent the higher order of hierarchy in an individual bedform. Their size can achieve the dimension of the ripple height, often combining several slip-face sets. Reactivation surfaces may provide a considerable rate of erosion, and their presence proves the unsteadiness of the stream lines of the water flow especially close to the sediment surface.

Ripple lamination in a small scale and cross-bedding in a larger scale also provides unconformities which are oriented more or less horizontally (Fig. 2). These are erosion features at the base of the individual ripple sets and are pro-

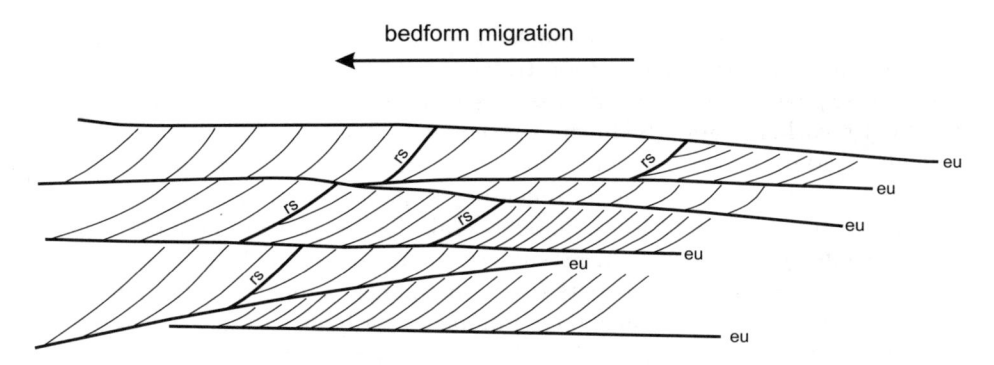

Fig. 2. Slip-faces from sandy aquatic bedforms form sets of downstream oriented concave-up inclined strata. They are subsequently eroded by the next younger generation of bedforms. Result is a sedimentary structure of either small-scale current ripple lamination or large-scale trough cross-bedding, depending on the size and form of the migrating bedforms, each of them separated by sub-horizontal erosion unconformities (eu). Minor deviations in the downstream water-flow and erosions of slip-face sets develop reactivation surfaces (rs), proving synchronous reworking of the migrating bedforms. Reactivation surfaces are the next lower hierarchy relative to the individual slip-face sets, followed by the erosion unconformities.

vided by the downstream migration of the bedforms. They are horizontal if no net sedimentation takes place, but may rise downstream if there is an increase in sediment volume, resulting in a climbing-ripple lamination (Leeder 1999). Trough-like cross-bed sets and current-ripple lamination will only be preserved to minor parts of the original height of the bedforms. Sediment layers subsequently following each other will erode the stratum of bedforms prior to the actual one. Thus, rippled resp. cross-bedded carpets of sands in a vertical section only represent about one third of the original bedform height or even less than this (Rubin 1987). The unconformities underneath the actual series of bedforms are syngenetic and include time breaks of a duration depending on the individual sub-environment where the formation of the bedforms takes place. The unconformities separate ripple sets or cross-bed sets and clearly are erosion surfaces. With respect to the slip-faces and their reactivation surfaces, they are of a lower order in the deposition hierarchy. They are near-horizontal bedding planes of depositional sequences with only minor differences between the individual grain-size populations. Erosion surfaces may also be syngenetic surfaces, while deposition and erosion is more or less at the same time.

2.2 Vertical Facies Successions

Vertically stacked facies successions form sedimentary sequences through layered sediments or sedimentary rocks. The sequences contain strata with bedding features, which reflect the hydrodynamic conditions during deposition (Walker 1984; Reading 1996; Leeder 1999), also forming the substrate for biota. Facies and facies successions of siliciclastic sediments describe processes of deposition by traction transport or suspension transport through the water column. Sedimentary sequences consist of individual strata of stacked depositional environments of past landscapes. Sedimentary sequences are uniform, if the conditions — which lead to their deposition — also are uniform. Yet, if for instance a cut-off of a meander loop occurs, the sedimentary sequence of the river plain will be changed from a channel facies to an oxbow-lake facies (Walker and James 1992).

Sedimentary sequences or facies succession motifs (sensu Cross, pers. comm.) describe depositional environments (e. g. a high-energy coast; Fig. 3). They contain the deposits of individual sub-environments verifying smaller reaches within them (e. g. shoreface, swash zone of the shore, dunes on the back shore). Each sub-environment alongshore has its specific grain-size range depending on the energy of the sea, its specific range of bedding-features, relevant to the paleoecology of species of plants and animals. Each sub-environment shows a different chance of preservation and a different rate of accumulation (Reineck and Singh 1980). These individual characteristics result in different vertical lengths of sedimentary sequences with different succession motifs of fining-up or coarsening-up features or in an interbedded style. To these sedimentary sequences, characteristic syngenetic unconformities and time gaps due to deposition, erosion, and re-deposition are included, only some of them being clearly visible. River channel scours in a fluvial plain commonly show specific evidence, multiple crevasse splays mostly will be obliterated. Tidal channel outlets at a high-energy beach being of equivalent grain-size range as the coast will not show up, whereas muddy tidal channels show clear evidence by point bars in mixed tidal flats.

Following this, sedimentary sequences are non-uniform build-ups, although they are used as idealised sections to describe depositional models (Walker 1984). A fining-up sequence of several metres in a cored well shows a grain-size suite from coarse to fine and is coupled with a set of sedimentary structures from trough cross-bedding to small-scaled current-ripple lamination, including high-energy lamination, wave ripples and the like, desiccation cracks, and plant roots in the overbank fines. This fining-up sequence may represent a continuous section through a meandering channel (Fig. 4). Nevertheless, it has to be considered that each sub-storey of this section consists of many sets of individual time spans, where deposition and reworking took place. The build-up of a sedimentary sequence in a meandering lowland river can be of a duration of half an hour or can be a matter of several years depending on climate, tectonic frame, or earth period. This can only be shown by means of Wheeler diagrams, if the time control is good (Wheeler 1964). The overall length of sedimentary sequences also is very variable, ranging from a metre to several tens of metres, rich in small- and/or large-scale stratal gaps. Although sedimentary sequences are by no means complete, they are widely used as ideal sections through depositional models. They are divided into units of grain-size trends related to the individual energy content of the depositional environment (Moslow 1984; Fielding 1987).

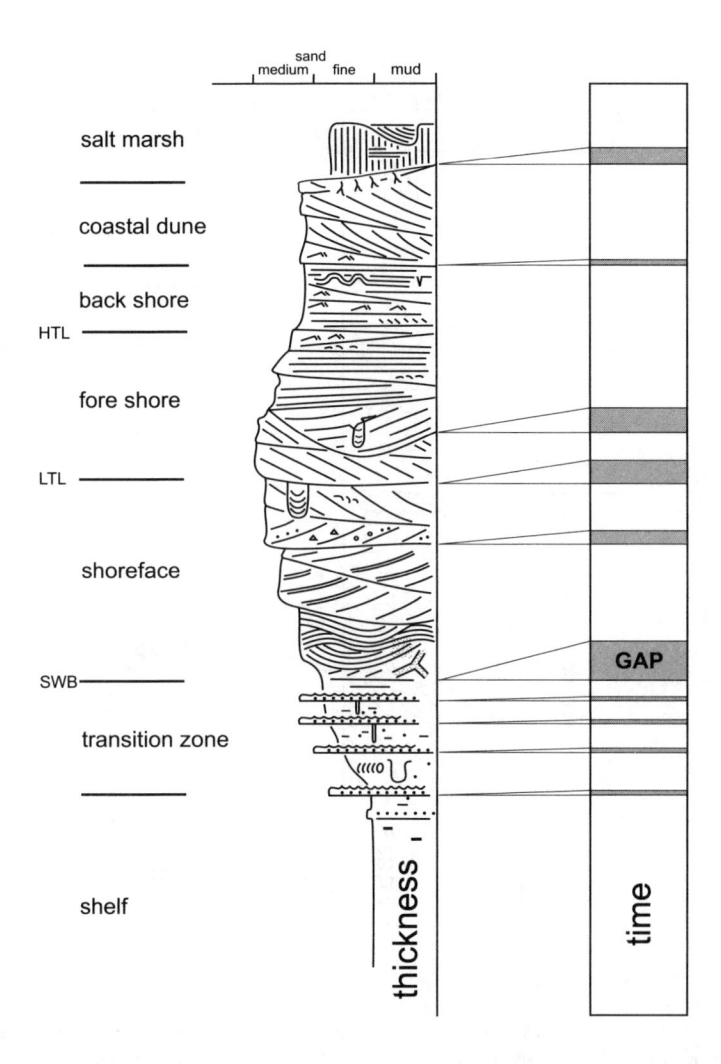

Fig. 3. This facies succession motif forms by sediment accumulation and reworking at a marine high-energy coast, being the most active coastal clinoform with respect to sediment accumulation and reworking. The paleogeographic range is from the shelf to the back shore, where peats accumulate in the salt marshes and elevated areas. The profile length is different with respect to the individual coast. The shoreface length depends on the exposition to the sea and energy of the wave fetch, causing different depths of the wave base. Moreover, the width of the fore shore is defined by the tidal range and can be rather different, from several tens of metres to several kilometres. Also, the width of the back shore with eolian sands and algal mats upon, vary considerably in area. The extension of the low-energy back-barrier tidal flats and the salt marshes along the high-tide line (HTL) of the coast together with lagoons and bays reflect the amount and quality of sediments provided from riverine input. The 'time' column reveals numerous stratigraphical gaps, yet only few of them show up in the 'thickness' column.

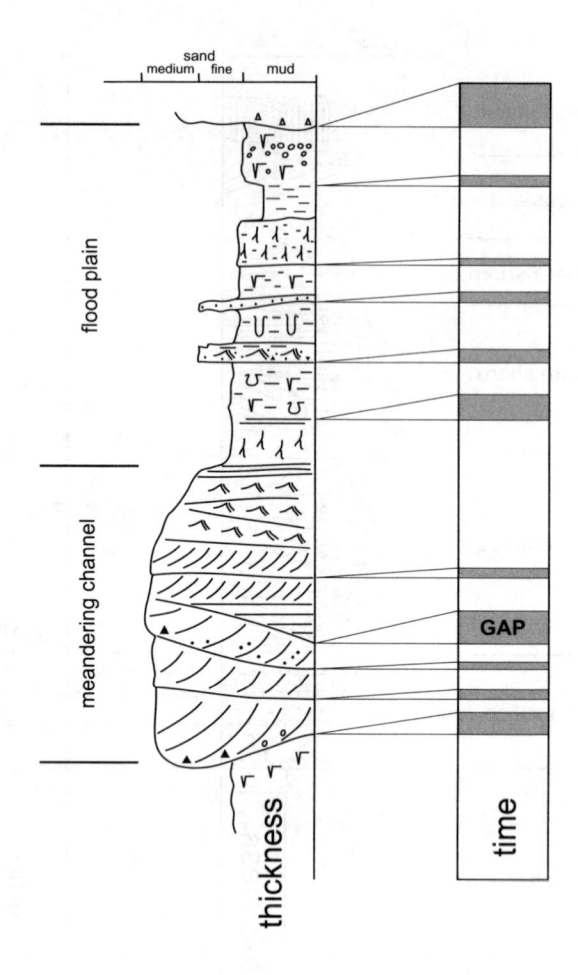

Fig. 4. A fluvial sedimentary sequence through a meandering channel shows a generalised section through its various storeys. The lower sandy part is the lateral accretion storey of the meandering channel where point bars accrete downstream and toward either sides of a channel in a lowland river. Individual downstream migrating megaripples form trough-cross-bed sets. They are embracketted by lateral accretion surfaces which can be understood as time surfaces and may be exemplified by major unconformities. Rapidly waning of the flood develop small-scaled sedimentary features, dominantly current-ripple lamination of fine-grained sands. In the vertical accretion storey muds are deposited on the flood plain. They derive from a low-energy flooding. Numerous paraconformities with soil horizons, rootlets, desiccation cracks, and a wide variety of concretion nodules will be preserved. Seldom and only close to the channels, a flooding causes crevasse splay, resulting both in a low-energy current-ripple lamination and a high-energy parallel-bedding of fine-grained sands, often resulting a soft deformation of muddy substrates underneath. The 'time' column reveals numerous stratigraphical gaps, yet only few of them show up in the 'thickness' column.

Sedimentary sequences contain stratal gaps, only some of them being well resolved in the ancient record. If these gaps are paraconformities in flood plains, the identification of the individual time gap is not possible. If these gaps are at the base of channels and form marked disconformities, it depends on the fluvial model, whether a time gap of several years or tens of years is evident due to the position of the river in a proximal or a distal part of the alluvial plain and, whether the base level rises or falls. Disconformities may prove the time gap that can range from tens to thousands of years, and easily reach geologic ages, if angular unconformities are revealed (Boggs 1995). As a result, the missing stratigraphical record in depositional sequences can be little, if they are within one depositional model. But stratigraphical gaps can be enormous, if the tectonic frame of the basin is responsible for disconformities or angular unconformities, when rivers cut their courses across a newly exposed alluvial plain.

2.3 Sequence Stratigraphy — Systems Tracts, Sequences, and Parasequences

Sequence stratigraphy provides a conceptual framework to coordinate the arrangement of sedimentary sequences in sedimentary basins (Vail et al. 1991; Posamentier and James 1993). Due to the rise and fall of the sea level, to the tectonics of the earth crust underneath, and to the input of sediment from an uplifted orogen in the hinterland, sedimentary sequences form in tight relation to the 'relative sea level'.

For the use of sequence stratigraphy, stratigraphic charts in regional geology are divided with a cyclicity of about 3 to 0.5 million years. This 3rd-order cyclicity (Haq et al. 1987; Wilgus et al. 1988) is related to the earth's stress field and to a large-scale climate development due to the earth's ice shield. The basic units are 'systems tracts'. They can be a lowstand, transgressive, and highstand systems tract, each of which consists of a series of different depositional environments, from the deep sea to the continental alluvium (Weimer and Posamentier 1993). The 'systems tracts' provide the stratigraphic correlation between continents across the oceans. Stratigraphical standards in sedimentary basins are of different scales, and they use sediment models to define sedimentary sequences. As stratigraphic approaches meanwhile differ somewhat from the initial concept, various 'systems tracts' exist today. Yet, measured sections in sedimentary basins are rather easily correlated with the 3rd-order cyclicity (Hardenbol et al. 1998).

Sequence stratigraphy also recognizes cyclicities higher than the 3rd order, which are the Milankovitch cycles. They depend on the geometry of the earth's orbit around the sun, the eccentricity (e) of the earth' circuit, the obliquity (o) of the earth' axis, and its precession (p) i.e. its wobble while turning (p) (Emery and Myers 1996). This provides four periods with different time spans: 413 ka and 112 ka (e), 41 ka (o), 21 ka (p). These measures of differing cyclicities allow subdivision of the 'systems tracts' (3rd order) into smaller units (4th, 5th, and 6th orders), so that sedimentary sequences of shorter time length can be used for geological field and subsurface work. Depending on the individual sedimentary basin, its time frame, and the sedimentary model of its basin-fill, in sequence stratigraphy 'parasequences' and 'sequences' are currently in use

(Fig. 5). 'Parasequences' are shoaling-up, progradational successions, separated from each other by 'marine flooding surfaces' being preferably used in coastal marine environments, whereas 'sequences' are separated from each other by subaerial unconformities, i. e. 'sequence boundaries'. These are preferably used in continental environments, considering that 'sequence boundaries' are surfaces with subaerial exposure (Fig. 6; Plint and Nummedal 2000).

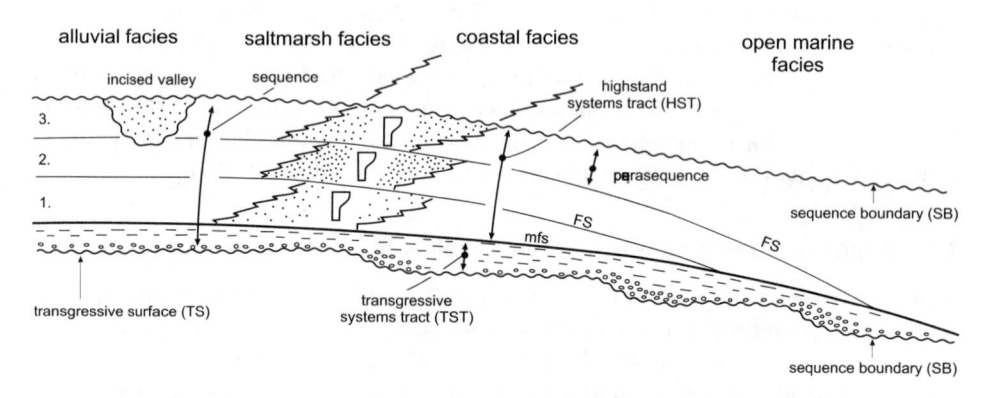

Fig. 5. A coastal scenery shows the relationship of 'systems tracts', 'sequences', and 'parasequences' in a sequence stratigraphical concept. Due to a relative sea-level highstand, 'parasequences' of a prograding coastal environment provide coarsening-up lithofacies sections. At the coastal clinoform (i. e. coastal facies), they are separated by 'flooding surfaces FS' (equivalent to 'transgressive surface TS'). The same time, in the alluvial reach of the coast, 'sequences' develop which are confined by erosive 'sequence boundaries SB'. The prograding 'parasequences' are combined to form a 'highstand systems tract HST' (modified after Walker 1990).

This sequence stratigraphy concept, since its origin more than 25 years ago, meanwhile has become the dominant stratigraphic paradigm to work in both siliciclastic and carbonate environments. Nevertheless, quite often it is difficult to overcome the restricted range of the outcrop size or basin geometry exposed. The stratigraphical lengths and gaps between 'systems tracts', 'sequences', and 'parasequences' in sequence stratigraphy may range widely. Therefore, they are not always ideal to individual facies studies. On the other hand, sequence stratigraphy has become established to use it in basin analysis, whether by seismic sections, well profiles, or measured sections in outcrop studies (Nummedal et al. 2001).

2.4 Base Levels

The concept of base level cyclicity (Barrell 1917; Cross and Lessenger 1997, 1998) is an approach with more practical use in field work. Transgressive surfaces and sequence boundaries still keep their stratigraphical meaning. Base levels are fixed to the storm wave base of the coastal clinoform forming an ideal surface

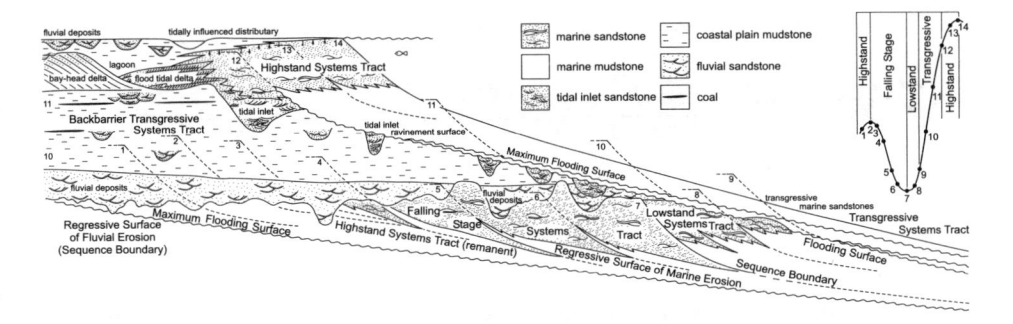

Fig. 6. This figure is redrawn from Plint and Nummedal (2000: Fig. 2), with respect to the original version the time section is omitted and the facies section slightly exaggerated in vertical height. The model (referring to a field example in Alberta, Canada) shows a ramp setting sequence constructed by shifting equilibrium profiles according to the relative sea-level curve at the right of the figure. Erosion and deposition happened along the axis of an incised valley, well verified during the falling stage of the relative sea-level, during regression. At the base of the section, a transgressive series sits on top of a maximum flooding surface of an older series, providing a Highstand Systems Tract. During regression, coastal sediments of the Falling Stage Systems Tract prograde toward the sea to the right. As regression continues, the proximal fluvial environment also prograde seaward and deeply erodes the coastal deposits. In the Lowstand Systems Tract, at the most basinward position, marine sediments are laid down. During subsequent transgression, the Transgressive Systems Tract follows with coastal plain muddy sediments of a backbarrier environment located behind shoaling sands and barrier islands. The Highstand Systems Tract causes frequent flooding, ravinement, and lastly a maximum flooding, the surfaces of all of them trending landward. On top of the Maximum Flooding Surface, the Highstand Systems Tract shows a variety of coastal environments which are fed from a river net widely extending to the hinterland.

of erosion, and can be traced from here in a landward direction (Fig. 7). The base level correlates with the height of the relative sea level. This is clear at the coast, but may be ambiguous in continental basins far off the sea, as the drainage system of major rivers also is guided by the structural model of the basin. Cross and Homewood (1997: p. 1624–1625) sum up the modern understanding of the base level concept, that will be cited partially in the following.

Barrell (1917): "Where base level is above the earth's surface, sediment will accumulate if sediment is available. Where base level is below the earth's surface, sediment is eroded and transported downhill to the next site where base level is above the earth's surface. These transits of base level ... produce a stratigraphic record ... as regular vertical successions of facies separated by surfaces of unconformity. ... time [is] continuous and ... is fully represented in the stratigraphic record by the combination of rocks plus surfaces of stratigraphic discontinuity."

Wheeler (1964) "considered stratigraphic base level as an abstract (nonphysical), nonhorizontal, undulatory, continuous surface that rises and falls with respect to the earth's surface. Stratigraphic base level is a descriptor of the interactions between processes that create and remove accommodation space.

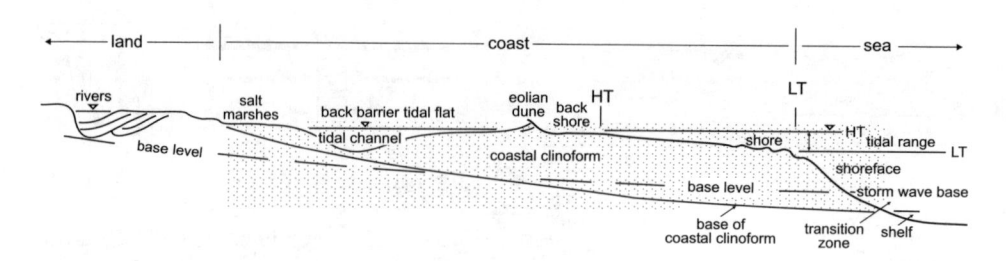

Fig. 7. The base level is fixed to the storm wave base at the coastal clinoform, separating the 'shoreface' from the 'transition zone'. During storms, especially winter storms, the coast can be reworked, also totally removed during a slow transgression. From here, the base level may be traced to continental environments in the coastal alluvium. Depending on the tendency 'transgression' or 'regression' of the sea, this surface of potential erosion shifts toward the land or toward the sea. It results in 'base level rise' or 'base level fall', each precisely depicted by a sequence of depositional features at any place of observation between the deep sea and continental alluvium (see text for more details).

... defined stratigraphic base level as a potentiometric energy surface that describes the energy required to move the Earth's surface up or down to a position where gradients, sediment supply, and accommodation are in equilibrium."

Wheeler's stratigraphic base level is the conceptual device that links several concepts (Cross and Homewood 1997):

– Gressley's facies and facies tracts,
– Wather's notion of the 'zig zag' uphill and downhill movement of facies tracts,
– Gressley's and Walther's identification that this movement is recorded as regular vertical facies successions that define stratigraphic cycles,
– Barrel's notion that base level cycles are the clock of stratigraphy (and therefore that correlations based upon physical stratigraphy are possible),
– Barrell's concept of sediment volume partitioning.

The base level method has been adopted since a while to sedimentological field research (Cross et al. 1993; Cross and Lessenger 1998). With some simplification, base level rise considers a deepening of the environment due to an increase of the A/S rate (accommodation space to sediment supply), whereas base level fall describes its shallowing due to a decrease of the A/S rate. Both half cycles are more or less equivalent to coastal onlap and coastal offlap of the sequence stratigraphy concept (Galloway 1989). The base level concept resolves a stratigraphic sequence in a much smaller scale, even in a 5th- and 6th-order cyclicity. The cyclic development of the depositional environment — whether it is in the deep sea, on the coastal clinoform, or far off the sea in a continental basin — provides a fining-up or a coarsening-up of the depositional sequence together with a shift of the individual energy content and the balance between the accommodation space of the sedimentary basin and the sediment supply to it. This observation can easily be made by sedimentological measures in field

Depositional Systems and Missing Depositional Records

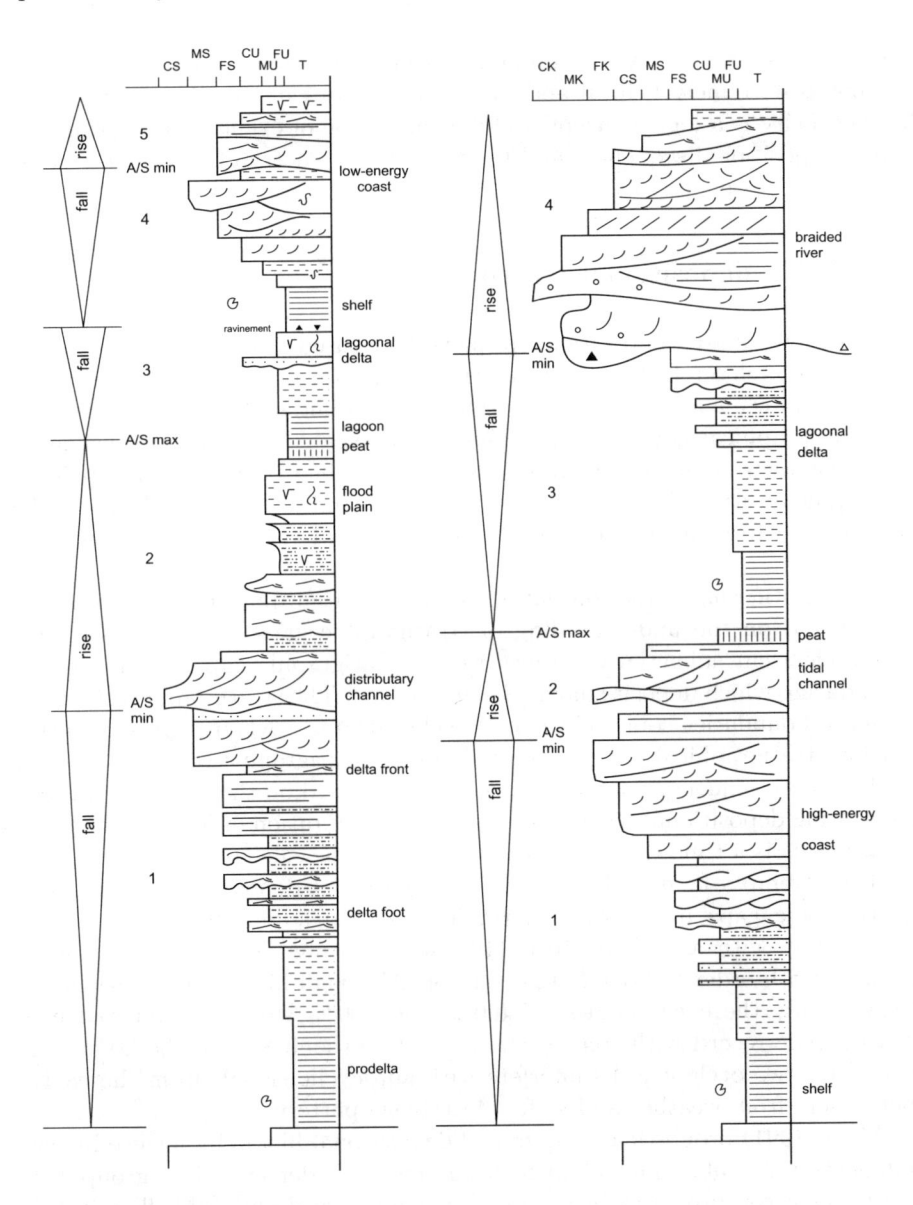

Fig. 8. Two successions of rocks are shown, each consisting depositional sequences which are 'measured' from ideal fluvial, deltaic, and coastal environments, each with the interpretation of the depositional environments and subenvironments alongside. In addition, the base level cyclicity is indicated by triangles pointing upward ('base level rise') or downward ('base level fall'), also the quotient A/S (accommodation space vs. sediment supply), reflecting the variation of the relative sea-level, the structural development of the sedimentary basin, and the sediment supply to it.

and on cores (Fig. 8). As the observation mostly is done in outcrop-scale, this concept is extremely helpful and preferably focuses on small-scale work. Base level cyclicity is a logic concept useful at any scale of cyclicity. It explains individual depositional sequences and the basin-fill geometry of sedimentary basins (Cross and Lessenger 1998).

3 Missing depositional record

As was shown above, depositional sequences are used as entities (cf. Figs. 3 and 4). Yet, at a closer look, numerous gaps of erosion or non-deposition in these sedimentary sequences occur (Dott 1983). Depending on the local environment, the stratigraphical gaps present different time spans. Sedimentological models prefer complete sedimentary sequences, ignoring their origin, stratigraphic, and geographic position. These models are ideal and tend to disregard the gaps which are provided from the environment, due to shifts of transport directions, tides, emersion, etc.

On the other hand, the concept of sequence stratigraphy and base level cyclicity is based on the understanding of stratigraphic gaps. Sequence boundaries with a clear-cut subaerial exposure, reveal stratigraphic gaps in thickness and time, during which deposits having been removed. The stratigraphic relevance of sequence boundaries comes clear in Wheeler diagrams when regressive surfaces provide stratigraphical gaps in the individual stratigraphic sequence (Wheeler 1964). As the sequence stratigraphic concept is based on the 3rd-order cyclicity, the missing depositional record can easily be of a duration of 0.5–3 million years. Clearly, this is a time span only relevant in a basin-scale context.

The attempts to increase the time resolution of depositional cycles and of stratigraphic gaps in between, normally leads to a high-frequency sequence stratigraphy with a cyclicity in a Milankovitch range of a 4th-order (500 ka–100 ka) or 5th-order (100 ka–20 ka) concept. Although high-order cycles are well defined, only the cycle lengths of 400 ka and 100 ka will be deciphered in the stratigraphic record with some certainty, not so the 40 ka and the 20 ka signal. The latter two cycle lengths interfere with autocyclic signals from the environment itself, from weather and shifts of drainage patterns.

Miall (1991) proposes a ten-grouped depositional hierarchy for use in clastic sediments that subdivides a range from the 3rd-order cyclicity (group 10) to burst-and-sweep processes in traction currents or grain-fall and -flow in eolian dunes (group 1). This range embraces a time-scale from 3 million years to a few seconds. The high-order groups 10 to 8 are established parallel to the 3rd-, 4th-, and 5th-order of the sequence stratigraphic concept, whereas below group 7, a 6th-order cyclic behaviour of depositional processes is revealed, both in continental and marine environments. The physical scale of the depositional units are of 14 orders of magnitude. Also, the sedimentation rates are from 10^5 m/ka (group 2) to 10^{-1} to 10^{-2} m/ka (groups 9 and 10). Bounding surfaces, surfaces of non-deposition, or erosional discordances are more frequent toward the higher-order groups. Sedimentation rates in low-order groups depend on

coastal sequence

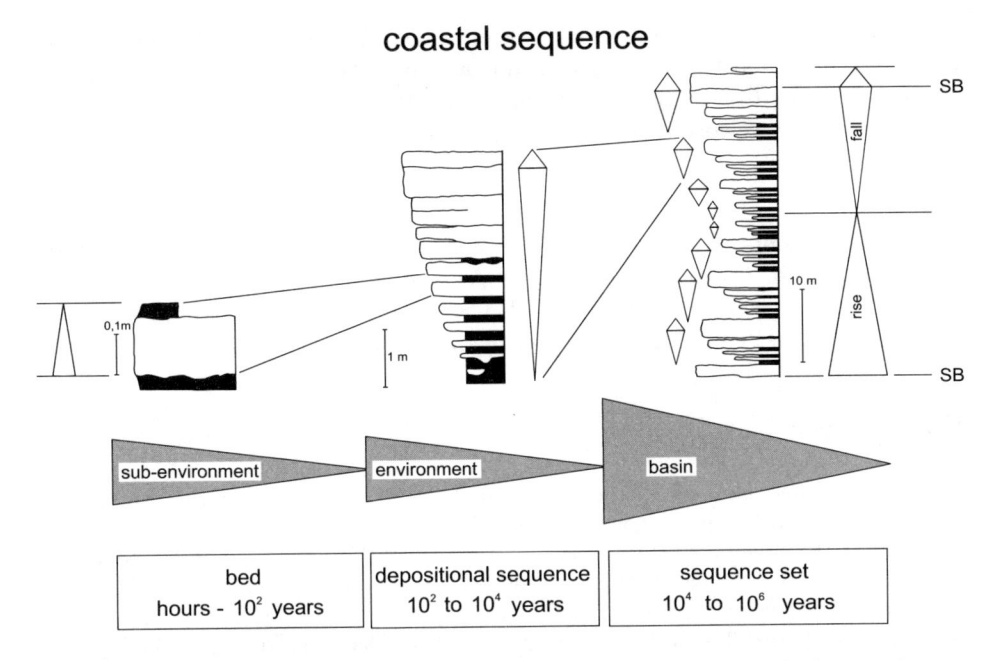

Fig. 9. Coastal depositional sequences consisting each of a set of beds, forming details of a full transgressive-regressive cycle as used in basin analysis (modified after Aigner et al. 1990, following the concept of Cross and Lessenger 1997). The base level cyclicity (verified as hemi-cycles) is indicated by triangles (base level rise and base level fall). Moreover, time ranges for the deposition, erosion, and redeposition of the individual beds, depositional sequences, and sequence sets are shown.

sedimentation/erosion processes in even one depositional cycle, whereas in high-order groups depositional processes are more due to the accommodation space of the environment, and lastly, of the entire sedimentary basin. As a consequence, "the sedimentation rate varies in inverse relation to the length of time over which the measurement is made".

Base level cyclicity enables to show more from the small-scaled variation of the environments (Cross and Lessenger 1997). In a continental environment, channel migration, floodplain emersion, multiple stacking of fluvial fan lobes etc. can be defined as base level rise cycles. These may be of a length of a 5th- or a 6th-order cyclicity. This environment-bound cyclicity defines flooding and erosion unconformities between depositional sequences. On the other hand, it enables to sum up small-scaled stacked channels to define a thick fluvial succession with relevance to the fill of a large-scaled sedimentary basin (Ramón and Cross 1997). Each, the individual environment and the basin-fill can well be verified. Missing depositional records provide precise surfaces on top of environments and paleogeographic settings. They form time lines which can be understood as interbasinal correlation surfaces. As an example, a set of coastal sequences show the hierarchy of sub-environments and environments for use in basin analysis (Fig. 9). The variability of scales of time used for deposition of these beds and their thicknesses may indicate that also gaps exist in which nothing was deposited or — more often in a coastal environment — in which beds were removed by reworking during winter storms or by variations of the relative sea level, in detail and in basin-scale.

Acknowledgements I appreciate the discussion that Tim A. Cross (Colorado School of Mines, Golden, Colorado, USA) offered to an early version of this paper. Also, the help of Agemar Siehl, Clemens Simmer, and Christian Derer (all from University of Bonn) is kindly acknowledged.

References

Aigner, Th., Bachmann, G. H. and Hagdorn, H. (1990): Zyklische Stratigraphie und Ablagerungsbedingungen von Hauptmuschelkalk, Lettenkeuper und Gipskeuper in Nordost-Württemberg. Jber. Mitt. oberrhein. geol. Ver., N. F. 72, 125–143.

Allen, J. R. L. (1979): Physical Processes of Sedimentation. 248 pp., 5th ed., (Allen and Unwin) London.

Allen, J. R. L. (1982): Sedimentary Structures. Their Character and Physical Basis. — volume I, 594 pp.; volume II, 644 pp., (Elsevier) Amsterdam, New York.

Ashley, G. M. (symposium chairperson) (1990): Classification of large-scale subaqueous bedforms: A new look at an old problem. J. Sedimentary Petrology, 60, 160–172.

Barrell, J. (1917): Rhythms and measurement of geologic time. — Geol. Soc. America, Bull., 28, 745–904.

Blum, M. D. and Tornqvist, T. E. (2000): Fluvial responses to climate and sea-level change: a review and look forward. Sedimentology, 47, Suppl. 1, 2–48.

Boggs, S. jr. (1995): Principles of Sedimentology and Stratigraphy. — 774 pp., 2nd ed., (Prentice-Hall) Englewood Cliffs, NJ.

Cross, T. A. et al. (1993): Application of high-resolution sequence stratigraphy to reservoir analysis. — in: Eschad, R. and Doligez, B. (eds.): Subsurface reservoir characterisation from outcrop observations. Proceedings of the 7th Exploration and Production Research Conference, Technip, 11–33, Paris.

Cross, T. A. and Homewood, P. W. (1997): Amanz Gressley's role in founding modern stratigraphy. — Geol. Soc. America, Bull., 109, 1617–1630.

Cross, T. A. and Lessenger, M. A. (1997): Correlation strategies for clastic wedges. — in: Coulson, E. B.; Osmond, J. C. and Williams, F. T. eds.): Innovative Applications of Petroleum Technology in the Rocky Mountain Area. Rocky Mountain Association of Geologists, 183–203.

Cross, T. A. and Lessenger, M. A. (1998): Sediment volume partitioning: rationale for stratigraphic model evaluation and high-reolution stratigraphic correlation. — in: Gradstein, F. M., Sandvik, K.O. and Milton, N. J., Sequence Stratigraphy — Concepts and Applications, NPF Spec. Publ., 8, 171–195.

De Mowbray, T. and Visser, M. J. (1984): Reactivation surfaces in subtidal channel deposits, Oosterschelde, Southwest Netherlands. J. Sediment. Petrol., 54, 811–824.

Dott, R. H. jr. (1983): 1982 SEPM Presidential Address: Episodic sedimentation — how normal is average? How rare is rare? Does it matter?, J. Sed. Petrol., 53, 1, 5–23.

Einsele, G. (2000): Sedimentary Basins. Evolution, Facies, and Sediment Budget. — 792 pp., 2nd ed., (Springer) Heidelberg, Berlin, New York.
Emery, D. and Myers, K. J. (ed. 1996): Sequence Stratigraphy. — 297 pp., (Blackwell) Oxford.
Fielding, C. R. (1987): Coal depositional models for deltaic and alluvial plain sequences. Geology, 15, 661–664.
Galloway, W. E. (1989): Genetic stratigraphic sequences in basin analysis I: architecture and genesis of flooding- bounded depositional units. Amer. Assoc. Petrol. Geol., Bull., 73, 125–142.
Haq, B. U.; Hardenbol, J. and Vail, P. R. (1987): Chronology of fluctuating sea levels since the Triassic. Science, 235, 1156–1166.
Hardenbol, J.; De Graciansky, C.; Jacquin, T. and Vail, P.R. (eds. 1998): Mesozoic and Cenozoic Sequence Stratigraphy of European Basins. Soc. Econ. Pal. Min., Spec.Publ., 60, 1–485.
Leeder, M. R. (1999): Sedimentology and Sedimentary Basins. From Turbulence to Tectonics. — 592 pp., (Blackwell) Oxford.
McCabe, P. J. and Jones, C. M. (1977): Formation of reactivation surfaces within superimposed deltas and bedforms. J. Sediment. Petrology, 47, 707–715.
Miall, A. D. (1991): Hierarchies of architectural units in terrigenous clastic rocks, and their relationship to sedimentation rate. — in: Miall, A. D. and Tyler, N. (eds.): The three-dimensional facies architecture of terrigenous clastic sediments and its implications for hydrocarbon discovery and recovery.- SEPM Concepts in Sedimentology and Paleontology, 3, 6–12, Tulsa.
Moslow, T. F. (1984): Depositional models of shelf and shoreline sandstones. Amer. Assoc. Petrol Geol., Contin. Educ. Course Note, Series 27, 102 pp.
Nummedal, D., Cole, R., Young, R., Shanley, K. and Boyles, M. (2001): Book Cliffs Sequence Stratigraphy: The Desert and Castlegate Sandstone. — 81 pp., GJGS/SEPM Fied Trip # 15, June 6–10, 2001 (in conjunction with the 2001 Annual Convention of the American Association of Petroleum Geologists); Denver, Colorado.
Pettijohn, F. J., Potter, P. E. and Siever, R. (1987): Sand and Sandstone. — 553 pp., 2nd ed., (Springer) Berlin, Heidelberg, New York.
Posamentier, H. W. and James, D. P. (1993): An overview of sequence-stratigraphic concepts: uses and abuses. Spec. Publs. Int. Ass. Sediment., 18, 3–18.
Ramón, J. C. and Cross, T. A. (1993): Characterization and prediction of reservoir architecture and petrophysical properties in fluvial channel sandstones, middle Magdalena Basin, Colombia. CT and F-Ciencia, Technología y Futuro, 1, 3, 19–46.
Reading, H. G. (1986): Sedimentary Environments and Facies. — 615 pp., 2nd ed., (Blackwell) Oxford.
Reineck, H.-E. (1960): Über Zeitlücken in rezenten Flachsee-Sedimenten. Geol. Rundschau, 49, 149–161.
Reineck, H.-E. and Singh, I. B. (1980): Depositional Sedimentary Environments. — 549 pp., 2nd ed., (Springer) Heidelberg, Berlin, New York.
Reineck, H.-E.; Singh, I. B. and Wunderlich, F. (1971): Einteilung der Rippeln und anderer mariner Sandkörper. Senckenbergiana maritima, 3, 93–101.
Rubin, D. M. (1987): Cross-bedding, bedforms, and paleocurrents. Concepts in Sedimentology and Paleontology, 1, 1–187, Tulsa.
Singh, I. B. and Kumar, S. (1974): Mega and giant ripples in the Ganga, Yamuna, and Son rivers, Uttar Pradesh, India. Sediment. Geol., 12, 53–66.
Southard, J. B. and Bogouchwal, L. A. (1990): Bed configurations in steady unidirectional water flows. Part 2: Synthesis of flume data. J. Sediment. Petrol., 60, 658–679.

Sundborg, A. (1967): Some aspects on fluvial sediments and fluvial morphology. I. General views and graphic methods. Geograf. Ann., 49, 333–343.

Tanner, W. F. (1978): Reynolds and Froude Numbers. — in: Fairbridge and Bourgeois (eds.): The Encyclopedia of Sedimentology. Encyclopedia of Earth Sciences Series, 6, 620–622.

Vail, P. R.; Audemard, F.; Bowman, S. A.; Eisner, P. N. and Perez-Cruz, C. (1991): The stratigraphic signatures of tectonics, eustasy and sedimentology — an overview. — in: Einsele, G.; Ricken, W. and Seilacher, A. (eds.): Cycles and Events in Stratigraphy. — 617–659, (Springer) Berlin, Heidelberg, New York.

Walker, R. G. (ed. 1984): Facies Models. Geoscience Canada Reprint Series 1, 317 pp., 2nd ed., (Geol. Assoc. Canada) St. Johns.

Walker, R. G. and James, N. P. (ed. 1992): Facies Models. Response to Sea Level Change. — 409 pp., (Geol. Assoc. Canada) St. Johns.

Walther, J. (1893/1894): Einleitung in die Geologie als historische Wissenschaft. Beobachtungen über die Bildung der Gesteine und ihrer organischen Einschlüsse. — Dritter Teil: Lithogenesis der Gegenwart. Beobachtungen über die Bildung der Gesteine an der heutigen Erdoberfläche. — 535–1055, (Gustav Fischer) Jena.

Wheeler, H. E. (1964): Baselevel, lithosphere surface, and time-stratigraphy. Geol. Soc. Amer., Bull., 75, 599–610.

Weimer, P. and Posamentier, H. W. (eds. 1993): Siliciclastic sequence stratigraphy. Amer. Assoc. Petrol. Geol., Memoir, 58, 1–492.

Wilgus, C. K.; Hastings, B. S.; Kendall, C. G. S. C.; Posamentier, H. W.; Ross, C. A. and Van Wagoner, J. C. (eds. 1988): Sea-level changes: An integrated approach. Soc. Econ. Pal. Min., Spec. Publ., 42, 1–407.

Multi-Scale Processes and the Reconstruction of Palaeoclimate

Christoph Gebhardt[1]*, Norbert Kühl[2], Andreas Hense[1], and Thomas Litt[2]

[1] Meteorological Institute, University of Bonn, Germany
[2] Institute for Palaeontology, University of Bonn, Germany

Abstract. In this section, we focus on multi-scale problems encountered when reconstructing palaeoclimate on the basis of terrestrial botanical proxy data (pollen and macro remains). Climate and the taxonomical composition of the vegetation are affected and linked by complex geo-processes acting on various spatial and temporal scales. We describe how the resolution of these data — both in space and time — limits the scale of features which can be reconstructed. We then present a Bayesian statistical transfer function relating vegetation and climate which is based on a Boolean approach ("taxon existent"–"taxon non-existent"). This transfer function allows for an effective avoidance of "no modern analogue"-situations and intents to reduce the influence of sub-scale processes (e. g. local effects) and of non-climatic factors on the reconstruction.

We present two types of transfer functions: a reconstruction model suitable for reconstructions of local palaeoclimate and a backward model which can be used to reconstruct spatially coherent climatological fields. The reconstruction model is given by probability density functions for the climate conditional on the existence of certain taxa. The backward model is deduced from the reconstruction model in a straightforward way using the Bayes theorem. For the spatially coherent reconstruction, the backward model is combined with a dynamical constraint in order to stabilise the reconstruction which is normally based on a relatively sparse network of palaeo-observations. We use a singular value decomposition of the dynamics to restrict the reconstruction to the dominant and resolvable scales.

1 Introduction

The reconstruction of palaeoclimate depends on the use of proxy data, because it is not directly measurable. The relation of proxy data to modern climate conditions can be converted into a transfer function which is applied back in time. The description of the underlying processes by the transfer functions can be based either on a statistical approach or on an explicit modelling.

Various types of data can serve as substitute for direct meteorological measurements, e. g. physical or biological (cf. Berglund 1986, Frenzel 1991, Schleser

---
* c.gebhardt@uni-bonn.de

et al. 1999). In the following, pollen and macro remains serve as an example for botanical proxy data. In suitable environments, they are conserved over long periods. The use of these data implies that a significant relation between the proxy data, i.e. the plants that produced the fossils, and the considered climate parameters exist. However, organisms are influenced by and interact with the environment in a very complex way. Moreover, the influence of climate as well as other factors differ on different scales, urging to address the appropriate scale.

Palaeo-conditions can be calculated quantitatively by transfer functions using today's dependences of the proxy data. For establishing the relationship between biological data and environmental variables multivariate numerical analyses were used based on multiple linear regression models (Birks and Birks 1980, Birks and Gordon 1985). In this respect climate estimates from botanical data and their application for palaeoclimatological reconstructions are important, but the quantification of the function relating vegetation and climate is problematic (see the classical study by Iversen 1944 using climatic indicator species). First numerical methods in pollen analysis were developed by Webb and Bryson (1972) to reconstruct palaeoclimate. This method was later adopted and modified (Guiot 1990; Huntley and Prentice 1988, Pons et al. 1992, Field et al. 1994). However, the lack of modern analogues for several fossil pollen spectra is a disadvantage of this method (as measured by the dissimilarity coefficients, see Overpeck et al. 1985).

The study of the plant — climate relations depends on the knowledge of the plant distributions (chorology) and the climate conditions in the studied geographical region. Fig. 2 shows an example of the combination of a gridded climatology and distribution area. To which extent factors influence plant distribution depends, amongst others, on the regarded scale. As can be learned from the contributions of the previous chapters, the methods suitable for the description of the dynamics of and the interactions between different earth systems (e.g. atmosphere, biosphere, hydrosphere, ...) depend on the considered scales. On the comparatively small scales, explicit modelling of the dynamical processes and the coupling between different scales is possible (Mendel et al. 2002). On larger scales, the numerical solution of the underlying differential equations is only feasible using a coarser resolution. The effects of the subgrid-scale processes and their interactions with the resolved scales can be parametrised in dependence of the large-scale features (Gross et al. 2002; Heinemann and Reudenbach 2002) or their mean behaviour is described using "effective parameters" (Diekkrüger 2002). Our contribution deals with the large-scale end of the processes presented in this book. On these large spatial and temporal scales, it is sufficient and more suitable to apply a statistical model which summarises the effects of the interaction between climate and vegetation using a few parameters. The atmospheric dynamics can be satisfactorily described as quasi-stationary processes acting on the broadest spatial scales. The quantification of the dependences of proxy data on climate is closely connected to the chosen scale. Therefore it is essential to find the suitable scales. In the following, we will demonstrate that the large scales are the important scales in the presented reconstruction problem.

Multi-Scale Processes and the Reconstruction of Palaeoclimate

2 Variability of data on different spatial and temporal scales

2.1 The meteorological data

We consider the case that the climatological data set is given as gridded fields of climatic near-surface variables. Such data are subject to various types of variability on different temporal and spatial scales. The spatial variability includes

- large-scale effects (O(10,000)–O(1,000)) km) due to solar insolation variations at the top of the atmosphere and land-sea contrasts
- meso-scale orographic and land-sea distribution effects (O(1,000)–O(100) km)
- and small-scale effects (< O(100) km) due to soil type, orography, and vegetation.

The temporal variability is composed of

- long timescale effects due to astronomic forcing factors (Milankovitch cycles) and changes in the oceanic thermohaline circulation
- decadal and interannual variations due to El Niño or long term changes of the North Atlantic Oscillation
- fast variations like the annual or even the daily cycle.

The palaeoclimate reconstruction is restricted to certain spatial and temporal scales which are basically determined by the characteristics specific to the given data — the recent climatological and botanical data as well as the palaeobotanical data —, the assumptions related to the chosen method of reconstruction, and the characteristics of atmospheric dynamics. First of all, the grid-spacing of the climatological data (e.g. 0.5° × 0.5° longitude-latitude grid (New et al. 1999)) determines a lower limit of the spatial scale that can be resolved by the reconstruction. However, due to the multi-scale variability inherent in the atmospheric dynamics, aliasing effects representing undersampled sub-scale variability might be misinterpreted as features on the grid scale.

Furthermore, a reconstruction based on terrestrial botanical proxy data assumes identical dependences on climate for both modern botanical observations and palaeovegetation. Additionally, the reconstruction of a high-dimensional field from a small set of palaeo-observations is a (seriously) ill-defined regression problem which can only be stabilised by additional information such as a given climatic reference field. In consequence of the last two points, the palaeoclimate to be reconstructed should be similar to an a priori chosen background state, e.g. the present climate. However, it is yet unclear how large the maximum deviation can be. A safe meteorological estimate would be to avoid deviations larger than about ±5 K which is about the internal variability scale of present-day natural climate fluctuations in Europe. Due to the characteristics of the atmospheric dynamics, such anomalies or deviations of climatic variables from the modern mean are basically large-scale features: If we assume that e.g. the rate of change of the temperature $\frac{dT}{dt}$ following the trajectory of an air parcel is controlled by a forcing function Q depending on space and time we get an equation which describes the temporal and spatial structure of the temperature field completely. This equation states that the temporal rate of change of the field structure can be described by the local temperature change and the horizontal advection in a given flow $\vec{v}$. We can write for two realisations $i = 1, 2$, e.g. the present day and the climate of the Last Interglaciation (Eemian),

$$\frac{dT}{dt} = \frac{\partial T_i}{\partial t} + \vec{\nabla}(\vec{v}_i T_i) = Q_i \qquad i = 1, 2 \qquad (1)$$

thereby assuming non-divergent flow conditions ($\vec{\nabla}\vec{v}_i = 0$).

As we investigate the structure of climate anomalies, we introduce the average temperature $\bar{T}$ and its anomaly T' as

$$\begin{aligned}\bar{T} &= \tfrac{1}{2}(T_1 + T_2) \\ T' &= T_1 - T_2.\end{aligned} \qquad (2)$$

Taking the difference of the process equations above we obtain an equation for the anomalies

$$\frac{\partial T'}{\partial t} + \vec{\nabla}(\bar{v}T') + \vec{\nabla}(\vec{v}'\bar{T}) = Q' \qquad (3)$$

where the averages and anomalies of the flow velocity and the forcing function are given as defined above for the temperature (Eq. 2).

Additionally it is assumed that the effects of the anomalous flow $\vec{v}'$ upon the anomalous temperature can be parametrised as a diffusion process (K being the diffusion coefficient) leading to an equation for the temperature anomaly T':

$$\frac{\partial T'}{\partial t} + \vec{\nabla}(\bar{v}T') - \vec{\nabla}(K\vec{\nabla}T') = Q'. \qquad (4)$$

Thus we have derived a linear advection-diffusion equation for the anomalies. The non-linearities are transformed into the equation for the average which needs not to be considered in the present context because anomalies from a background climate are studied. As Eq. (4) is solved on a regular spatial grid with N grid points it has to be discretised. Let $\vec{T}$ be an N-dimensional vector with the grid values of T' as components. In consequence, the local rate of change and the linear advection-diffusion operator can be written as an $N \times N$ matrices $\mathbf{D}$ and $\mathbf{M}$, respectively. Thus the discretised equation is given by

$$\mathbf{D}\vec{T} + \mathbf{M}\vec{T} = \vec{Q} \qquad (5)$$

where $\vec{Q}$ is defined as the vector of forcing anomalies analogous to $\vec{T}$.

If we further assume a quasi-stationary anomaly field $\mathbf{D}\vec{T} \approx 0$, the solution of the remaining equation can be discussed best within the framework of dynamical modes of the advection-diffusion operator $\mathbf{M}$. The dynamical modes can be found by performing a singular value decomposition (SVD) of the operator matrix. The SVD of the matrix $\mathbf{M}$ is given by $\mathbf{M} = \mathbf{U}\mathbf{\Lambda}\mathbf{V}^\mathsf{T}$ (Press et al. 1992) where the superscript T denotes the transpose of a matrix. The matrix $\mathbf{U}$ contains the left singular vectors $\vec{u}_n$ in its columns ($\mathbf{U}^T\mathbf{U} = \mathbf{1}$), $\mathbf{\Lambda}$ is a diagonal matrix with the singular values λ_n, $n = 1, \ldots, N$ as components ($\lambda_n \geq 0$). The right singular vectors $\vec{v}_n$ are the columns of $\mathbf{V}$ and build a complete, orthogonal basis of the N-dimensional space ($\mathbf{V}^T\mathbf{V} = \mathbf{V}\mathbf{V}^T = \mathbf{1}$). It is convenient to take these vectors as dynamical modes of the advection-diffusion operator. The temperature anomalies can be expressed in terms of these modes as $\vec{T} = \sum_{n=1}^{N} \tau_n \vec{v}_n$ with each coefficient given by $\tau_n = \vec{v}_n^T \cdot \vec{T}$. Writing the τ_n into a vector $\vec{\tau}$ and inserting the SVD in the stationary form of Eq. (5) leads to

$$\mathbf{M}\vec{T} = \mathbf{U}\mathbf{\Lambda}\mathbf{V}^T\vec{T} = \mathbf{U}\mathbf{\Lambda}\vec{\tau} = \vec{Q}. \qquad (6)$$

The solution is thus given by $\vec{\tau} = \mathbf{\Lambda}^{-1}\mathbf{U}^T\vec{Q}$ and can be transformed back to $\vec{T} = \mathbf{V}\vec{\tau}$. Note that the forcing anomalies $\vec{Q}$ are unknown but not zero. In case this forcing has a non zero projection upon the left singular vectors $\vec{u}_n$, the stationary solution for $\vec{T}$ is strongly affected by those dynamical modes which have the smallest singular values λ_n. Experience shows that these modes possess the largest spatial scales which means that small-scale features of the considered variable are subjected to a larger "friction" than the large-scale deviations. Thus large-scale deviations have a longer life time than small-scale anomalies thereby determining the stationary solution. This is exemplified in Fig. 1 which shows for the study area Mid-Europe the two gravest modes (b,c) for a discretised advection-diffusion operator using as a mean flow $\vec{v}$ the climatological mean for January (a) at about 1,500 m height above sea level (pressure height 850 hPa). The modes resemble Fourier modes subjected to modifications by advection, geometry and boundary conditions describing in (b) a positive definite but non-uniform contribution and in (c) a zonally uniform meridional gradient with negative contribution in the northern half of this area. This example shows that the large-scale features emerging from the atmospheric dynamics are the dominant scales of spatial variability. Thus non-representative small-scale variability in the climatological as well as in the botanical data is effectively treated by constraining the reconstruction to the large scales of the dominant modes of an atmospheric model such as the advection-diffusion model.

Furthermore, using only a small number of large-scale dynamical modes in the reconstruction reduces the dimension of the climatic phase space which helps to further stabilise the reconstruction based on a small number of palaeo-observations. It will be shown in Section 3 how the dynamical constraints can be incorporated in the reconstruction.

The time scale of the modern climate is set by the averaging period of the data set which usually ranges over several decades (e. g. 1961–1990 (New et al. 1999)). When considering temporal scales, it has to be kept in mind that transfer functions derived from vegetation data rely on the assumption of a stationary biome model where the species distribution is in equilibrium with the forcing parameters. This assumption is not valid for periods with highly dynamical climate that results in dynamic, short-term deviations in vegetation patterns during transition periods. It is clear that under rapid transient conditions which were probably observed at the end of the last glacial period, equilibrium biome models

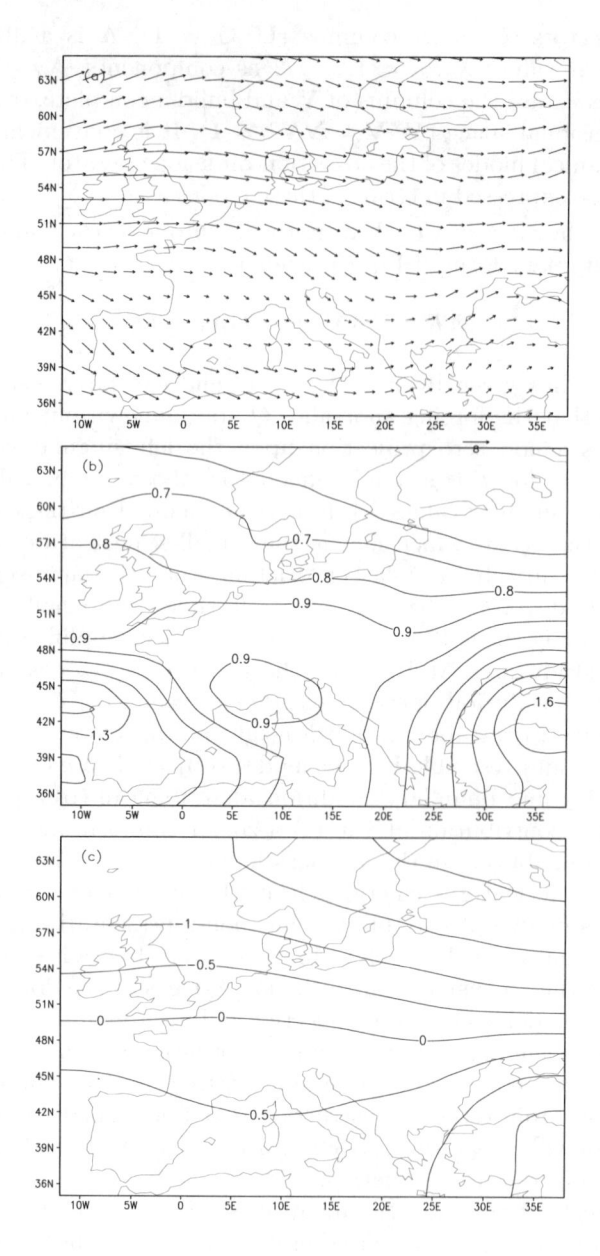

Fig. 1. Example for the structure of dynamical modes of a linear advection-diffusion model to describe the temperature field for palaeo-reconstruction over Mid Europe. (a) average flow field $\bar{\vec{v}}$ climatological mean at pressure level 850 hPa ($\sim$ 1,500 m above sea level) for January, (b) and (c) the two gravest dynamical modes of a discrete advection diffusion operator Eq. (3)

can not be used directly for reconstruction. Furthermore, if the sampling interval of the palaeo-observations from the stratigraphy is larger than the response time of the vegetation in case of a transient climate variation aliasing effects could result in mimicking slow temporal variability through undersampled fast processes.

2.2 The botanical data

The distribution of plants changes over time for various reasons, of which climatic changes are a main driving force on all scales. However, species distributions depend on other factors as well. The impact of these factors on the occurrence and abundance of a taxon varies on different spatial and temporal scales. This impact has to be recognised and eliminated as far as possible when plants are used as proxy data solely for macroclimate. Major factors on a large scale include the anthropogenic impact, the atmospheric composition (e. g. CO_2 content) and competition as well as the vegetation history. The relief, soil properties and soil development, ecological factors (e. g. fire and diseases) and hydrology have a pronounced influence on plant distribution on a somewhat smaller scale.

Today the anthropogenic impact plays a crucial role for the composition of the vegetation. Without anthropogenic help several taxa would not have enough competitive strength to be abundant in plant communities which otherwise would be dominated by stronger competitors. The modern chorology should include the present knowledge about anthropogenic changes of the vegetation and provide maps of the natural species distributions, e. g. maps of Meusel et al. (1964), Meusel et al. (1978), Meusel and Jäger (1992).

Competition as a natural source of interference has a major influence on plant distribution on small and large spatial as well as temporal scales. Competition leads to a difference between the physiological optimum of a plant and its ecological optimum (Ellenberg 1996) which depends on the interaction with other plants and is therefore hard to determine. For example, although climatic conditions would be favourable for gymnosperms such as *Pinus* (pine) to grow all over Europe, its modern natural habitat is limited to poor soils or quite cold climatic conditions. Only in these areas, the absence of stronger competitors allows *Pinus* to survive. Ancient plant societies may differ from today, resulting in a possible error by assuming present-day plant dependence on climate under today's competitive situation being the same as it was under palaeo-conditions. Since little is known about the complex modifications of distributions, this assumption has to be made in fact. However, the problem can be minimised by addressing the presence and absence of a taxon (Boolean approach) rather than its abundance. When macro climate conditions are favourable for a plant to grow at all, its abundance is also subject to other factors than climate. However, the less favourable the overall climate conditions are, the more likely it is that other factors dominate up to the extent that the species does not occur. How this fact can be used for a statistical approach when determining the transfer function is shown in the next chapter.

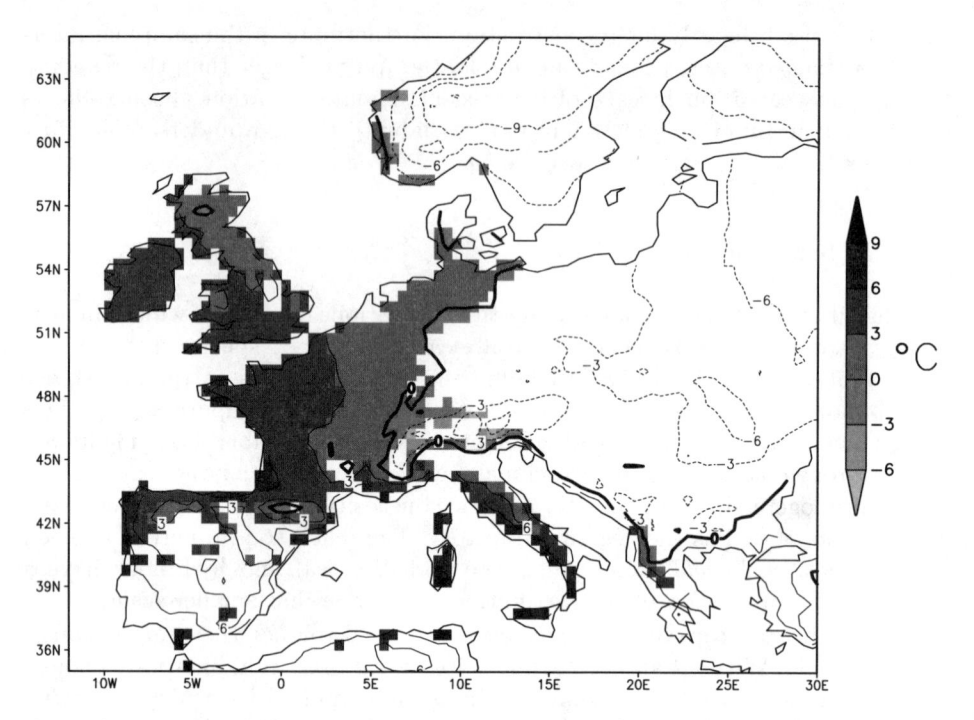

Fig. 2. Combination of the distribution area of *Ilex aquifolium* (coloured areas) with January temperature (in °C, as indicated by colours and isolines)

The use of a relatively large resolution has the major advantage of reducing the influence of factors with impact on a smaller scale. Among these factors are — besides the ones already described — the geology of the substratum as well as the impact that animals have on vegetation (grazing, dispersion). The geology determines hydrological and soil features, but also the relief which can cause severe differences between micro and macro climate conditions. Comparing plant distribution and climate on a large scale, it is unlikely for climate values in areas with a strong relief to represent the spot where the plant actually occurs. For that reason, those data have to be omitted, as is the case e. g. for the Alpes (compare Fig. 2). In addition to the chosen scale, the use of the information about the presence and absence of taxa (Boolean approach) rules out the danger of a great portion of systematic errors, because species abundance is subject to small-scale influences to a much greater extent than just species occurrence.

3 The botanical-climatological transfer functions

In general, two kinds of transfer functions can be established: a reconstruction model where climate is described in dependence of the proxy data and a backward model which assigns a pattern of proxies to a given climatic state. These models can be given either by an explicit model of the underlying processes or by a statistical approach which summarises the effects of these processes. The suggested form of a statistical reconstruction model is based on a Bayesian approach: The reconstruction model R is defined as the conditional probability density function (pdf) for the climate given the co-existence of one or several taxa $R := \text{pdf}(climate|vegetation)$. Such an approach transforms the processes controlling the interdependence between vegetation and climate to an expression which assigns limiting and optimal climatic states to a given spectrum of plants. The pdf can be estimated from the recent botanical and climatological data by either a non-parametric (e. g., Kernel) or a parametric (e. g., Gaussian) density function (Fig. 3).

In the first case, the smoothness of the estimated pdf and therefore the approximation level of the statistically summarised processes can be controlled easily. With a parametric estimation, an a priori assumption about the relevant processes is made and their effect is described by the parameters of the pdf. The

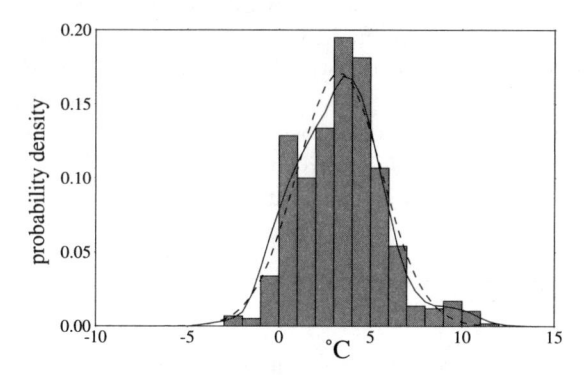

Fig. 3. Probability density functions describing the probability of January temperature, given the occurrence of *Ilex aquifolium*: histogram, Kernel-approximation (—) and Gaussian function (—)

pdf(*climate|vegetation*) is suitable for local reconstructions of the palaeoclimate which is gained from the pdf by the most probable climate and its uncertainty range given the taxa of the palaeorecord at a specific site. Due to the statistical nature of the transfer function, the inclusion of all the co-existing taxa using a common pdf results in a reduction of the uncertainty range. This follows from the gain of information which is provided by the existence of each taxon.

For a spatially coherent reconstruction of a climatological field, the usually small number of fossil sites is insufficient. Thus auxiliary conditions such as constraints C on the reconstructed climate have to be used to overcome these problems. As these constraints are normally given as functions of the climatic state, the backward model is chosen for the reconstruction. A statistical backward model describes the probability $B = \text{Prob}(V|climate)$ of the existence of a certain taxon V given a specific climatic state and can be deduced from the pdf(*climate*|V) with help of the Bayes theorem:

$$\text{Prob}(V|\textit{climate}) = \frac{\text{pdf}(\textit{climate}|V) \cdot \text{Prob}(V)}{\text{pdf}(\textit{climate})}$$

The a priori probability Prob(V) corresponds to an assumption about the existence of taxon V which has to be based on any knowledge other than climate. If there is no prior knowledge, Prob(V) = 0.5 can be chosen. The a posteriori probability Prob(V|*climate*) represents the effect on the taxon V given a specific climatic state according to the processes linking climate and vegetation.

The inclusion of information provided by observational data into the determination of a transfer function can lead to more reliable results. The implementation of a reconstruction model is straightforward as it can be applied to the fossil proxy data directly. A reconstruction on the basis of a backward model is slightly more challenging from a mathematical point of view but it has the advantage that any constraints on the climate (e. g., dynamical processes and their multi-scale variability) can be taken into account easily. Let $\vec{k}$ be the climatic state vector at a given point in space where the dimension of this vector corresponds to the number of reconstructed variables. Then $B(\vec{k})$ is the proxy-information related to a given climate according to the backward model. The constraints on the climate are described by $C(\vec{k}) \approx 0$. A cost function $\mathcal{J}(\vec{k})$ integrates the deviations of $B(\vec{k}_i)$ from the fossil proxy data $B_o(\vec{r}_i)$ at the fossil sites $\vec{r}_i$ and the violation of the constraints over a geographical area Ω:

$$\mathcal{J}(\vec{k}) = \sum_{i=1}^{I} ||B_o(\vec{r}_i) - B(\vec{k}_i)||_B + \gamma \cdot \int_\Omega ||C(\vec{k})||_C \, d\Omega$$

where γ is a weighting factor used to control the relative importance of the two terms, I is the number of fossil sites, and $||\cdot||_B$ and $||\cdot||_C$ are any suitable norms. This cost function summarises the processes 'proxies ⇌ climate' and the physical processes which are assumed to be valid for the climate. The climatic state which minimises the cost function represents the reconstructed palaeoclimate and is found by variational methods.

The reconstruction method using a cost function is adjusted to this special statistical backward model by replacing $B(\vec{k}_i)$ with $\text{Prob}(V|\vec{k}_i)$ and $B_o(\vec{r}_i)$ with 1 (i. e. Taxon V does occur in the palaeorecord) in the cost function. A reasonable reconstruction of a climatological field can only be done on spatial scales at least as coarse as the resolution of the input data, because no information about the smaller scales is available.

4 Summary

The knowledge of plant dependences on climate allows their use for climate reconstructions based on transfer functions. However, the complexity of interaction between vegetation and its environment on various spatial and temporal scales complicates the identification and quantification of transfer functions. This problem can be widely overcome by restricting these functions to the information "taxon existent" instead of the more detailed frequency approach and by applying it to the most preferable scales. The use of relatively large scales and the Boolean approach leads to a reduction of local effects and to the minimisation (but not exclusion) of other impacts than climate. We discussed ways of generating transfer functions (backward model and reconstruction model) on the basis of a quantitative description of the underlying processes using Bayesian statistics. This results in a robust botanical climatological transfer function for the reconstruction of palaeoclimate which is relatively insensitive to sample errors, particularly to those in the modern data. Another major advantage of the Boolean approach is the effective avoidance of "no modern analogue situations". These situations occur in periods with very dissimilar climate compared to present climate and therefore dissimilar vegetation composition.

The presented approaches can serve as tools for climate reconstructions that result in a reconstructed most probable climate on the basis of a given palaeovegetation, including a quantification of a reconstruction uncertainty.

References

Atkinson, T. C., Briffa, K. R. and Coope, G. R. (1987): Seasonal temperatures in Britain during the past 22,000 years, reconstructed using beetle remains. Nature 325:587–582.

Bartlein, P. J., Lipsitz, B. B. and Thompson, R. S. (1986): Climatic response surfaces from pollen data for some eastern North American taxa. Journal of Biogeography 13:35–37.

Bartlein, P. J., Anderson, K. H., Anderson, P. M., Edwards, M. E., Mock, C. J., Thompson, R. S., Webb, R. S., Webb, T., III and Whitlock, C. (1998): Paleoclimate Simulations for North America Over the Past 21,000 Years: Features of the Simulated Climate and Comparisons with Paleoenvironmental Data. Quaternary Science Reviews 17/6–7:549–585.

Berglund, B. E. (ed.) (1986): Handbook of Holocene Palaeoecology and Palaeohydrology. Wiley and Sons Ltd. pp. 407–422.

Birks, H. J. B. and Birks, H. H. (1980): Quaternary Palaeoecology. Arnold, London. 289 pp.

Birks, H. J. B. and Gordon, A. D. (1985): Numerical Methods in Quaternary Pollen Analysis. Academic Press, London and other. 317 pp.

Diekkrüger, B. (2002): Upscaling of hydrological models by means of parameter aggregation technique. This volume.

Ellenberg, H. (1996): Vegetation Mitteleuropas mit den Alpen. 5. Aufl., 1095 pp.

Field, M. H., Huntley, B. and Müller, H. (1994): Eemian climate fluctuations observed in a European pollen record. Nature 371:779–783.

Frenzel, B. (ed.) (1991): Evaluation of climate proxy data in relation to the European Holocene. Gustav Fischer Verlag, Stuttgart and other, 309 pp.

Gross, P., Simmer, C. and Raschendorfer, M. (2002): Parameterization of turbulent transport in the atmosphere. This volume.

Guiot, J. (1990): Methodology of the last climatic cycle reconstruction in France from pollen data. Palaeogeography, Palaeoclimatology, Palaeoecology 80:49–69.

Heinemann, G. and Reudenbach, C. (2002): Precipitation dynamics. This volume.
Huntley, B. and Prentice, J.C. (1988): July Temperatures in Europe from Pollen data, 6000 Years Before Present. Science 241:687–690.
Iversen, J. (1944): *Viscum, Hedera* and *Ilex* as climate indicators. Geol. Fören. Förhandl. 66:463–483.
Mendel, M., Hergarten, S. and Neugebauer, H. J. (2002): Water uptake by plant roots — a multi-scale approach. This volume.
Meusel, H., Jäger, E. and Weinert, E. (1964): Vergleichende Chorologie der zentraleuropäischen Flora -Karten-. Band I, 258 pp.
Meusel, H., Jäger, E., Rauschert, S. and Weinert, E. (1978): Vergleichende Chorologie der zentraleuropäischen Flora -Karten-. Band II, 421 pp.
Meusel, H. and Jäger, E. (1992): Vergleichende Chorologie der zentraleuropäischen Flora-Karten. Band III, 688 pp.
New, M. G., Hulme, M. and Jones, P. D. (1999): Representing 20th century space-time climate variability. In: Development of a 1961–1990 mean monthly terrestrial climatology. J. Climate 12:829–856.
Overpeck, J. T., Webb III, T. and Prentice, I. C. (1985): Quantitative Interpretation of Fossil Pollen Spectra: Dissimilarity Coefficients and the Method of Modern Analogs. Quaternary Research 23:87–108.
Pons, A., Guiot, J., De Beaulieu, L. and Reille, M. (1992): Recent Contributions to the Climatology of the last Glacial-Interglacial Cycle based on French Pollen Sequences. Quaternary Science Reviews 11:439–448.
Press, W. H., Vetterling, W. T., Teukolsky, S. A. and Flannery, B. P. (1992): Numerical Recipes in C: the art of scientific computing. Second edition. Cambridge University Press, 994 pp.
Schleser, G. H., Helle, G., Lücke, A., Vos, H. (1999): Isotope signals as climate proxies: the role of transfer functions in the study of terrestrial archives. Quaternary Science Reviews 18(7):927–943.
Thompson, R. S., Anderson, K. H. and Bartlein, P. J. (1999): Atlas of Relations Between Climatic Parameters and Distributions of Important Trees and Shrubs in North America. U.S. Geological Survey Professional Paper 1650–A+B.
Webb, T., III and Bryson, R. A. (1972): Late- and post-glacial climatic change in the northern Midwest, USA: quantitative estimates derived from fossil pollen spectra by multivariate statistical analysis. Quaternary Research 2:70–115.
Webb, T., III (1980): The reconstruction of climatic sequences from botanical data. Journal of Interdisciplinary history 10:749–772.

A Statistical-Dynamic Analysis of Precipitation Data with High Temporal Resolution

Hildegard Steinhorst, Clemens Simmer*, and Heinz-Dieter Schilling

Meteorological Institute, University of Bonn, Germany

Abstract. The analysis of precipitation data time series with high temporal resolution is carried out. A first task is the definition and separation of independent rainfall events in the time series. Based on the empirical probability density functions of the precipitation features models are fitted to the data. Two classes of models were used, each in the original version and then in a modified version optimised for the high resolution measurements. For the first class of models each rainfall event is described by a Poisson process. The second class of models additionally considers the clustering of rainfall events: the individual events again are generated by a Poisson process, but each event can give rise to a whole cluster of rain cells. The comparison of the simulated autocorrelation functions with the empirical ones shows the advantage of the modified models for the investigated scale.

1 Introduction

Precipitation is the most important component of the hydrological cycle. Information on precipitation is necessary for hydrological investigations like the calculations of runoff and discharge or the prediction of floods. Measured precipitation time series contain either the cumulative rain amount or the precipitation intensity for subsequent disjoint time intervals. These time series exhibit always the intermittent character of precipitation: rainfall of highly variable intensity is always followed by completely rain-free periods. In addition to the measured precipitation records also simulated time series, generated with suitable models, can be used for this purposes. Simulated time series are interesting for example when the measured time series are too short or incomplete, or when the resolution of the record does not match the actual demands. For precipitation data in the range of hours or days a lot of models do exist, but only few investigations about the structure and the statistics of precipitation data in the range of minutes are known.

We analyse precipitation records of a dense city precipitation gauge network (Cologne), which has been recording in five minutes intervals. We first check the feasibility of existing precipitation models for this high temporal scale. The

---
* csimmer@uni-bonn.de

models assume for each precipitation descriptor (for example the duration of the rain events or the total precipitation amount) certain statistical distributions. So we first check for all descriptors the applicability of the assumed empirical distribution on the five minutes scale. We find that the inclusion of the high-frequency scales requires some modifications of the distributions (Chapter 2 and 4).

In the literature the construction of models which can statistically match all observed scales is an important task (Rodriguez-Iturbe et al., 1987). This implies the validity of the main statistical characteristics and also the model parameters over a range of levels of aggregation. An advantage of this would be the possibility to calculate the statistical structure of the hourly scale for example, even if observations are made only in the daily scale.

2 Precipitation models

A precipitation model is a mathematical framework for representing the continuous rainfall process (Rodriguez-Iturbe et al., 1984). The measured, discrete values, however, are in fact already integrals over this process. The kind of precipitation model, we consider for the representation of rainfall time series consists of a set of mathematical equations for the first order statistics (mean, variance, covariances with a lag of minutes to hours) of the discrete, cumulative precipitation amount during the time interval T. This interval can be hours, days, or weeks. With the model equations a direct comparison of the theoretical moments (simulated with the model equations) and the empirical moments (obtained from the measurement record) is possible. This comparison is used here for the determination of the optimal model parameters (Chapter 3.c) and for the final evaluation of the models (Chapter 4). In the models, which are valid so far for hourly or daily records, some assumptions about the descriptors of a precipitation event and about their distributions are made. The parameter values of these distributions are considered unknown a priori and will be determined in a separate step with the help of the model equations (Chapter 3.c).

In general the precipitation models under consideration here are based on the theory of stochastic point processes. In these models the rainfall intensity process $\xi(t)$ is used for the mathematical description of the rainfall process. This continuous process $\xi(t)$ contains the basic rainfall characteristics, like the frequency and occurrence of the events (storms). Also the duration and the intensity of the precipitation events is determined by the rainfall intensity process.

2.1 Rectangular pulses models

Precipitation models have different dependence structures referring to the behaviour of events and cells. Mostly used are rectangular pulse models. In these models each event is associated with a period of rainfall of random duration and constant intensity. An example is the rectangular pulse model of Rodriguez-Iturbe et al. (1984) with a Markov dependence structure (RPM). A Poisson process with a parameter λ determines the chance of occurrence of rainfall pulses.

A Statistical-Dynamic Analysis of Precipitation Data

T_1 marks the beginning of the first pulse, T_2 the beginning of the second and so on (Fig. 1). With each occurrence time T_n is connected the duration t_r^n and the intensity i_r^n. Both are assumed to be independent random variables. Different pulses may overlap in this kind of model. Then the total intensity is the sum of all active contributions (Rodriguez-Iturbe et al., 1987). In Fig. 1 the pulses No. 2 and No. 3 overlap and form one event. The distributions of duration and intensity are assumed to follow an exponential function with the parameters η and μ. The size of the interval T in the original model equations may vary from one hour, to 6 or 12 hours, or even to a day or a month.

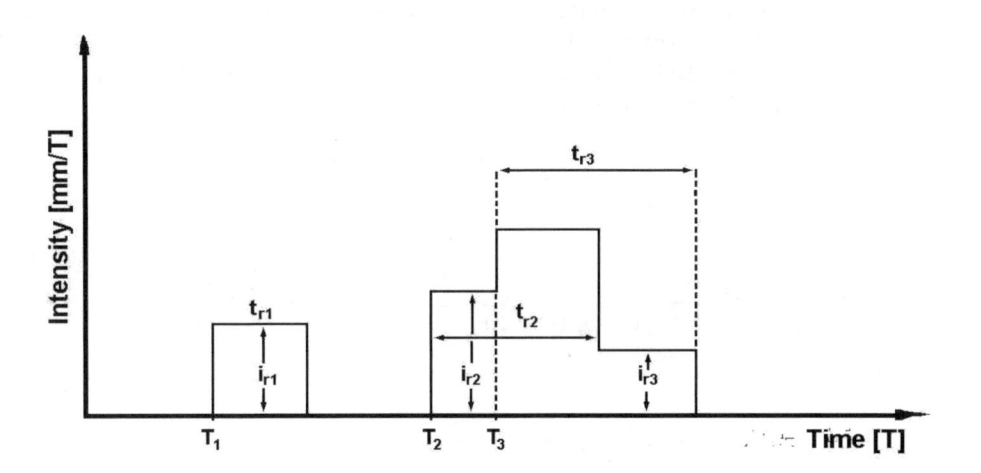

Fig. 1. Schematic description of the rectangular pulses model (after Rodriguez-Iturbe et al., 1984)

The equations for the mean, variance and lag-covariances of the cumulative precipitation amount $Y(T)$ in the interval T of the original rectangular pulse model (RPM) follow from the form of the distributions and are dependent on the a priori unknown distribution parameters λ, η and μ:

Mean:
$$E[Y(T)] = \frac{T\lambda}{\mu\eta} \quad (1)$$

Covariance (lag k-1):
$$\mathrm{Cov}[Y_1(T), Y_k(T)] = \frac{2\lambda}{\eta^3\mu^2}\left(1 - e^{-\eta T}\right)^2 e^{-\eta T(k-2)} \quad k \geq 2 \quad (2)$$

Variance (k=1):
$$\mathrm{Var}[Y(T)] = \frac{4\lambda}{\eta^3\mu^2}\left(\eta T - 1 + e^{-\eta T}\right) \quad (3)$$

It seems obvious that a record with a finer resolution of five minutes should be more suitable for the representation of the real rainfall intensity process than e.g. an hourly record. Thus we tested the goodness of fit for the frequency distributions of the high resolution time series. The result was, that for the high resolution data the distribution forms assumed for the larger time intervals did not give always the best fit. In the modified rectangular pulse model (MRPM) for the five minutes resolution data, the occurrence of the events is better described by a geometrical distribution with the parameter p. For the intensity distributions the Pareto distribution

$$f(x, x_0, \alpha) = \frac{\alpha}{x_0}\left(\frac{x_0}{x}\right)^{\alpha+1}$$

with two free parameters α and x_0 leads to a better fit than the exponential distribution. The lower threshold of the density function is given by x_0. For the duration, the exponential distribution with the parameter η is still acceptable. Thus for the equations of the MRPM follows:

Mean:
$$E[Y(T)] = \frac{p}{(1-p)} \frac{\alpha x_0}{(\alpha-1)} \frac{T}{\eta} \qquad (4)$$

Covariance (lag k-1):
$$\mathrm{Cov}[Y_1(T), Y_k(T)] = \frac{p}{(1-p)} \frac{\alpha x_0^2}{(\alpha-2)} \frac{e^{-\eta T(k-2)}}{\eta^3} \left(1 - e^{-\eta T}\right)^2 \qquad k \geq 2 \qquad (5)$$

Variance (k=1):
$$\mathrm{Var}[Y(T)] = \frac{p}{1-p} \frac{\alpha x_0^2}{(\alpha-2)} \frac{e^{\eta T}}{\eta^3} (1 - e^{-\eta T})^2 \qquad (6)$$

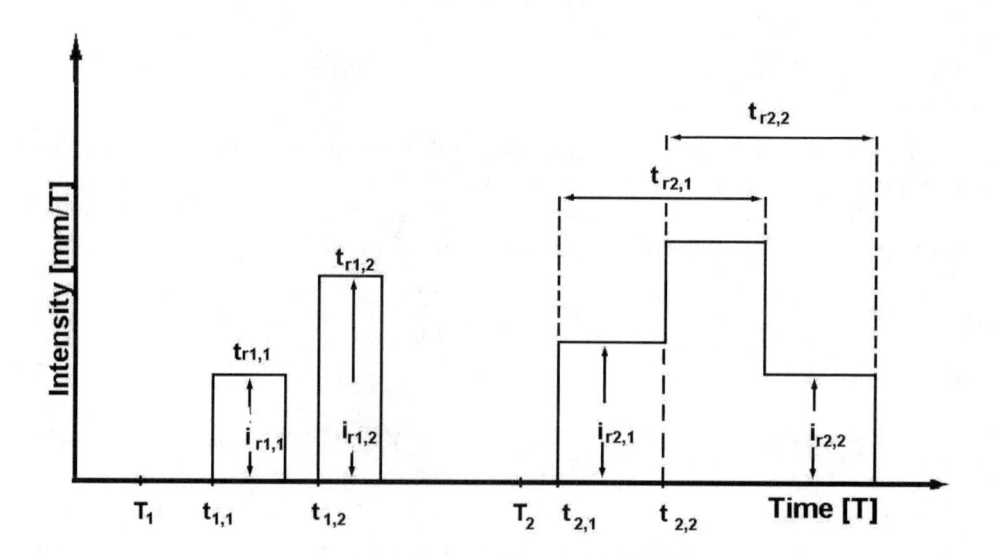

Fig. 2. Schematic description of the Neyman-Scott rectangular pulses model

2.2 Cluster models

More complicated but also more close to the reality are cell cluster models (Kavvas and Delleur, 1981). The Neyman-Scott cluster model (NSM, see Fig. 2) belongs to this category (Rodriguez-Iturbe et al., 1987; Entekhabi et al., 1989). The rainfall intensity is now the result of two independent Poisson processes. The first process controls the occurrence of the cluster origins and determines the rain events with the parameter λ. In Fig. 2 T_i indicates the beginning of the rain event number i. Each rain event may consist of one or more cells. The second Poisson process, with the parameter μ_c controls the occurrence of cells belonging to an event. $t_{1,1}$ and $t_{1,2}$ indicate the beginning of the two cells belonging to the first event. In the same way the numbering of the intensity and duration of the cells is indicated. Overlap of pulses is allowed, even if they belong to different events. In the case of overlap the intensities are given by the sum of the individual pulse intensities (for example event No. 2 in Fig. 2). The distance of the cell start from the beginning of the storm follows an exponential distribution with the parameter β. The duration and the intensity of the cells is exponentially distributed with the parameters η and μ, respectively. After Entekhabi et al. (1989) the equations of the original Neyman-Scott cluster model NSM are:

Mean:
$$E[Y(T)] = \frac{T\lambda\mu_c}{\mu\eta} \tag{7}$$

Covariance (lag k):
$$\mathrm{Cov}[Y_i(T), Y_{i+k}(T)] = \frac{\lambda}{\eta^3}(1 - e^{-\eta T})^2 e^{-\eta T(k-1)} \left\{ \frac{2\mu_c}{\mu^2} + \frac{(\mu_c^2 + 2\mu_c)}{\mu^2} \frac{\beta^2}{2(\beta^2 - \eta^2)} \right\}$$
$$-\lambda(1 - e^{-\beta T})^2 e^{-\beta T(k-1)} \frac{(\mu_c^2 + 2\mu_c)}{2\mu^2\beta(\beta^2 - \eta^2)} \qquad k \geq 1 \tag{8}$$

Variance:
$$\mathrm{Var}[Y(T)] = \frac{\lambda}{\eta^3}\left(\eta T - 1 + e^{-\eta T}\right) \left\{ \frac{4\mu_c}{\mu^2} + \frac{(\mu_c^2 + 2\mu_c)}{\mu^2} \frac{\beta^2}{(\beta^2 - \eta^2)} \right\}$$
$$-\lambda\left(\beta T - 1 + e^{-\beta T}\right) \frac{(\mu_c^2 + 2\mu_c)}{\mu^2\beta(\beta^2 - \eta^2)} \tag{9}$$

As in the case of the rectangular pulse model the distributions of the model characteristics were analysed for both the aggregated daily and the five minutes records. Again modifications of the model proved to be more suitable for the fine resolution data, resulting in the modified Neyman-Scott rectangular pulse model (MNSM).

Mean:
$$E[Y(T)] = \frac{p}{(1-p)} \frac{c}{(1-c)} \frac{\alpha x_0}{(\alpha - 1)} \frac{T}{\eta} \tag{10}$$

Covariance (lag k):

$$\text{Cov}[Y_i(T), Y_{i+k}(T)] = \frac{p}{(1-p)} \frac{c}{(1-c)} \frac{e^{-\eta T(k-1)}}{\eta^3} \left(1 - e^{-\eta T}\right)^2$$

$$\left\{ \frac{\alpha x_0^2}{(\alpha-2)} + \left[\frac{c}{(1-c)} - 1\right]\left(\frac{\alpha x_0}{\alpha-1}\right)^2 \frac{\beta^2}{(\beta^2-\eta^2)} \right\}$$

$$- \frac{p}{(1-p)} \frac{c}{(1-c)} \left[\frac{c}{(1-c)} - 1\right]\left(\frac{\alpha x_0}{\alpha-1}\right)^2$$

$$\frac{1}{(\beta^2-\eta^2)} \left\{ \frac{e^{-\beta T(k-1)}}{\beta} \left(1 - e^{-\beta T}\right)^2 \right\} \quad k \geq 1 \quad (11)$$

Variance:

$$\text{Var}[Y(T)] = \frac{p}{(1-p)} \frac{c}{(1-c)} \frac{e^{\eta T}}{\eta^3} \left(1 - e^{-\eta T}\right)^2$$

$$\left\{ \frac{\alpha x_0^2}{(\alpha-2)} + \left[\frac{c}{(1-c)} - 1\right]\left(\frac{\alpha x_0}{\alpha-1}\right)^2 \frac{\beta^2}{(\beta^2-\eta^2)} \right\}$$

$$- \frac{p}{(1-p)} \frac{c}{(1-c)} \left[\frac{c}{(1-c)} - 1\right]\left(\frac{\alpha x_0}{\alpha-1}\right)^2$$

$$\frac{1}{(\beta^2-\eta^2)} \left\{ \frac{e^{\beta T}}{\beta} \left(1 - e^{-\beta T}\right)^2 \right\} \quad (12)$$

The occurrence times of both the events and the cells is better represented by geometrical distributions with parameters p and c. The intensity of the cells is better fitted by a Pareto distribution with parameters α and x_0. For the distance of the cells from the origin of the event and the duration of the cells the exponential distribution is maintained with parameters β and η. For the cluster models NSM and MNSM it is possible also to aggregate or disaggregate the size of the time interval T.

3 Theory for the selection of suitable model parameters

The analysis, so far, was only directed towards finding the most appropriate statistical distributions for the models. The explicit values of the parameters of the distributions, however, are considered unknown. Even in the case of the modified models, where the empirical and the theoretical distributions were calculated and compared with a Kolmogorov-Smirnov test, the values of the empirical distribution parameters can not be used as model parameters. The main reason is the different definition of an event or of a cell in the statistics of the empirical data and the definition of the pulses which are the base of the model events and model cells: In the models an overlap of pulses is allowed which makes no sense in the real precipitation process. The problems arising with the empirical data to this respect are presented in the following two sub-chapters. For all cases the

A Statistical-Dynamic Analysis of Precipitation Data 343

optimum model parameters will be calculated in an separate step based directly on the model equations and presented in the third sub-chapter. This procedure ensures that the parameter values harmonise both with each other and with the model. The procedure is then applied to the rain gauge date of the city of Cologne (last sub-chapter).

3.1 Identification of independent rainfall events

A first analysis task is the definition and separation of independent rainfall events. Such a separation has to take into consideration both the prevailing weather conditions and the investigated scale. In a single synoptic disturbance can be embedded multiple meso-scale disturbances and each meso-scale disturbance may produce several cells with precipitation, followed by time periods without rainfall (Restrepo-Posada and Eagleson, 1982). In the analysis it is assumed that rain events are independent from each other in case the rain-free periods between successive rain periods are large compared to the investigated scale. If the break between two rain periods is small, they probably form different cells of one event. Thus rain events are independent and may contain a single cell or several cells. In the five minutes time series rainfall events show in general many cells with rather short duration, interrupted by breaks in the same range of five to ten minutes. For larger scales generally a smoothing effect can be observed.

Restrepo-Posada and Eagleson (1982) present a practical criterion to separate independent rain events based on the duration of the rain-free periods between the rain events. It is assumed that a Poisson process describes the random arrival of precipitation events and that the events are mutually independent. Further, they suppose that in this case the duration of the rain-free periods t_b is distributed exponentially and that the duration of the rain events is small in comparison with the duration of the rain-free periods. For the exponential distribution the coefficient of variation $CV[t_b]$, which is defined as the ratio of the mean and the standard deviation has the value one:

$$CV[t_b] = \frac{\sigma[t_b]}{E[t_b]} = 1 \qquad (13)$$

Thus CV can be used as an identification criterion for independent events: First the CV of the rain-free time periods is calculated for a range of different t_b. From the results the value t_{bo} which satisfies Eq. 13 is found, which is then chosen as the criterion for the separation of independent events. With this methodology t_{bo} can be calculated for different values of minimum rain-free periods in a single processing of the original data.

We find, that his criterion works well for precipitation records in the range of hours or days. For higher temporal resolutions, however, the resulting independent rain events and the durations of the rain-free periods exceed by far the investigated scale. The mean duration of the rain-free period t_b between independent events analysed for the rain gauge network of the city of Cologne (for details see below) using the above criterion is 14.7 hours. Obviously, this criterion only separates events of the synoptic scale in this data set. If this criterion would be used for the five minutes resolution records, the specific characteristics of this range would be lost.

To represent and maintain the characteristic Al features of the five minutes records another empirical criterion — the PQ-criterion — is tested. The motivation for this criterion comes from the inherent scale independent characteristic of rainfall: if, for example, precipitation originates from a large, synoptical disturbance, the duration of precipitation is in general longer (in the range of some hours) and similar characteristics hold for the rain-free periods. If convective conditions prevail, the duration of both the precipitation events and the rain-free periods are in general smaller. The dependence of statistical properties of precipitation records on the duration of the event has been proposed by Koutsoyiannis and Foufoula-Georgiou (1993). In hydrological design, for example. the concept of mass curves (normalised cumulative storm depth versus normalised cumulative time since the beginning of the event) is such a relative and extensively used measure.

To follow this concept in a first step rain-free periods were sorted out, when they are smaller than a given threshold P_{gr}. This results in a provisional clustering of cells (see Fig. 3 a and b.). In a second step, the ratio of the duration of the rain-free period (following a cell cluster) and the mean duration of this cluster is calculated. The mean duration means, that the mean value of the duration of the last and the last but one cell cluster is used. If this ratio is smaller than a given threshold Q_{gr}, it is assumed, that it separates only cells of the same event and that the cell clusters before and the cluster after belong together (Fig. 3c). Empirical results show that for the majority of the cell clusters this ratio is smaller or equal to four. To determine the optimum values of the thresholds P_{gr} and Q_{gr} different values are tested. In every case the empirical cumulative frequency distribution of the number of events per day is determined. This is compared with a geometrical distribution under the null hypothesis, namely, that the distributions do not differ significantly (using the Kolmogorov-Smirnov test). If the null hypothesis is not rejected, it is assumed that the given thresholds separate independent events. Independent events means in this context only, that they can be regarded as independent in the scale of five minutes. The minimum values, which satisfy this condition are chosen as the thresholds P_{gr} and Q_{gr}. For the data of Cologne the distribution of the original cells is rejected. After the application of the above criterion using the thresholds $P_{gr} = 2$ and $Q_{gr} = 3$ the resulting rain events meet the requirements satisfactorily (Fig. 4).

3.2 Checking model assumptions by the parameter distributions

To check the model assumptions over the range of hours, days, and months the five minutes records were aggregated to the respective sums and their empirical frequency distributions were compared with the expected, theoretical distribution of the models. In these aggregated scales the given assumptions can all be accepted. For the five minutes resolution data also the distributions of the occurrence of events per day, the number of cells per event and their temporal distribution related to the beginning of the event, of the duration of cells and

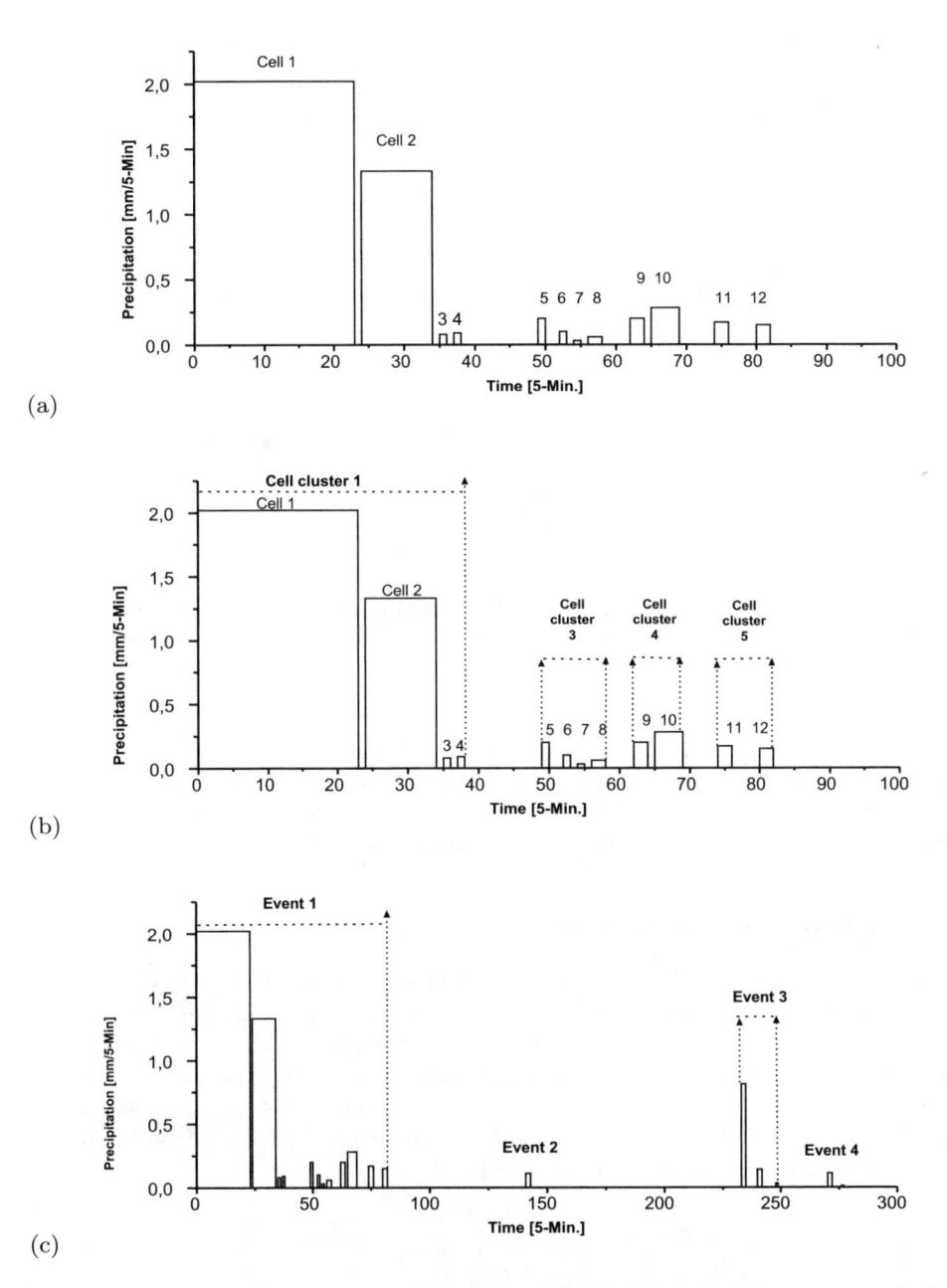

A Statistical-Dynamic Analysis of Precipitation Data 345

Fig. 3. Schematic application of the PQ-criterion: **(a)** Original precipitation record, the numbers specify the cells. **(b)** The cells are grouped into provisional cell clusters (after the application of the first step). The figure shows the same time interval as above. **(c)** After the second step the cell clusters are grouped into independent events. To show more than one event, the time domain is greater than above. The full time domain of (a) and (b) represents here event No. 1.

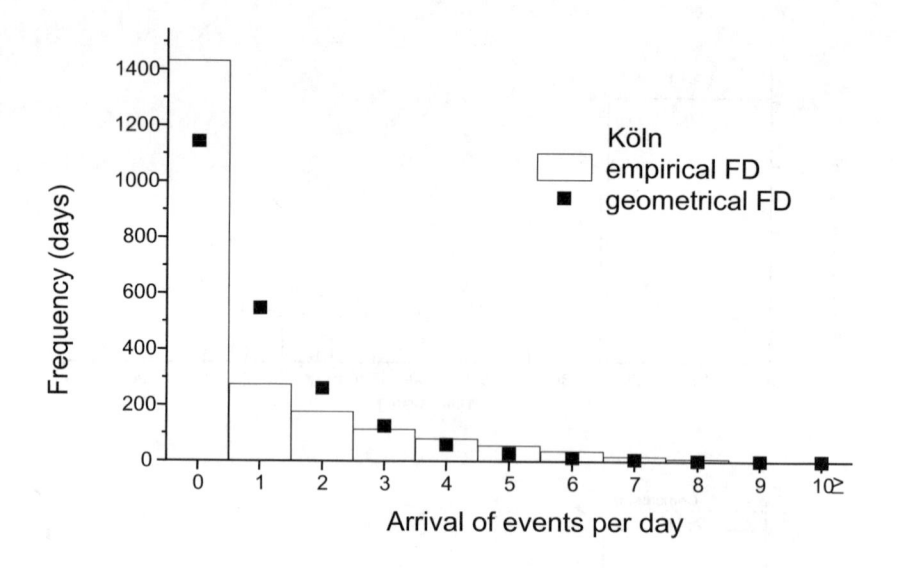

Fig. 4. Empirical frequency distribution of the number of arrivals of events per day after the application of the PQ-criterion and a geometrical distribution for the precipitation data of Cologne.

the amount of the precipitation data are checked. We had to reject in some cases the theoretical distributions used for the hourly or daily scales. Like we found for the optimal description of the models above the geometrical distribution replaces here the Poisson distribution. For the precipitation amount the Pareto distribution matches better than the exponential distribution.

3.3 Estimation of model parameters

The model equations contain a set of m unknown parameters, namely the parameters of the distributions of the statistics of the events and cells (arrival, duration, rain amount, ...). The geometrical, exponential and Poisson distribution each need only one parameter. For the Pareto distribution two parameters are necessary. To estimate the optimum parameter values a minimum least square techniques is used (Entekhabi et al., 1989; Islam et al., 1990), e.g. the minimum of the following expression is determined.

$$\left(\frac{F_1(X)}{\theta_1 - 1}\right)^2 + \left(\frac{F_2(X)}{\theta_2 - 1}\right)^2 + \cdots + \left(\frac{F_m(X)}{\theta_m - 1}\right)^2 \tag{14}$$

Here the set of parameters is determined by the vector X. The empirical statistical moments are given by θ_1 to θ_m, and $F_1(X)$ to $F_m(X)$ indicate their theoretical values (calculated with the model equations). The parameter set for which expression (14) reaches a minimum, defines the optimum values for the model parameters.

A Statistical-Dynamic Analysis of Precipitation Data 347

The parameter set estimation is performed by fitting empirical moments of the cumulative rain process at different levels of aggregation (Rodriguez-Iturbe et al., 1987). The choice of the fitted empirical moments is free. Important is only, that the number of the given empirical parameters is equal to the number m of the unknown model parameters. It is possible to use only moments of a single level of aggregation and to fit then the model to this scale. For some complex models, which need the specification of more parameters, it is also possible to fit the model to moments of a range of scales simultaneously. This indicates, that it might be impossible to determine for each model a single "true" parameter set.

In the literature the final evaluation of a model is made by testing how good it fits the empirical autocorrelation function. Interesting is the behaviour of the simulated autocorrelation function for greater lags. Usually, these values are not directly specified in the model. So the correspondence between model and measured time series finally proves the appropriateness of the model assumptions. Also the autocorrelation function for another level of aggregation is interesting. Here the model can show or not show its ability to represent and simulate the statistical characteristics of different scales with a single parameter set.

3.4 Application to the rain gauge network of Cologne

The city of Cologne maintains a dense network of precipitation gauges with one minute temporal resolution. The average distance between the gauges is about five km. To obtain representative empirical values we included all stations independently in the statistics being aware, of course, of some dependence between the time series due to common events. To test the possible influence of the weather situation our analysis is done for three samples. The first sample contains the complete five minute record (the one minute data has been summed up accordingly to reduce some artefacts inherent to the measuring method) of four summer months, namely June to August 1995 and June 1996. Also two reduced samples were used, either only the data from days with thunderstorms or its complement. For all cases the theoretical autocorrelation is determined with the model equations and then compared to the empirical autocorrelation function of the measurements. For the three samples and two temporal scales (five minutes and one hour) the following precipitation models were fitted:

1. Original rectangular pulses model (RPM): This model has three model parameters and requires therefore the declaration of three empirical moments. For the fitting of each scale the mean value, the variance and the covariance (lag=1) of the respective scale are used in expression (14).
2. Modified rectangular pulse model (MRPM): This model has four model parameters. For the necessary additional parameter we use the autocorrelation (lag=1) of the fitted scale.
3. Original Neyman-Scott model (NSM): This model with five model parameters requires another additional information, which we provide with the autocorrelation (lag=2) of the fitted scale.

4. Modified Neyman-Scott model (MNSM): This model has six model parameters. Additional to the five parameters mentioned above (mean value, the variance and the covariance (lag=1), autocorrelation (lag=1), autocorrelation (lag=2)) the autocorrelation (lag=1) of the other (aggregated or disaggregated, respectively) scale will be used.

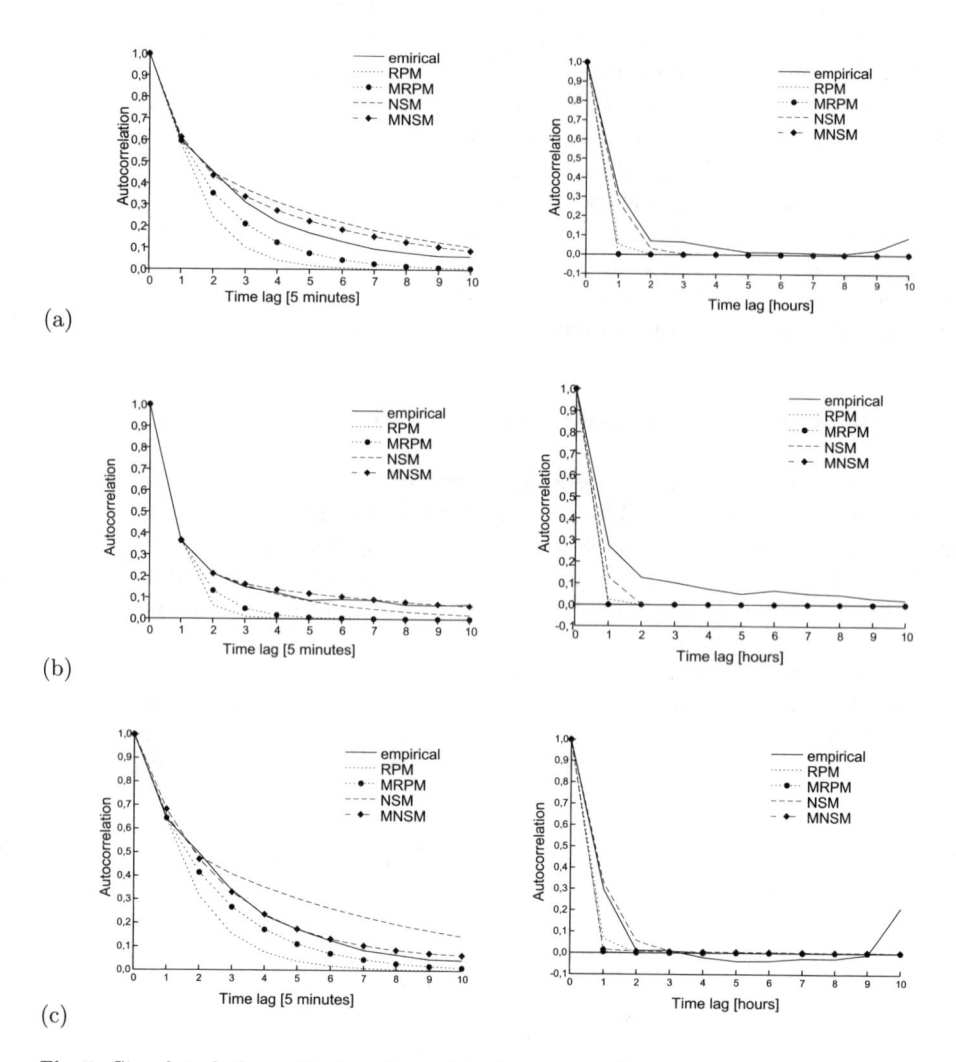

Fig. 5. Simulated theoretical and empirical autocorrelation of the precipitation data of the city of Cologne. The model parameters were fitted to the 5 minutes scale: (a) complete sample, (b) days without thunderstorms, (c) days with thunderstorms.

Fig. 5 represents the results for Cologne with the model parameters fitted to the five minutes scale and Fig. 6 the results for the hourly scale. When the model parameters are fitted to the statistical moments of the five minutes scale

A Statistical-Dynamic Analysis of Precipitation Data

the resulting simulated autocorrelation for this scale is poor for the simple pulses models. The modified pulse model (MRPM) shows some better results as the original RPM. The best results, however, show the cluster models. In general the modified version (MNSM) is better than the original one (NSM), regardless which data set is used. The simulated autocorrelation function of the aggregated hourly scale, however, falls too quickly to zero in all cases.

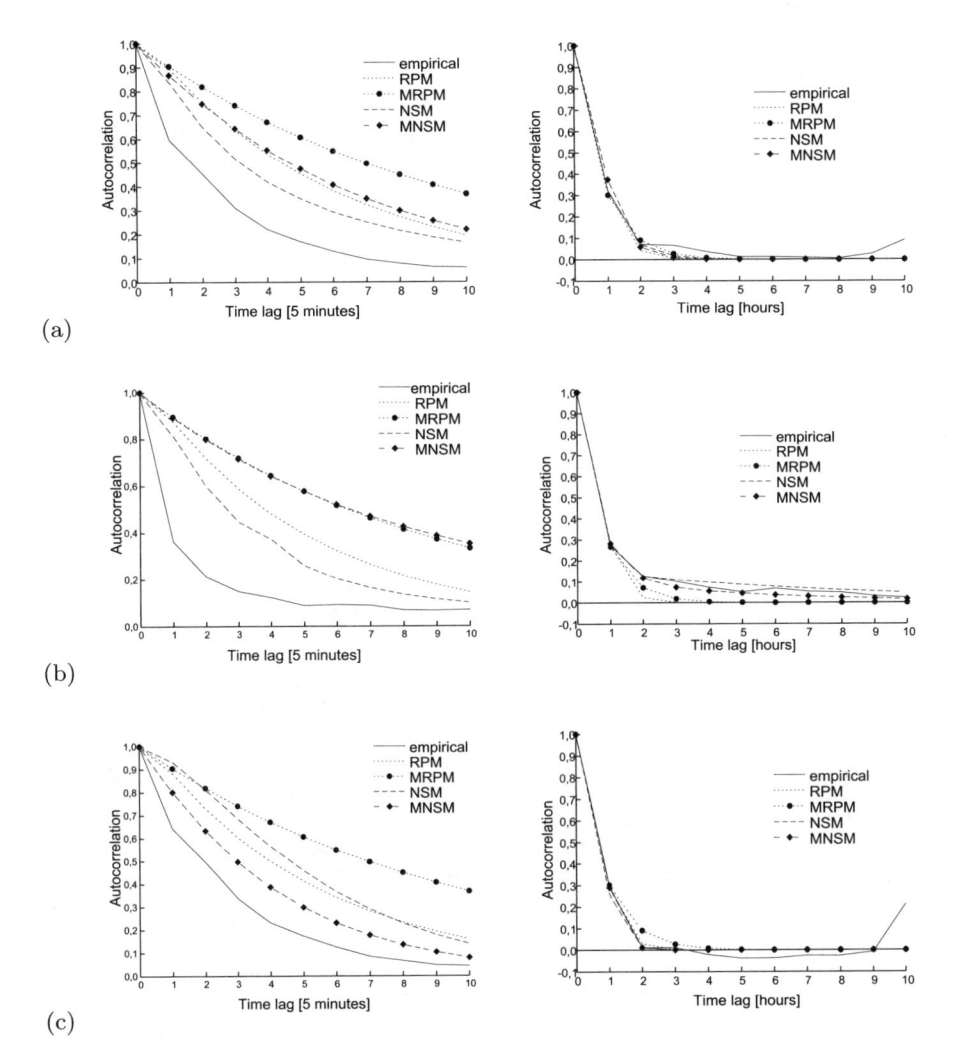

Fig. 6. Simulated theoretical and empirical autocorrelation of the precipitation data of the city of Cologne. The model parameters were fitted to the hourly scale: (a) complete sample, (b) days without thunderstorms, (c) days with thunderstorms

When the model parameters are fitted to the statistical moments of the hourly data for the simulated autocorrelation function of the five minutes data the results are in general poor; the simulated autocorrelation is always too large. For larger time lags both original models show somewhat more similarity than the modified versions. The simulation of the autocorrelation of the hourly data with the modified cluster model (MNSM) shows the best results with all data sets (all days, thunderstorm days, days without thunderstorms).

4 Discussion

As expected the results of the cluster models with more parameters to fit to were better than the simple pulses models. This underlines the cluster characteristics of the precipitation process. If the models should be valid for different scales, the original models lead to more stable results. Thus the modifications undertaken lead to improvements only for the representation of scale specific characteristics of rainfall. For the original models too, the results are better, if for each scale and data set a new parameter set is determined. But the improvements were not comparable to the improvements obtained for the modified models.

References

Entekhabi, D., Rodriguez-Iturbe, I., and Eagleson, P. S., 1989: Probabilistic Representation of the Temporal Rainfall Process by a Modified Neyman-Scott Rectangular Pulses Model: Parameter Estimation and Validation, Water Resour. Res., Vol. 25, No. 2, 295–302.

Islam, S., Entekhabi, D., Bras, R. L. and Rodriguez-Iturbe, I., 1990: Parameter Estimation and Sensitivity Analysis for the Modified Bartlett-Lewis Rectangular Pulses Model of Rainfall. J. Geophys. Res., Vol. 95 (D3), 2093–2100.

Koutsoyiannis, D. and Foufoula-Georgiou, E., 1993: A Scaling Model Of A Storm Hyetograph. Water Resour. Res., Vol. 29, No. 7, 2345–2361.

Restrepo-Posada, P. and Eagleson, P., 1982: Identification Of Independent Rainstorms, Journal of Hydrology, 55, 1982, 303–319.

Rodriguez-Iturbe, I., Febres De Power, B. and Valdès, J. B., 1987: Rectangular Pulses Point Process for Rainfall: Analysis of Empirical Data, Journal of Geophysical Research, Vol. 92, 9645–9656.

Rodriguez-Iturbe, I., Gupta, V. K. and Waymire, E., 1984: Scale Considerations in the Modelling of Temporal Rainfall, Water Resour. Res., Vol. 20, No. 11, 1611–1619.

Index

A/S rate, 318
accommodation space, 318
accuracy, 6, 7, 11
advection term, 171
aerosol, 169
aggregated parameters, 159
aliasing, 7, 8, 327
anisotropy
– smoothing, 24
approximation
– boundary layer, 176
– Boussinesq, 171, 175
– free atmosphere, 175
attractor, 254
Austausch theory, 173
auto regressive models, 34
autocorrelation, 347, 349
– function, 337, 339
average
– statistical, 60

balanced volume-controlled 3D-restoration, 105
balancing operations, 109
basin evolution, 294
Bayes theorem, 334
Bayesian statistics, 335
bedding plane, 311
bedforms, 309
bias, 6, 7
boundary layer
– atmospheric, 171
– nocturnal, 177
box counting, 242
Brownian motion, 267
bubbling pressure, 221
buoyancy, 173, 174

causability, 53
causality, 57, 91
cellular automaton, 51, 284
change, 42
– characteristics, 57, 59
– fluctuating, 54, 56
– management of, 60
– permanent, 58

– random, 59
– stationary, 55
– sudden, 59
channel network
– optimal, 280
chorology, 326
closure
– assumption, 172
– first order, 172–173
– level 2.0, 176
– level 2.5, 176
– problem, 169–172
– second order, 172–176
cloud
– microphysics, 186
– model, 189
– processes, 172
– water, 173
cluster model, 341–342
coastline clinoform, 313
coefficient
– of turbulent diffusion, 173
cohesion theory, 216
Common Object Request Broker Architecture, 114
complexity, 52, 56, 59
condition, 127
– convective, 344
conditional relations, 57, 59
conductivity curve, 147
conservation law, 55
conservation law
– for momentum, 168
– for inner energy, 168
– for mass, 168
consistency, 131
constant flux layer, 177
constitutive relations, 56
convection, 133
convective
– conditions, 173
– systems, 186
convergence rate, 126
CORBA, 114
CORBA/ODBMS integration layer, 115
Coriolis force, 168, 175

correlation, 171
- triple, 171
cost function, 334
counter gradient flux, 173
critical state, 254
criticality
- self-organised, 250
-- in earth surface processes, 272
crustal heterogeneities, 258
cyclicities, 315

DAG, 113
Darcy velocity, 220
Darcy's law, 146
data
- analysis, 82
- management
-- consistent spatial, 109
- noisy, 83
- proxy, 325
- storage, 83
- transmission, 83
database
- management systems
-- object-oriented (ODBMS), 107
-- relational (RDBMS), 107
- queries
-- spatial, 109
- view, 115
density
- air, 167
- of momentum, 168
depocentre, 308
depositional
- model, 308
- system, 307
diffusion, 133
- coefficient, 177
- molecular, 173
- problem, 137
dimension
- Euclidean, 264
- fractal, 239, 258
directed acyclic graph, 113
discharge, 337
disconformity, 315
discretisation, 124
distribution
- exponential, 339–343, 346
- fractal, 59, 239

- geometric, 346
- geometrical, 340, 342
- of drainage areas, 273
- of landslides, 281
- Pareto, 340, 342, 346
- Poisson, 346
- power law, 239
disturbance
- meso-scale, 343
- synoptic, 343, 344
double porosity, 136, 138
drag production, 174
drainage
- area, 272
- network, 272
drop size
- distribution, 187
dynamic
- fractal, 255
- range, 6, 7
dynamical
- constraints, 329
- modes, 328

Earth
- rotation vector, 168
- surface processes, 271
- system, 5–7, 13, 14
earthquake, 241
- magnitude of, 241
- rupture area of, 241
eddy, 171
- diffusivity, 173
Eemian, 328
effective
- equations, 138
- operator, 135, 138
- parameter, 135, 145
- size, 173
equation
- advection-diffusion, 328
- budget, 167
-- filtered, 170
-- for covariances, 171
-- for liquid water, 168
-- for water vapour, 168
-- scalar, 172
- integro-differential, 140
- master, 203
- of continuity, 168

– of motion, 168
– of state, 168
– – ideal gas, 167
– of temperature tendency, 168
– partial differential, 124
– prognostic, 169
– thermo-hydrodynamic, 168, 171, 172
erosion, 272
error, 6
extraction function, 221

facies tract, 307
fault
– displacement, 50
– dynamics, 48
– model, 43
– slip, 49
faulting
– episodes, 51
Fick's law, 173
filter
– box filter, 23
– Gaussian, 24
filtering, 83
finite elements, 70, 80
floods, 337
fluctuations, 51
flux
– density
– – turbulent, 171
– gradient parameterisation, 173
– turbulent, 173
form
– element, 111
– facet, 111
fossil, 326
Fourier
– analysis, 70
– modes, 329
fractal, 238
– dimension, 59
– dynamic, 255
– geometry, 258
– static, 255
fragmentation, 249
frequency-magnitude relation
– of landslides, 281
Froude number, 309
function
– lognormal, 58

– power law, 58

gas constant
– specific, 167
– – for dry air, 170
– – for water-vapour, 170
Gaussian, 333
generalisation, 104
– of sub-classes, 104
geometry model
– database support, 296
– multi-resolution, 304
– time-dependent, 298
geophysics, 69
GeoStore, 105, 110
GeoToolKit, 106, 110
Global-Modell, 176
GOCAD, 296
gradient approach, 175
graphical database query languages, 103
gravity, 168
– of Earth, 168
Gutenberg-Richter law, 241

Hack's law, 272
heat
– capacity
– – specific, 168
– source, 168
– specific
– – of water fusion, 174
heating
– adiabatic, 172
– radiative, 168, 174
homogenisation, 135, 136
Hurst exponent, 264
hydraulic lift, 215

inheritance, 104
integration of heterogeneous data sources, 114
interoperability, 114
Interoperable Object References (IORs), 114
iterative methods, 126

Java, 115

K-theory, 173
Kell-factor, 22

kernel, 333
kinetic energy
- of turbulence, 175
Kolmogorov hypothesis, 175
Kolmogorov-Smirnov test, 342, 344
KTB, 258

landform, 111
- modelling
-- multi-scale, 111
- objects, 111
- phenomena, 111
landsliding, 281
Last Interglaciation, 328
Latin Hypercube, 148
least square technique, 346
length scale
- turbulent, 175
liquid water, 168
Lokal-Modell, 172, 176–191

macro remains, 326
magnitude
- of earthquakes, 241
manifold, 81
map
- logistic, 37
- plane restoration, 296
- scale, 19
marine flooding surface, 316
Markov process, 338
mass
- balance equation, 201
- curve, 344
mediators, 114
memory term, 136, 140
meta-information, 115
Meteosat, 192
middleware platform, 114
Milankovitch cycles, 315
mixing length theory, 173
model, 18
- backward, 332
- categories, 54
- climate, 168, 176
- cloud, 169
- complexity, 50
- conceptual, 43
- consistence, 48, 55
- description, 18

- deterministic, 55
- dynamical, 49
- generic, 18
- geometrical, 47
- kinematic, 105, 295
- large-scale, 107
- mathematical, 18
- non-linear, 36
- object
-- large-scale, 105
- omni-scale, 38
- pulse, 338–340
- small-scale, 107
D-, 105
- specific, 18
- weather prediction, 168, 176
modelling
- concepts, 52
- interactive geological, 109
- inverse, 153
- large-scale geological, 105
- small-scale geometric-kinetic, 105
moment
- seismic, 246
- statistic, 346
momentum transport
- sub-grid scale, 174
Monin-Obukhov similarity, 177
Monte Carlo method, 149
multi-resolution
- analysis, 75
- objects, 109
multigrid, 129, 130
- algebraic, 132

Nabla operator, 167
network
- of rain gauges, 347
noise
- $1/f$, 254, 257
- flicker, 254
- pink, 254
- scaling, 264
- white, 257

object
- causation, 53
- function, 53
- manifestation, 53
- quality, 53

octree, 22
optimal channel network, 280

palaeobotany, 327
palaeoclimate, 325
– reconstruction, 325
palinspastic restoration, 294
paraconformity, 315
parameterisation
– sub-grid scale, 167–185
parasequence, 316
parcel method, 173
perception
– spatial, 45
performance criterion, 154
petrophysical
– logs, 261
– properties, 258
phase change
– of water, 168
pixel, 7
plant physiology, 216
Poisson process, 337, 338, 341, 343
pollen, 326
– analysis, 326
– spectrum, 326
pore size index, 221
porous medium, 138
power law, 13, 14, 239
precipitation, 84, 337–350
– cumulative, 338, 339
– models, 337–350
precision, 6–8
preconditioning, 128
prediction, 60
pressure
– atmospheric, 167
– covariance, 174
– gradient, 168, 174
– work, 168
primitive equations, 168
principal curvature, 92
probability
– density function, 333, 337
– distribution, 58
process
– adiabatic, 174
– turbulent, 172

quadtree, 22

R*-tree, 109
R-tree, 109
radiation, 172
rain gauge networks, 196
random walk, 273
reactivation surface, 310
reality
– objective, 61
– relative, 57
reconstruction model, 332
recovery, 83
recursivity, 91
relative sea level, 315
relief classification, 93
rescaled range (or R/S-) analyses, 263
resolution, 5–8, 11–14, 20
retention curve, 147
return-to-isotropy, 174
Reynolds
– approximation, 170
– number, 309
– stress tensor, 172
Richards
– equation, 146, 218
Richardson number
– gradient, 176
ripple, 309
– formation, 206
robustness, 132
rolling
– of sediment grains, 201
Rotta hypothesis, 175
Rouby's method, 296
roughness length, 177
runoff, 337
rupture area
– of earthquakes, 241

saltation, 203
sampling, 5–8, 10, 12, 13, 59
– theorem, 22
sandpile, 284
satellite retrieval, 193
scalability, 69
scale, 69
– anisotropic, 19
– average wavelength, 25, 29
– concept, 42
– data internal, 27
– definition, 42

- hierarchy, 60
- in AR-models, 34
- in maps, 18
- in models, 34
- independence, 58
- inhomogeneous, 19
- interaction, 171
- invariance, 237, 258
- invariant, 59
- level of abstraction, 33
- level of aggregation, 33
- local, 51
- map scale, 29
- map scale number, 29
- meso-β, 176
- multiple scales, 36
- non-linear models, 36
- number, 19
- parameter, 137
- properties, 57, 61
- relations, 32
- relative smoothness, 27, 29
- representative, 58, 61
- resolution, 29
- resolved, 326
- scale matrix, 24
- smoothness, 27
- space, 90
-- Gaussian, 90, 91
- spatial, 104, 326, 327
- statistics, 57, 60
- subgrid, 326
- synoptic, 343
- temporal, 104, 106, 326, 327
- transitions, 191
- window size, 29
scaling, 145
self-organisation, 56
self-organised
- criticality, 61, 250
-- in earth surface processes, 272
separation
- of time scales, 203
sequence
- boundary, 316
- sedimentary, 311
- stratigraphy, 315
singular value decomposition, 328
size
- effective, 173

slip-face, 309
smoothing, 91, 93, 95, 129
- anisotropic, 24
soil erosion, 204
space
- limitation, 55
spatial
- access methods, 109
- anisotropy, 301
- consistency, 291
- locality, 69
specialisation, 104
- of a super-class, 104
spectral exponent, 258
splines, 80
stability
- functions, 176, 177
- of atmosphere, 173
- static, 173
static fractal, 255
statistic, 6, 12, 14
statistical
- moment
-- second order, 171
- upscaling, 196
storm wave base, 318
stratal gap, 312
stratification
- neutral, 173
stream
- density, 201
- velocities
-- rapid, 309
-- tranquil, 309
subgrid variability, 162
supercell storm, 186
surface
- fluxes, 177
- heating, 171
- runoff, 205
suspended load, 202
sustainability, 60
systems tracts, 315

temperature
- air, 167, 168
- potential, 172, 174
-- virtual, 170
terrain, 38
- profile, 38

thermodynamics
– first law of, 168
– statistical, 169
threshold, 83
– critical, 56
thunderstorm, 187, 347, 350
time
– historical, 46, 55
– model, 56
– scales
– – separation of, 203
– scene
– – tree, 295
– universal, 46
Tissot's indicatrix, 20
transfer
– coefficient
– – turbulent, 177
– function, 332
transpiration, 216
triangle networks, 105
triangular irregular networks, 22
turbulence
– atmospheric, 167–185
turbulent waterflow, 309
type change, 140

uncertainty range, 333
unconformity, 315
undersampling, 7, 8, 13, 14
upscaling, 136, 145

van Genuchten parameters, 221
variables
– conservative, 174
variation coefficient, 343
variational methods, 334
velocity
– effective, 173
– scale
– – turbulent, 175
viscosity
– kinematic, 168

water vapour, 168, 172
wavelet, 69, 71, 80, 131
– transform, 73, 77
weather
– prediction
– – numerical, 175
– radar, 187
Wheeler diagram, 308
wind
– shear, 171
– – vertical, 173
– vector, 167
window size, 23

XML, 115

Z/R relation, 191

Lecture Notes in Earth Sciences

For information about Vols. 1–22
please contact your bookseller or Springer-Verlag

Vol. 23: K. B. Föllmi. Evolution of the Mid-Cretaceous Triad. VII, 153 pages. 1989.

Vol. 24: B. Knipping, Basalt Intrusions in Evaporites. VI, 132 pages. 1989.

Vol. 25: F. Sansò, R. Rummel (Eds.). Theory of Satellite Geodesy and Gravity Field Theory. XII, 491 pages. 1989.

Vol. 26: R. D. Stoll, Sediment Acoustics. V. 155 pages. 1989.

Vol. 27: G.-P. Merkler, H. Militzer, H. Hötzl, H. Armbruster, J. Brauns (Eds.). Detection of Subsurface Flow Phenomena. IX. 514 pages. 1989.

Vol. 28: V. Mosbrugger. The Tree Habit in Land Plants. V. 161 pages. 1990.

Vol. 29: F. K. Brunner, C. Rizos (Eds.). Developments in Four-Dimensional Geodesy. X. 264 pages. 1990.

Vol. 30: E. G. Kauffmann, O. H. Walliser (Eds.), Extinction Events in Earth History. VI. 432 pages. 1990.

Vol. 31: K.-R. Koch, Bayesian Inference with Geodetic Applications. IX, 198 pages. 1990.

Vol. 32: B. Lehmann, Metallogeny of Tin. VIII, 211 pages. 1990.

Vol. 33: B. Allard, H. Borén, A. Grimvall (Eds.), Humic Substances in the Aquatic and Terrestrial Environment. VIII, 514 pages. 1991.

Vol. 34: R. Stein, Accumulation of Organic Carbon in Marine Sediments. XIII, 217 pages. 1991.

Vol. 35: L. Håkanson, Ecometric and Dynamic Modelling. VI, 158 pages. 1991.

Vol. 36: D. Shangguan, Cellular Growth of Crystals. XV, 209 pages. 1991.

Vol. 37: A. Armanini, G. Di Silvio (Eds.), Fluvial Hydraulics of Mountain Regions. X, 468 pages. 1991.

Vol. 38: W. Smykatz-Kloss, S. St. J. Warne, Thermal Analysis in the Geosciences. XII, 379 pages. 1991.

Vol. 39: S.-E. Hjelt, Pragmatic Inversion of Geophysical Data. IX, 262 pages. 1992.

Vol. 40: S. W. Petters, Regional Geology of Africa. XXIII, 722 pages. 1991.

Vol. 41: R. Pflug, J. W. Harbaugh (Eds.), Computer Graphics in Geology. XVII, 298 pages. 1992.

Vol. 42: A. Cendrero, G. Lüttig, F. Chr. Wolff (Eds.), Planning the Use of the Earth's Surface. IX, 556 pages. 1992.

Vol. 43: N. Clauer, S. Chaudhuri (Eds.), Isotopic Signatures and Sedimentary Records. VIII, 529 pages. 1992.

Vol. 44: D. A. Edwards, Turbidity Currents: Dynamics, Deposits and Reversals. XIII, 175 pages. 1993.

Vol. 45: A. G. Herrmann, B. Knipping, Waste Disposal and Evaporites. XII, 193 pages. 1993.

Vol. 46: G. Galli, Temporal and Spatial Patterns in Carbonate Platforms. IX, 325 pages. 1993.

Vol. 47: R. L. Littke, Deposition, Diagenesis and Weathering of Organic Matter-Rich Sediments. IX, 216 pages. 1993.

Vol. 48: B. R. Roberts, Water Management in Desert Environments. XVII, 337 pages. 1993.

Vol. 49: J. F. W. Negendank, B. Zolitschka (Eds.), Paleolimnology of European Maar Lakes. IX, 513 pages. 1993.

Vol. 50: R. Rummel, F. Sansò (Eds.), Satellite Altimetry in Geodesy and Oceanography. XII, 479 pages. 1993.

Vol. 51: W. Ricken, Sedimentation as a Three-Component System. XII, 211 pages. 1993.

Vol. 52: P. Ergenzinger, K.-H. Schmidt (Eds.), Dynamics and Geomorphology of Mountain Rivers. VIII, 326 pages. 1994.

Vol. 53: F. Scherbaum, Basic Concepts in Digital Signal Processing for Seismologists. X, 158 pages. 1994.

Vol. 54: J. J. P. Zijlstra, The Sedimentology of Chalk. IX, 194 pages. 1995.

Vol. 55: J. A. Scales, Theory of Seismic Imaging. XV, 291 pages. 1995.

Vol. 56: D. Müller, D. I. Groves, Potassic Igneous Rocks and Associated Gold-Copper Mineralization. 2nd updated and enlarged Edition. XIII, 238 pages. 1997.

Vol. 57: E. Lallier-Vergès, N.-P. Tribovillard, P. Bertrand (Eds.), Organic Matter Accumulation. VIII, 187 pages. 1995.

Vol. 58: G. Sarwar, G. M. Friedman, Post-Devonian Sediment Cover over New York State. VIII, 113 pages. 1995.

Vol. 59: A. C. Kibblewhite, C. Y. Wu, Wave Interactions As a Seismo-acoustic Source. XIX, 313 pages. 1996.

Vol. 60: A. Kleusberg, P. J. G. Teunissen (Eds.), GPS for Geodesy. VII, 407 pages. 1996.

Vol. 61: M. Breunig, Integration of Spatial Information for Geo-Information Systems. XI, 171 pages. 1996.

Vol. 62: H. V. Lyatsky, Continental-Crust Structures on the Continental Margin of Western North America. XIX, 352 pages. 1996.

Vol. 63: B. H. Jacobsen, K. Mosegaard, P. Sibani (Eds.), Inverse Methods. XVI, 341 pages, 1996.

Vol. 64: A. Armanini, M. Michiue (Eds.), Recent Developments on Debris Flows. X, 226 pages. 1997.

Vol. 65: F. Sansò, R. Rummel (Eds.), Geodetic Boundary Value Problems in View of the One Centimeter Geoid. XIX, 592 pages. 1997.

Vol. 66: H. Wilhelm, W. Zürn, H.-G. Wenzel (Eds.), Tidal Phenomena. VII, 398 pages. 1997.

Vol. 67: S. L. Webb, Silicate Melts. VIII. 74 pages. 1997.

Vol. 68: P. Stille, G. Shields, Radiogenetic Isotope Geochemistry of Sedimentary and Aquatic Systems. XI, 217 pages. 1997.

Vol. 69: S. P. Singal (Ed.), Acoustic Remote Sensing Applications. XIII, 585 pages. 1997.

Vol. 70: R. H. Charlier, C. P. De Meyer, Coastal Erosion – Response and Management. XVI, 343 pages. 1998.

Vol. 71: T. M. Will, Phase Equilibria in Metamorphic Rocks. XIV, 315 pages. 1998.

Vol. 72: J. C. Wasserman, E. V. Silva-Filho, R. Villas-Boas (Eds.), Environmental Geochemistry in the Tropics. XIV, 305 pages. 1998.

Vol. 73: Z. Martinec, Boundary-Value Problems for Gravimetric Determination of a Precise Geoid. XII, 223 pages. 1998.

Vol. 74: M. Beniston, J. L. Innes (Eds.), The Impacts of Climate Variability on Forests. XIV, 329 pages. 1998.

Vol. 75: H. Westphal, Carbonate Platform Slopes – A Record of Changing Conditions. XI, 197 pages. 1998.

Vol. 76: J. Trappe, Phanerozoic Phosphorite Depositional Systems. XII, 316 pages. 1998.

Vol. 77: C. Goltz, Fractal and Chaotic Properties of Earthquakes. XIII, 178 pages. 1998.

Vol. 78: S. Hergarten, H. J. Neugebauer (Eds.), Process Modelling and Landform Evolution. X, 305 pages. 1999.

Vol. 79: G. H. Dutton, A Hierarchical Coordinate System for Geoprocessing and Cartography. XVIII, 231 pages. 1999.

Vol. 80: S. A. Shapiro, P. Hubral, Elastic Waves in Random Media. XIV, 191 pages. 1999.

Vol. 81: Y. Song, G. Müller, Sediment-Water Interactions in Anoxic Freshwater Sediments. VI, 111 pages. 1999.

Vol. 82: T. M. Løseth, Submarine Massflow Sedimentation. IX, 156 pages. 1999.

Vol. 83: K. K. Roy, S. K. Verma, K. Mallick (Eds.), Deep Electromagnetic Exploration. X, 652 pages. 1999.

Vol. 84: H. V. Lyatsky, G. M. Friedman, V. B. Lyatsky. Principles of Practical Tectonic Analysis of Cratonic Regions. XX, 369 pages. 1999.

Vol. 85: C. Clauser, Thermal Signatures of Heat Transfer Processes in the Earth's Crust. X, 111 pages. 1999.

Vol. 86: H. V. Lyatsky, V. B. Lyatsky, The Cordilleran Miogeosyncline in North America. XX, 384 pages. 1999.

Vol. 87: M. Tiefelsdorf, Modelling Spatial Processes. XVIII, 167 pages. 2000.

Vol. 88: S. Rodrigues-Filho, G. Müller, A Holocene Sedimentary Record from Lake Silvana, SE Brazil. XII, 96 pages. 1999.

Vol. 89: M. Makhous, The Formation of Hydrocarbon Deposits in the North African Basins. XII, 329 pages. 2000.

Vol. 90: R. Klees, R. Haagmans (Eds.), Wavelets in the Geosciences. XVIII, 241 pages. 2000.

Vol. 91: I. Gilmour, C. Koeberl (Eds.), Impacts and the Early Earth. XVIII, 445 pages. 2000.

Vol. 92: P. C. Hansen, B. H. Jacobsen, K. Mosegaard (Eds.), Methods and Applications of Inversion. X, 304 pages. 2000.

Vol. 93: A. Montanari, Ch. Koeberl, Impact Stratigraphy. XIII, 364 pages. 2000.

Vol. 94: M. Breunig, On the Way to Component-Based 3D/4D Geoinformation Systems. XI, 199 pages. 2001.

Vol. 95: A. Dermanis, A. Grün, F. Sansò (Eds.), Geomatic Methods for the Analysis of Data in the Earth Sciences. XII, 265 pages. 2000.

Vol. 96: J. Cobbing, The Geology and Mapping of Granite Batholiths. XII, 141 pages. 2000.

Vol. 97: H. J. Neugebauer, C. Simmer, Dynamics of Multiscale Earth Systems. XIII, 357 pages. 2003.

Vol. 98: T. Takanami, G. Kitagawa, Methods and Applications of Signal Processing in Seismic Network Operations. XIV, 266 pages. 2003.